UNDERSTANDING ENZYMES

Second Edition

ELLIS HORWOOD SERIES IN
BIOLOGICAL CHEMISTRY AND BIOTECHNOLOGY
Series Editor: Dr. ALAN WISEMAN, University of Surrey

Ambrose, E. J.	The Nature and Origin of the Biological World
Berkely, R. C. W., et al.	Microbial Adhesion to Surfaces
Blackburn, F. and Knapp, J. S.	Agricultural Microbiology*
Butt, W.R.	Topics in Hormone Chemistry Vol. 1
Dean, A. C. R., et al.	Continuous Culture Vols. 6 & 8
Dunhill, P., Wiseman, A. and Blakeborough, N.	Enzymic and Non-enzymic Catalysis
Horobin, R. W.	Chemical and Physical Basis of Biological Staining*
Kennedy, J. F., et al.	Cellulose and its Derivatives
Kennedy, J. F. and White, C. A.	Bioactive Carbohydrates
Kricka, L. J. and Clark, P. M. S.	Biochemistry of Alcohol and Alcoholism
Palmer, T.	Understanding Enzymes 2nd Edition
Reid, E.	Methodological Surveys in Biochemistry Vols. 6–11
Rennie, M. J.	Biochemistry of Muscle and Exercise*
Roe, F. J. C.	Microbiological Standardisation of Laboratory Animals
Sammes, P. G.	Topics in Antibiotic Chemistry Vols. 1–6
Sikyta, B. and Moss, M. O.	Methods in Industrial Microbiology
Verrall, M. S.	Discovery and Isolation of Microbial Products
Wiseman, A.	Handbook of Enzyme Biotechnology 2nd Edition
Wiseman, A.	Topics in Enzyme and Fermentation Biotechnology Vols. 1–10

In preparation

UNDERSTANDING ENZYMES

Second Edition

TREVOR PALMER, B.A., Ph.D.
Principal Lecturer in Biochemistry
Trent Polytechnic
Nottingham

ELLIS HORWOOD LIMITED
Publishers · Chichester

Halsted Press: a division of
JOHN WILEY & SONS
New York · Chichester · Brisbane · Toronto

This Second Edition first published in 1985 by

ELLIS HORWOOD LIMITED

Market Cross House, Cooper Street, Chichester, West Sussex, PO19 1EB, England

The publisher's colophon is reproduced from James Gillison's drawing of the ancient Market Cross, Chichester.

Distributors:

Australia, New Zealand, South-east Asia:
Jacaranda-Wiley Ltd., Jacaranda Press,
JOHN WILEY & SONS INC.,
GPO Box 859, Brisbane, Queensland 4001, Australia

Canada:
JOHN WILEY & SONS CANADA LIMITED
22 Worcester Road, Rexdale, Ontario, Canada.

Europe, Africa:
JOHN WILEY & SONS LIMITED
Baffins Lane, Chichester, West Sussex, England.

North and South America and the rest of the world:
Halsted Press: a division of
JOHN WILEY & SONS
605 Third Avenue, New York, NY 10158, USA

© 1985 T. Palmer/Ellis Horwood Limited

British Library Cataloguing in Publication Data
Palmer, Trevor
Understanding enzymes. — 2nd ed.
1. Enzymes
I. Title
547.7'58   QP601
ISBN 0-85312-873-1 (Ellis Horwood Limited — Library Edn.)
ISBN 0-85312-874-X (Ellis Horwood Limited — Student Edn.)
ISBN 0-470-20173-8 (Halsted Press)

Typeset by Ellis Horwood Limited
Printed in Great Britain by R. J. Acford, Chichester

Table of Contents

Part 2 : Kinetic and Chemical Mechanisms of Enzyme-Catalysed Reactions

Chapter 6   An Introduction to Bioenergetics, Catalysis and Kinetics

Chapter 7   Kinetics of Single-Substrate Enzyme-Catalysed Reactions

For Jan, James and Caroline

Author's Preface

This book was written, as all student text-books should be, with the requirements of the student firmly in mind. It is intended to provide an introduction to enzymology, and to give a balanced, reasonably-detailed account of all the various theoretical and applied aspects of the subject which are likely to be included in a course (strangely enough, something rarely attempted in enzymology books at this level). Furthermore, some of the later chapters may serve as a bridge to more advanced textbooks for students wishing to proceed further in this area of biochemistry.

The book is intended mainly for students taking degree courses which have a substantial biochemistry component. In addition, large portions may be of value to students on non-degree courses (e.g. in applied biology or medical laboratory sciences), or even those on MSc or other advanced courses who are approaching the subject of enzymology for the first time (or for the first time in many years).

No previous knowledge of biochemistry, and little of chemistry, is assumed; most scientific terms are defined and placed in context when they first appear. Enzymology inevitably involves a certain amount of elementary mathematics, and some of the equations which are derived may appear somewhat complicated at first sight; however, once the initial biochemical assumptions have been understood, the derivations usually follow on the basis of simple logic, without involving any difficult mathematical manipulations. Numerical and other problems (with answers) are included, to test the student's grasp of certain points. These problems use hypothetical data, although the results are sometimes based on findings reported in the biochemical literature.

If the size of a book is to be kept reasonable, some things of value have to be left out. The chief aim of this particular book is to help the student understand the concepts involved in enzymology (hence the title!); it is not a reference book for practising enzymologists, so no comprehensive tables of data or long, finely-detailed accounts are included. Instead, an attempt has been made to give a perspective of each topic, and examples are quoted where appropriate. Credit has been given wherever possible to those responsible for the development of the subject, but many names deserving of mention have been excluded for reasons of space. Individual scientific papers have not been referred to, but at the end of each chapter is a list of relevant books and review articles, from which references to the original papers may be obtained.

As with any book at this level, certain topics have been presented in a simplified (possibly even over-simplified) form. However, an effort has been made to avoid giving a distorted account of any topic. It is hoped that this book can provide a foundation for those wishing to pursue more advanced studies, and that nothing learned from it will have to be 'un-learned' later. A slight exception to this may be in the use of symbols: experience has shown that symbols v_0 and V_{max} help students to understand some of the basic concepts, so these are adopted here, whereas the Enzyme Commission recommended v and V for general use.

I must record my immense debt to my teachers: in particular to Dr (later Professor) Malcolm Dixon, for awakening my interest in enzymology, and to Dr Peter Sykes, for demonstrating that organic chemistry is a science and not just a list of reactions. I am also grateful to Dr's Barnett Levin, Victor Oberholzer and Ann Burgess, who introduced me, albeit indirectly, to the applications of enzymology in medicine. My thanks are due to Dr Walter Morris and the late Mr Gerald Leadbeater, for giving me the opportunity to teach enzymology, and to my predecessors in my present post, whose legacy of notes proved useful when I began teaching. I am indebted to my colleagues, Dr's Terry Vickers, Clive Williams, Richard Olsson, Martin Griffin and Bernard Scanlon, and my research student Mohammed Ameen, whose advice has been of great value in the preparation of this book. For the same reason, thanks are due to Dr Alan Wiseman and the publishers. Finally, I am extremely grateful to my wife, Jan, who has been involved throughout and who typed the bulk of the manuscript.

Any errors of fact or interpretation which may have inadvertently crept into the book are, of course, entirely my own responsibility, and I would be obliged if I could be informed about them.

1980   T.P.

For this second edition I have corrected errors and revised and updated the text and illustrations. In particular, more has been included about oligomeric enzymes, HPLC, nmr and immunoassay techniques, while reference to some obsolete procedures has been omitted. Several extra problems have also been added. In addition to those people acknowledged above, I would like to thank the following for their help and encouragement: my Head of Department, Professor Keith Short; Deputy Head, Dr Clive Mercer; colleague, Dr Ken Pallett; and research associates Susan Price and Ian Hunter. I am also grateful to the staff of the Trent Polytechnic Science Library.

1985   T.P.

Part 1

Structure and Function of Enzymes

CHAPTER 1

An Introduction to Enzymes

1.1 WHAT ARE ENZYMES?

Enzymes are biological catalysts. They increase the rate of chemical reactions taking place within living cells without themselves suffering any overall change. The reactants of enzyme-catalysed reactions are termed **substrates** and each enzyme is quite specific in character, acting on a particular substrate or substrates to produce a particular product or products.

All enzymes are proteins. However, without the presence of a non-protein component called a **cofactor**, many enzyme proteins lack catalytic activity. When this is the case, the inactive protein component of an enzyme is termed the **apoenzyme**, and the active enzyme, including cofactor, the **holoenzyme**. The cofactor may be an organic molecule, when it is known as a **coenzyme**, or it may be a metal ion. Some enzymes bind cofactors more tightly than others. When a cofactor is bound so tightly that it is difficult to remove without damaging the enzyme it is sometimes called a **prosthetic group**.

To summarise diagramatically:

$$
\text{ENZYME} \begin{cases} \text{INACTIVE PROTEIN (APOENZYME) + COFACTOR} \\ \xleftarrow{\hspace{2cm}} \text{(HOLOENZYME)} \xrightarrow{\hspace{2cm}} \\ \text{ACTIVE PROTEIN} \end{cases} \begin{cases} \text{ORGANIC MOLECULE} \\ \text{(COENZYME)} \\ \\ \text{METAL ION} \end{cases}
$$

As we shall see later, both the protein and cofactor components may be directly involved in the catalytic processes taking place.

1.2 A BRIEF HISTORY OF ENZYMES

Until the nineteenth century, it was considered that processes such as the souring of milk and the fermentation of sugar to alcohol could only take place through the action of a living organism. In 1833, the active agent breaking down the sugar was partially isolated and given the name **diastase** (now known as **amylase**). A little later, a substance which digested dietary protein was extracted from gastric juice and called **pepsin**. These and other active preparations were given the general name **ferments**. Liebig recognised that these ferments could be non-living materials obtained from living cells, but Pasteur and others still maintained that ferments must contain living material.

While this dispute contined, the term ferment was gradually replaced by the name **enzyme**. This was first proposed by Kühne in 1878, and comes from the Greek, enzumé (ἐνζυμη) meaning "in yeast". Appropriately, it was in yeast that a factor was discovered which settled the argument in favour of the inanimate theory of catalysis: the Büchners, in 1897, showed that sugar fermentation could take place when a yeast cell extract was added even though no living cells were present.

In 1926, Sumner crystallised urease from Jack-bean extracts, and in the next few years many other enzymes were purified and crystallised. Once pure enzymes were available, their structure and properties could be determined, and the findings form the material for most of this book.

Today, enzymes still form a major subject for academic research. They are investigated in hospitals as an aid to diagnosis and, because of their specificity of action, are of great value as analytical reagents. Enzymes are still widely used in industry, continuing and extending many processes which have been used since the dawn of history.

1.3 THE NAMING AND CLASSIFICATION OF ENZYMES

1.3.1 Why Classify Enzymes?

As we have seen in the previous section, there is a long tradition of giving enzymes names ending in "-ase". The only major exceptions to this are the **proteolytic enzymes**, whose names usually end with "-in", e.g. **trypsin**.

The names of enzymes usually indicate the substrate involved. Thus, **lactase** catalyses the hydrolysis of the disaccharide **lactose** to its component monosaccharides **glucose** and **galactose**:

$$C_{12}H_{22}O_{11} + H_2O \rightleftharpoons C_6H_{12}O_6 + C_6H_{12}O_6$$
$$\text{lactose} \qquad\qquad \text{glucose} \qquad \text{galactose}$$

The name lactase is a contraction of the clumsy, but more precise, lactosase. The former is used because it sounds better but it introduces a possible trap for the unwary because it could easily suggest an enzyme acting on the substrate lactate.

There is nothing in the name of this enzyme or many others to indicate the type of reaction being catalysed. **Fumarase**, for example, by analogy with lactase might be supposed to catalyse a hydrolytic reaction; but, in fact, it *hydrates* **fumarate** to form **malate**:

$$^-O_2C.CH = CH.CO_2^- + H_2O \rightleftharpoons {}^-O_2C.CHOH.CH_2CO_2^-$$

$$\text{fumarate} \qquad\qquad\qquad \text{malate}$$

The names of other enzymes, e.g. **transcarboxylase**, indicate the nature of the reaction without specifying the substrates (which in the case of transcarboxylase are methylmalonyl-CoA and pyruvate). Some names, such as **catalase**, indicate neither the substrate nor the reaction (catalase mediates the decomposition of hydrogen peroxide).

Needless to say, whenever a new enzyme has been characterised, great care has usually been taken not to give it exactly the same name as an enzyme catalysing a different reaction. Also, the names of many enzymes make clear the substrate and the nature of the reaction being catalysed. For example, there is little ambiguity about the reaction catalysed by **malate dehydrogenase**: this enzyme mediates the removal of hydrogen from malate to produce oxaloacetate:

$$^-O_2C.\underset{\underset{OH}{|}}{CH}.CH_2.CO_2^- + NAD^+ \rightleftharpoons {}^-O_2C.\underset{\underset{O}{\|}}{C}.CH_2.CO_2^- + NADH + H^+$$

$$\qquad\qquad \text{malate} \qquad\qquad\qquad\qquad \text{oxaloacetate}$$

However, malate dehydrogenase, like many other enzymes, has been known by more than one name.

So, because of the lack of consistency in the nomenclature, it became apparent as the list of known enzymes rapidly grew that there was a need for a systematic way of naming and classifying enzymes. A commission was appointed by the International Union of Biochemistry, and its report, published in 1964 and updated in 1972 and 1978, forms the basis of the present accepted system.

1.3.2 The Enzyme Commission's System of Classification

The Enzyme Commission divided enzymes into six main classes, on the basis of the total reaction catalyzed. Each enzyme was assigned a code number, consisting of four elements, separated by dots. The first digit shows to which of the main classes the enzyme belongs, as follows:

First digit	Enzyme class	Type of reaction catalyzed
1	Oxidoreductases	Oxidation/reduction reactions
2	Transferases	Transfer of an atom or group between two molecules (excluding reactions in other classes)
3	Hydrolases	Hydrolysis reactions

First digit	Enzyme class	Type of reaction catalysed
4	Lyases	Removal of a group from substrate (not by hydrolysis)
5	Isomerases	Isomerisation reactions
6	Ligases	The synthetic joining of two molecules, coupled with the breakdown of a pyrophosphate bond in a nucleoside triphosphate.

The second and third digit in the code further describe the kind of reaction being catalysed. There is no general rule, because the meanings of these digits are defined separately for each of the main classes. Some examples are given later in this chapter.

Enzymes catalysing very similar but non-identical reactions, e.g. the hydrolysis of different carboxylic acid esters, will have the same first three digits in their code. The fourth digit distinguishes between them by defining the actual substrate, e.g. the actual carboxylic acid ester being hydrolysed.

However, it should be noted that **isoenzymes**, that is to say, different enzymes catalysing identical reactions, will have the same four figure classification. There are, for example, five different isoenzymes of **lactate dehydrogenase** within the human body and these will have an identical code. The classification, therefore, provides only the basis for a unique identification of an enzyme: the particular isoenzyme and its source still have to be specified.

It should also be noted that all reactions catalysed by enzymes are reversible to some degree and the classification which would be given to the enzyme for the catalysis of the forward reaction would not be the same as that for the reverse reaction. The classification used is that of the most important direction from the biochemical point of view or according to some convention defined by the Commission. For example, for oxidation/reduction involving the interconversion of NADH and NAD^+ (see Chapter 11.5.2) the classification is usually based on the direction where NAD^+ is the electron acceptor rather than that where NADH is the electron donor.

Some problems are given at the end of this chapter to help the student become familiar with this system of classification.

1.3.3 The Enzyme Commission's Recommendations on Nomenclature

The Commission assigned to each enzyme a systematic name in addition to its existing trivial name. This systematic name includes the name of the substrate or substrates in full and a word ending in "-ase" indicating the nature of the process catalysed. This word is either one of the six main classes of enzymes or a subdivision of one of them. When a reaction involves two types of overall change, e.g. oxidation and decarboxylation, the second function is indicated in brackets, e.g. oxidoreductase (decarboxylating). Examples are given below.

The systematic name and the Enzyme Commission (E.C.) classification number unambiguously describe the reaction catalysed by an enzyme and should always be included in a report of an investigation of an enzyme, together with the source of the enzyme, e.g. rat liver mitochondria.

However, these names are likely to be long and unwieldy. Trivial names may, therefore, be used in a communication, once they have been introduced and defined in terms of the systematic name and E.C. number. Trivial names are also inevitably used in everyday situations in the laboratory. The Enzyme Commission made recommendations as to which trivial names were acceptable, altering those which were considered vague or misleading. Thus, "fumarase", mentioned above, was considered unsatisfactory and was replaced by "fumarate hydratase".

1.3.4   The Six Main Classes of Enzymes
Main Class 1: Oxidoreductases

These enzymes catalyse the transfer of H atoms, O atoms or electrons from one substrate to another. The second digit in the code number of oxidoreductases indicates the donor of the reducing equivalents (hydrogen or electrons) involved in the reaction. For example,

Second digit	*Hydrogen or Electron Donor*
1	alcohol ($>$CHOH)
2	aldehyde or ketone ($>$C=O)
3	$-$CH.CH$-$
4	primary amine ($-$CHNH$_2$ or $-$CHNH$_3^+$)
5	secondary amine ($>$CHNH$-$)
6	NADH or NADPH (only where some other redox catalyst is the acceptor).

The third digit refers to the hydrogen or electron acceptor, as follows:

Third digit	*Hydrogen or Electron Acceptor*
1	NAD$^+$ or NADP$^+$
2	Fe^{3+} (e.g. cytochromes)
3	O$_2$
99	An otherwise unclassified acceptor.

Trivial names of oxidoreductases include oxidases (transfer of H to O$_2$) and dehydrogenases (transfer of H to an acceptor other than O$_2$). These often indicate the identity of the donor and/or acceptor.

Here are some examples:

L-lactate: NAD$^+$ oxidoreductose (E.C. 1.1.1.27) (trivial name lactate dehydrogenase) catalyses:

$$CH_3.\underset{\underset{OH}{|}}{CH}.CO_2^- + NAD^+ \rightleftharpoons CH_3.\underset{\underset{O}{\|}}{C}.CO_2^- + NADH + H^+$$

$$\text{L-lactate} \qquad\qquad\qquad \text{pyruvate}$$

Note that it is the alcohol group of lactate, rather than the carboxyl group, which is involved in the reaction and this is indicated in the classification.

threo-D_s isocitrate: NAD^+ oxidoreductase (decarboxylating) (E.C. 1.1.1.41) (trivial name isocitrate dehydrogenase) catalyses:

$$^-O_2C.CH_2.\underset{\overset{|}{^-O_2C}}{\underset{|}{CH}}.\underset{\overset{|}{OH}}{\underset{|}{CH}}.CO_2^- + NAD^+ \rightleftharpoons {}^-O_2C.CH_2.CH_2.\underset{\overset{||}{O}}{C}.CO_2^- + NADH + H^+ + CO_2$$

threo-D_s-isocitrate   2-oxoglutarate

D-amino-acid: oxygen oxidoreductase (E.C. 1.4.3.3) (trivial name D-amino-acid oxidase) catalyses:

$$R.\underset{\overset{|}{^+NH_3}}{\underset{|}{CH}}.CO_2^- + H_2O + O_2 \rightleftharpoons R.\underset{\overset{||}{O}}{C}.CO_2^- + {}^+NH_4 + H_2O_2$$

D-amino-acid   oxo acid

Note that this enzyme is less specific than most and will act on any D-amino-acid.

Main Class 2: Transferases

These catalyse reactions of the type:

$$AX + B \rightleftharpoons BX + A,$$

but specifically exclude oxidoreductase and hydrolase reactions. In general, the Enzyme Commission recommends that the names of transferases should end "X-transferase", where X is the group transferred, although a name ending "trans-X-ase" is an acceptable alternative. The second digit in the classification describes the type of group transferred. For example:

Second digit	*Group transferred*
1	1-carbon group
2	aldehyde or ketone group ($>C=O$)
3	acyl group ($-\underset{\overset{||}{O}}{C}-R$)
4	glycosyl (carbohydrate) group
7	phosphate group.

In general, the third digit further describes the group transferred. Thus,

E.C. 2.1.1 enzymes are methyltransferases (transfer $-CH_3$) whereas
E.C. 2.1.2 enzymes are hydroxymethyltransferases (transfer $-CH_2OH$) and
E.C. 2.1.3 enzymes are carboxyl- or carbamoyl-transferases (transfer $-\underset{\overset{||}{O}}{C}-OH$ or $-\underset{\overset{||}{O}}{C}-NH_2$).

Similarly, E.C. 2.4.1 enzymes are hexosyltransferases (transfer hexose units) and E.C. 2.4.4 enzymes are pentosyltransferases (transfer pentose units).

The exception to this general rule for transferases is where there is transfer of phosphate groups: these cannot be described further, so there is opportunity to indicate the acceptor.

E.C. 2.7.1 enzymes are phosphotransferases with an alcohol group as acceptor,

E.C. 2.7.2 enzymes are phosphotransferases with a carboxyl group as acceptor and

E.C. 2.7.3 enzymes are phosphotransferases with a nitrogenous group as acceptor.

Phosphotransferases usually have a trivial name ending in "-kinase".

Some examples of transferases are:

Methylmalonyl-CoA: pyruvate carboxyltransferase (E.C. 2.1.3.1) (trivial name methylmalonyl-CoA carboxyltransferase, formerly transcarboxylase) which catalyses the transfer of a carboxyl group from methylmalonyl-CoA to pyruvate:

$$CH_3.CH.COSCoA + CH_3.CO.CO_2^- \rightleftharpoons CH_3.CH_2.COSCoA + CH_2.CO.CO_2^-$$
$$\qquad | \qquad\qquad\qquad\qquad\qquad\qquad\qquad\qquad\qquad\qquad\qquad |$$
$$\qquad CO_2^- \qquad\qquad\qquad\qquad\qquad\qquad\qquad\qquad\qquad\qquad CO_2^-$$

methylmalonyl-CoA   pyruvate   propionyl-CoA   oxaloacetate

ATP: D-hexose-6-phosphotransferase (E.C. 2.7.1.1) (trivial name hexokinase) which catalyses:

$$C_5H_9O_5.CH_2OH + ATP \rightleftharpoons C_5H_9O_5.CH_2OPO_3^{2-} + ADP$$

D-hexose   D-hexose-6-phosphate

This enzyme will transfer phosphate to a variety of D-hexoses.

Main Class 3: Hydrolases

These enzymes catalyse hydrolytic reactions of the form:

$$A-X + H_2O \rightleftharpoons X-OH + HA.$$

They are classified according to the type of bond hydrolysed. For example:

Second digit	Bond hydrolysed
1	ester
2	glycosidic (linking carbohydrate units)
4	peptide ($-\overset{\text{H}}{\underset{\text{O}}{\overset{|}{\underset{\|}{C}}}}-N-$)
5	C—N bonds other than peptides.

The third digit further describes the type of bond hydrolysed. Thus,

E.C. 3.1.1 enzymes are carboxylic ester ($-\overset{O}{\overset{\|}{C}}-O-$) hydrolases,

E.C. 3.1.2 enzymes are thiol ester ($-\overset{\overset{\text{O}}{\|}}{\text{C}}-\text{S}-$) hydrolases,

E.C. 3.1.3 enzymes are phosphoric monoester ($-\text{O}-\text{PO}_3^{2-}$) hydrolases and

E.C. 3.1.4 enzymes are phosphoric diester ($-\text{O}-\overset{\overset{-\text{O}}{|}}{\underset{\underset{\text{O}}{\|}}{\text{P}}}-\text{O}-$) hydrolases.

For example, orthophosphoric monoester phosphohydrolase (E.C. 3.1.3.1) (alkaline phosphatase) catalyses:

$$R-O-\overset{\overset{-\text{O}}{|}}{\underset{\underset{\text{O}}{\|}}{\text{P}}}-O^- + H_2O \rightleftharpoons R-OH + HO-\overset{\overset{-\text{O}}{|}}{\underset{\underset{\text{O}}{\|}}{\text{P}}}-O^-$$

            organic phosphate   inorganic phosphate

Alkaline phosphatases are relatively non-specific, and act on a variety of substrates at alkaline pH.

The trivial names of hydrolases are recommended to be the only ones to consist simply of the name of the substrate plus "-ase".

Main Class 4: Lyases

These enzymes catalyse the non-hydrolytic removal of groups from substrates, often leaving double bonds.

The second digit in the classification indicates the bond broken, for example,

Second digit	*Bond broken*
1	C—C
2	C—O
3	C—N
4	C—S.

The third digit refers to the type of group removed. Thus, for the C—C lyases:

Third digit	*Group removed*
1	carboxyl group (i.e. CO_2)
2	aldehyde group ($-CH{=}O$)
3	ketoacid group ($-\overset{\overset{\,}{|}}{\underset{\underset{\text{O}}{\|}}{\text{C}}}.\text{CO}_2^-$).

For example, L-histidine carboxy-lyase (E.C. 4.1.1.22) (trivial name histidine decarboxylase) catalyses:

$$C_3N_2H_3.CH_2.\overset{}{\underset{\underset{\text{CO}_2^-}{|}}{\text{CH}}}.NH_3^+ \rightleftharpoons C_3N_2H_3.CH_2.CH_2.\overset{+}{N}H_3 + CO_2$$

        histidine   histamine

(Note the importance in the hyphen in the systematic name, because carboxy-lyase and carboxylase do not necessarily mean the same thing: carboxylase simply refers to the involvement of CO_2 in a reaction without being specific.)

Also classified as lyases are enzymes catalysing reactions whose biochemically important direction is the reverse of the above, i.e. addition across double bonds. These may have the trivial name **synthase** or, if water is added across the double bond, **hydratase**, as discussed earlier in the example of fumarate hydratase (fumarase); the systematic name of this particular enzyme is L-malate hydro-lyase (E.C. 4.2.1.2).

Main Class 5: Isomerases
Enzymes catalysing isomerisation reactions are classified according to the type of reaction involved. For example:

Second digit	Type of reaction
1	Racemisation or epimerisation (inversion at an asymmetric carbon atom)
2	cis-trans isomerisation
3	intramolecular oxidoreductases
4	intramolecular transfer reaction.

The third digit describes the type of molecule undergoing isomerisation. Thus, for racemases and epimerases:

Third digit	Substrate
1	amino-acids
2	hydroxy acids
3	carbohydrates.

An example is alanine racemase (E.C. 5.1.1.1) which catalyses:

L-alanine $\rightleftharpoons$ D-alanine.

Main Class 6: Ligases
These enzymes catalyse the synthesis of new bonds, coupled to the breakdown of ATP or other nucleoside triphosphates. The reactions are of the form:

$$X + Y + ATP \rightleftharpoons X - Y + ADP + P_i$$

or $$X + Y + ATP \rightleftharpoons X - Y + AMP + (PP)_i$$

The second digit in the code indicates the type of bond synthesised. For example:

Second digit   *Bond synthesised*

1	C–O
2	C–S
3	C–N
4	C–C.

The third digit further describes the bond being formed. Thus,

E.C. 6.3.1 enzymes are acid–ammonia ligases (amide, $-\overset{\overset{\displaystyle O}{\|}}{C}-NH_2$, synthetases) and

E.C. 6.3.2 enzymes are acid–amino-acid ligases (peptide, $-\overset{\overset{\displaystyle O}{\|}}{C}-\underset{H}{N}-$, synthetases).

An example is L-glutamate:ammonia ligase (E.C. 6.3.1.2) (trivial name glutamine synthetase) which catalyses:

$$O=C.CH_2.CH_2.CH.CO_2^- + ATP + NH_3 \rightleftharpoons O=C.CH_2.CH_2.CH.CO_2^- + ADP + P_i$$

$$\underset{\text{L-glutamate}}{\overset{|}{{}^-O}\qquad\overset{|}{{}^+NH_3}} \qquad\qquad \underset{\text{L-glutamine}}{\overset{|}{NH_2}\qquad\overset{|}{{}^+NH_3}}$$

SUMMARY OF CHAPTER 1

Enzymes are proteins which catalyse, in a highly specific way, chemical reactions taking place within the living cell. Often a further, non-protein, component called a cofactor is required before an enzyme has catalytic activity.

Enzymes have been used for many centuries, although their true nature has only become known relatively recently, and they are still of great importance in scientific research, clinical diagnosis and industry.

Because of the lack of consistency and occasional lack of clarity in the names of enzymes, an Enzyme Commission appointed by the International Union of Biochemistry has given all known enzymes a systematic name and a four-figure classification. These, together with the source of the enzyme concerned, should be quoted in any report.

FURTHER READING

Barman, T. E. (1969), *Enzyme Handbook*, Springer-Verlag.

Dixon, M., Webb, E. C., Thorne, C. J. R. and Tipton, K. F. (1979), *Enzymes*, third edition, Longman (Chapters 1 and 5).

Enzyme Nomenclature. Recommendations (1978) of the Nomenclature Committee of the International Union of Biochemistry. Published 1979 by Academic Press. Corrections and additions listed in *European Journal of Biochemistry* (1982), **125** (pages 1-13).

PROBLEMS

1.1 Give the systematic names and the first three digits in the E.C. classifications of the enzymes catalysing the following reactions:
(Note: it should be possible to deduce the classification and make a reasonable attempt at the systematic name from the information given in Chapter 1.)

(a) $R.\overset{\shortmid}{\underset{O}{C}}O.CH_2.CH_2.\overset{+}{N}(CH_3)_3 + H_2O \rightleftharpoons R.\overset{\shortmid}{\underset{O}{C}}O^- + HOCH_2.CH_2\overset{+}{N}(CH_3)_3$

acyl choline   acid anion   choline

(b) $H_2N.\overset{\shortmid}{\underset{O}{C}}OPO_3^{2-} + H_3\overset{+}{N}(CH_2)_3.\underset{+NH_3}{\overset{\shortmid}{C}H}.CO_2^- \rightleftharpoons H_2N.\overset{\shortmid}{\underset{O}{C}}.NH.(CH_2)_3.\underset{+NH_3}{\overset{\shortmid}{C}H}.CO_2^- + P_i$

carbamoyl   L-ornithine   citrulline
phosphate

(c) $ATP + H_3\overset{+}{N}.\underset{CH_3}{\overset{\shortmid}{C}H}.CO_2^- + H_3\overset{+}{N}.\underset{CH_3}{\overset{\shortmid}{C}H}.CO_2^- \rightleftharpoons H_3\overset{+}{N}.\underset{CH_3O}{\overset{\shortmid}{C}H}.\overset{H}{\underset{\shortmid}{\overset{\shortmid}{C}}}.N.\underset{CH_3}{\overset{\shortmid}{C}H}.CO_2^- + ADP + P_i$

D-alanine   D-alanine   D-alanyl-alanine

(d)
$$\begin{array}{ccc}
CH_2OH & & CH_2OH \\
| & & | \\
C=O & + NADH + H^+ \rightleftharpoons & CHOH & + NAD^+ \\
| & & | \\
CH_2OH & & CH_2OH
\end{array}$$

dihydroxyacetone   glycerol

(e)
$$\begin{array}{l}
CH_2OPO_3^{2-} \\
| \\
C=O \\
| \\
CHOH \\
| \\
CHOH \\
| \\
CHOH \\
| \\
CH_2OPO_3^{2-}
\end{array}
\rightleftharpoons
\begin{array}{l}
CH_2OPO_3^{2-} \\
| \\
C=O \\
| \\
CH_2OH
\end{array}
+
\begin{array}{l}
HC=O \\
| \\
CHOH \\
| \\
CH_2OPO_3^{2-}
\end{array}$$

D-glyceraldehyde-3-phosphate
dihydroxyacetone phosphate

D-fructose-1,6-bisphosphate

(f) $NADH + 2$ ferricytochrome $b_5 \rightleftharpoons NAD^+ + 2$ ferrocytochrome b_5

(g) UDP-galactose $\rightleftharpoons$ UDP-glucose
(glucose and galactose are aldohexoses differing in configuration at C4).

1.2 Give the E.C. classification of the enzyme catalysing the following reactions:
(This question has been designed to encourage the student to become familiar with the Enzyme Commission's report and can only be answered satisfactorily by reference to this report or to a detailed account of it.)

(a) $CH_3.\underset{\underset{O}{\Vert}}{C}.CO_2^- + ATP + CO_2 + H_2O \rightleftharpoons {}^-O_2C.CH_2.\underset{\underset{O}{\Vert}}{C}.CO_2^- + ADP + P_i$

pyruvate — oxaloacetate

(b) $H_2S + 3NADP^+ + 3H_2O \rightleftharpoons$ sulphite $+ 3NADPH$

(c) $ATP + AMP \rightleftharpoons ADP + ADP$

(d)
$$\begin{array}{c} HC{=}O \\ | \\ CHOH \\ | \\ CH_2OPO_3^{2-} \end{array} \rightleftharpoons \begin{array}{c} CH_2OH \\ | \\ C{=}O \\ | \\ CH_2OPO_3^{2-} \end{array}$$

D-glyceraldehyde-3-P

dihydroxyacetone-P

(e) $H_3\overset{+}{N}.CH_2.\underset{\underset{O}{\Vert}}{C}.\overset{H}{N}.CH_2.CO_2^- + H_2O \rightleftharpoons H_3\overset{+}{N}.CH_2.CO_2^- + H_3\overset{+}{N}.CH_2.CO_2^-$

glycylglycine — glycine — glycine

(f) Endohydrolysis of α-1,4 glucan links in polysaccharides containing 3 or more α-1,4 linked D-glucose units.

(g)
$$H_3\overset{+}{N}.\underset{\underset{CO_2^-}{|}}{CH}.(CH_2)_3.NH.\underset{\underset{{}^+NH_2}{\Vert}}{C}.NH.\overset{\overset{CO_2^-}{|}}{\underset{\underset{\underset{CO_2^-}{|}}{CH_2}}{\underset{|}{CH}}} + H_2O \rightleftharpoons H_3\overset{+}{N}.\underset{\underset{CO_2^-}{|}}{CH}.(CH_2)_3.NH.\underset{\underset{{}^+NH_2}{\Vert}}{C}.NH + \begin{array}{c} H \quad CO_2^- \\ \diagdown\ \diagup \\ C \\ \Vert \\ C \\ \diagup\ \diagdown \\ {}^-O_2C \quad H \end{array}$$

L-argininosuccinate — L-arginine — fumarate

CHAPTER 2

The Structure of Proteins

2.1 INTRODUCTION

Since all enzymes are proteins, a knowledge of protein structure is clearly a prerequisite to any understanding of enzymes.

Proteins are **macromolecules** (i.e. large molecules) with molecular weights of at least several thousand daltons. They are found in abundance in living organisms, making up more than half the dry weight of cells. Two distinct types are known: fibrous and globular proteins.

Fibrous proteins are insoluble in water and are physically tough, which enables them to play a structural role. Examples include α–keratin (a component of hair, nails and feathers) and collagen (the main fibrous element of skin, bone and tendon). In contrast, **globular proteins** are generally soluble in water and may be crystallised from solution. They have a functional role in living organisms, all enzymes being globular proteins.

Unlike polysaccharides and lipids, which may be hoarded by cells solely as a store of fuel, each protein in a cell has some precise purpose which is related to its shape and structure. Nevertheless, should the need arise, proteins may be broken down, either to provide energy or to supply raw materials for the synthesis of other macromolecules.

All proteins consist of **amino-acid** units, joined in series. The sequence of amino-acids in a protein is specific, being determined by the structure of the genetic material of the cell (see Chapter 3.1), and this gives each protein unique properties. Some proteins are composed entirely of these amino-acid building blocks and are termed **simple proteins**. Others, called **conjugated proteins**, contain extra material, which is firmly bound to one or more of the amino-acid units. For example:

conjugated protein	*extra component present*
nucleoprotein	a nucleic acid
lipoprotein	a lipid
glycoprotein	an oligosaccharide
haemoprotein	an iron protoporphyrin
flavoprotein	a flavin nucleotide
metalloprotein	a metal.

As we have already see (Chapter 1.1), enzymes may be either simple or conjugated proteins.

2.2 AMINO-ACIDS, THE BUILDING BLOCKS OF PROTEINS

2.2.1 Structure and Classification of Amino-Acids

Amino-acids, by definition, are organic compounds which contain within the same molecule an amino group ($-NH_2$ or $>NH$) and a carboxyl group ($-C\overset{\displaystyle O}{\underset{\displaystyle OH}{}}$). Thus they have properties of both bases and acids.

The amino group in all but one of the twenty amino-acids commonly found in proteins is a primary one ($-NH_2$), the exception being proline, which contains a secondary amino group ($>NH$). The carbon atoms of organic molecules containing a carboxyl group may be identified with Greek letters as follows:

$$\gamma\text{-carbon atom}$$

$$\alpha\text{-carbon atom}$$

$$-C-C-C-C-(-CO_2H)$$

$$\beta\text{-carbon atom}$$

$$\delta\text{-carbon atom} \qquad \text{carboxyl group.}$$

All the amino-acids commonly found in proteins are **α-amino-acids**, since the amino group is on the α-carbon atom. The general formula is:

$$H_2N-\overset{\displaystyle H}{\underset{\displaystyle R}{C}}-C\overset{\displaystyle O}{\underset{\displaystyle OH}{}}\text{, or more conveniently, } H_2N.CHR.CO_2H.$$

The symbol R represents the rest of the molecule, often called the **side chain**. The amino and carboxyl groups attached to the α-carbon atom are termed the α-amino and α-carboxyl groups, to distinguish them from similar groups which may be present as part of the side chain. Proline, whose α-amino group forms part of an imino ring, is an **imino acid** with a formula slightly different from the general one given above (see Fig. 2.1).

The α-amino-acids may have polar or non-polar side chains. A **polar** molecule or group has a degree of ionic character and is **hydrophilic**, i.e. it is quite soluble in water because its structure may be stabilised by hydrogen bonding in aqueous solution. Polar groups may be acidic, basic or neutral. A **non-polar** molecule or group is entirely covalent in character and is **hydrophobic**, i.e. it is relatively insoluble in aqueous solvents but more soluble in organic solvents such as diethyl ether. The side chains of the amino-acids commonly found in proteins, classified according to their polar or non-polar characteristics, are shown in Fig. 2.1.

Non-polar side chains		Polar side chains	
—R	Amino-acid	—R	Amino-acid

Non-polar side chains		Polar side chains	
—CH$_3$	Alanine (Ala)	**Negative charge at pH 7**	
—CH.CH$_3$ with CH$_3$	Valine (Val)	—CH$_2$C(=O)(O$^-$)	Aspartic Acid (Asp) or Aspartate
—CH$_2$CH.CH$_3$ with CH$_3$	Leucine (Leu)	—CH$_2$CH$_2$C(=O)(O$^-$)	Glutamic Acid (Glu) or Glutamate
—CH.CH$_2$CH$_3$ with CH$_3$	Isoleucine (Ile)	**Positive charge at pH 7**	
		—(CH$_2$)$_4$NH$_3^+$	Lysine (Lys)
—CH$_2$—⟨ring⟩	Phenylalanine (Phe)	—(CH$_2$)$_3$NHC.NH$_2$ with =NH$_2^+$	Arginine (Arg)
		Uncharged at pH 7	
—CH$_2$—(indole ring, N, H)	Tryptophan (Trp)	—H	Glycine (Gly)
		—CH$_2$OH	Serine (Ser)
—CH$_2$CH$_2$—S—CH$_3$	Methionine (Met)	—CH.CH$_3$ with OH	Threonine (Thr)
		—CH$_2$SH	Cysteine (Cys)
(pyrrolidine ring) CH.CO$_2^-$, N$^+$H$_2$	Proline (Pro) (complete structure)	—CH$_2$—⟨ring⟩—OH	Tyrosine (Tyr)
		—CH$_2$C(=O)(NH$_2$)	Asparagine (Aspn)
		—CH$_2$CH$_2$C(=O)(NH$_2$)	Glutamine (Glun)
		—CH$_2$—(imidazole ring HN, N)	Histidine (His)

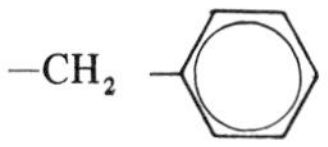

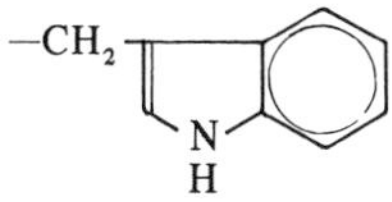

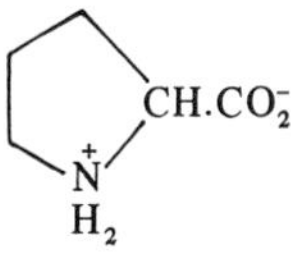

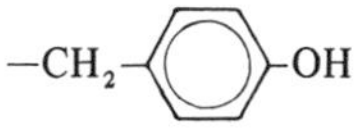

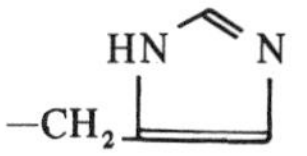

Fig. 2.1 The side chains of the twenty amino-acids commonly found in the proteins. (Several polar side chains contain ionisable group, the degree of ionisation being pH-dependent (see Chapter 2.3.2). Only the form which predominates at pH 7 is shown in the figure.)

It will be seen that the side chain of **histidine** contains an **imidazole** ring, while that of **tryptophan** includes a double-ringed structure called an **indole**. One of these rings is an aromatic benzene ring, so tryptophan, in common with phenylalanine and tyrosine, may be called an **aromatic** amino-acid. In **tyrosine** the aromatic ring is linked to −OH to form a **phenolic** group. **Glutamic acid** and **aspartic acid** contain a **carboxyl** group in their side chains, which is converted to an **amide** group in **glutamine** and **asparagine**. The side chains of **lysine** and **arginine** contain **amino** groups, which in the case of **arginine** forms part of a **guanidine** structure. The R groups of **valine**, **leucine** and **isoleucine** have a **branched-chain aliphatic hydrocarbon** structure while **proline**, as mentioned previously, is an **imino acid**. **Methionine** and **cysteine** contain **sulphur**, which in the case of **cysteine** is present as part of a **sulphydryl** (−SH) group. Cysteine is readily oxidised to form the dimeric compound **cystine**, the two component cysteine units being linked by a **disulphide bridge**.

$$H_2N.CH.CO_2H$$
$$|$$
$$CH_2$$
$$|$$
$$S$$
$$|$$ cystine
$$S$$
$$|$$
$$CH_2$$
$$|$$
$$H_2N.CH.CO_2H$$

Thus, amino-acids with a considerable variety of side chain characteristics are found in proteins. As we shall see later, this explains the range of properties shown by these macromolecules.

2.2.2 Stereochemistry of Amino-Acids

Each carbon atom in a molecule can form 4 single covalent bonds with other atoms. These are often represented at right angles to each other on a single plane, as in Chapter 2.2.1. However it must be realised that this is done entirely for convenience, since a page of a book is two-dimensional and thus lends itself to a two-dimensional representation of structure. In fact the four bonds are evenly distributed in three-dimensional space, which means they point to the four corners of a regular tetrahedron, each bond forming an angle of 109° with each of the other bonds.

If we consider the bonds involving the α−carbon of an amino-acid, we see that two different spatial arrangements, or **stereoisomeric forms**, are possible: the structure depicted in Fig. 2.2(a) cannot be superimposed on that in Fig. 2.2(b) by rotation of the molecules. The α−carbon atom is covalently linked to four different atoms or groups, so it is **asymmetric**; no plane drawn through this carbon atom can divide the molecule into two parts in such a way that each half is the exact mirror image of the other. As a consequence of this, two mirror image forms of the complete molecule can exist. Such forms are termed **optical isomers**, since one will usually rotate the plane of polarised light passing through it to the right, and the other to the left.

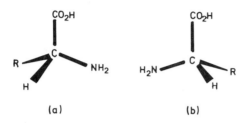

(a)   (b)

Fig. 2.2   Three-dimensional arrangements about the α–carbon atom for (a) a D-amino-acid and (b) an L-amino-acid. A bond coming out of the plane of the page towards the reader is indicated by a thickening of the line; one going away from the reader is represented by a narrowing of the line.

The molecule shown in Fig. 2.2(a) is *defined* as a **D-amino-acid** and the one in Fig. 2.2(b) as an **L-amino-acid**. This says nothing about how each isomer will affect the plane of polarised light, a property which has to be determined by experiment. Thus L-alanine is found to rotate polarised light to the right, but L-leucine rotates it to the left.

All the common amino-acids, with the exception of glycine, exist as optical isomers. Glycine does not have an asymmetric carbon atom since in this case there are two hydrogen atoms attached to the α-carbon (R=H). Threonine and isoleucine possess two asymmetric carbon atoms, but this extra complication need not concern us here.

If amino-acids are synthesised by an uncatalysed chemical process, a **racemic mixture** (one containing equal amounts of L- and D- isomers) is produced, the isomeric forms being almost indistinguishable from a chemical point of view. However, proteins are built almost exclusively of L-amino-acids, and most naturally ocurring amino-acids are in this same isomeric form. The explanation is that protein biosynthesis (see Chapter 3.1) and most other metabolic processes are mediated by enzymes which are specific for a particular isomeric form of the substrate (see Chapter 4.1); this is essential for ensuring the high degree of three-dimensional organisation which is found in structures within cells. It is presumably evolutionary chance which has determined that life as we know it is based on L- rather than D- amino-acids.

2.3   THE BASIS OF PROTEIN STRUCTURE

2.3.1   Levels of Protein Structure

Four separate levels of protein structure can be determined: these are the primary, secondary, tertiary and quaternary structures.

The **primary structure** is the sequence of amino-acids making up the protein: a **peptide bond** connects the α-carboxyl group of each amino-acid to the α-amino group of the next in the chain.

$$H_2N.CHR'.CO_2H + H_2N.CHR''.CO_2H \rightarrow H_2N.CHR'.\overset{\overset{\textstyle O}{\textstyle \|}}{C}\underset{\underset{\textstyle H}{\textstyle |}}{-}N.CHR''.CO_2H + H_2O$$

peptide
bond

Since a molecule of water is lost when two free amino-acid molecules undergo this reaction, only their **residues** are linked. A molecule consisting of two amino-acid residues joined by a peptide bond is called a **dipeptide**. Several residues linked in this way form an **oligopeptide**, while a chain of many amino-acid residues is termed a **polypeptide**. The **covalent backbone** of such a structure consists of α-carbon atoms linked by peptide bonds, the R groups sticking out from the chain. Each peptide chain has one free amino end (the **N-terminus**) and one free carboxyl end (the **C-terminus**); all the other α-amino and α-carboxyl groups present are involved in peptide bonds. For example:

N-terminus   C-terminus

$$H_2N.CHR'.CONH.CHR''.CONH.CHR'''.CONH.CHR''''.CO_2H$$

N-terminal   C-terminal
amino-acid residue   amino-acid residue

Proteins may contain one or more polypeptide chains, each one having a specific primary structure.

Although a two-dimensional representation of a polypeptide chain can give the impression that the backbone is linear, it should be understood that this is not so. The even distribution in three-dimensional space of the single covalent bonds about the carbon and nitrogen atoms in the backbone means that no two bonds emerging from the same atom will be diametrically opposite each other (see Chapter 2.2.2). Molecules may rotate freely about single covalent bonds, so an unlimited number of arrangements of a polypeptide chain in space are possible. However, some of these will be more stable than others, so are more likely to exist. **Secondary structure** refers to regular, repeating patterns formed by the backbone of at least part of a polypeptide chain and stabilised by hydrogen bonding.

Certain amino-acids cannot be accommodated into these regular arrangements, so the secondary structure is disrupted wherever they occur. Again the possibility of free rotation about a bond at each point of disruption suggests that a great number of different structures could result, but in fact each polypeptide chain is found to have a single, characteristic, three-dimensional structure. This is termed the **tertiary structure** and, once formed, it may be stabilised by bonding between amino-acids which find themselves in close proximity. It should be noted that amino-acids which are widely separated in the primary structure may be close together in space, because of the twists of the polypeptide chain.

Several identical or non-identical polypeptide chains may then be linked together to form the actual protein. The complete three-dimensional structure, including the interactions between the component polypeptide chains, is termed the **quaternary structure**.

2.3.2 Bonds involved in the Maintenance of Protein Structure

A **single covalent bond** is formed by the sharing of a pair of electrons between two atoms, each atom contributing one electron to the pair. By means of such a sharing arrangement, involving one or more covalent bonds, an atom can achieve an arrangement of electrons round its nucleus identical to that of an inert gas and thus become chemically stable. Two atoms of given identity, when linked by a single covalent bond, are located a characteristic distance apart, this distance being known as the **bond length**. If two atoms share *two* pairs of electrons between them, then a **double covalent bond** is formed. In this case the bond length is less than for the equivalent single bond. Molecules can rotate about single covalent bonds but not about double covalent bonds, which are more rigid.

The primary structure of a protein consists of amino-acid residues linked by covalent peptide bonds. Covalent disulphide bridges (—S—S—), linking cysteine residues, are often involved in the maintenance of tertiary structure. In a very few instances, disulphide bridges may also link the separate polypeptide components of a protein (see Chapters 3.1.4 and 5.1.2).

An alternative way by which an atom might achieve stability by obtaining the same electron structure as an inert gas is for it to gain or lose a number of electrons, i.e. to form an **ion**. Ions which are formed by loss of electrons from an atom will have a net positive charge and are called **cations** while those formed by the addition of electrons will have a negative charge and are termed **anions**. The magnitude of the charge will depend on the number of electrons transferred.

An **electrostatic interaction** occurs between each pair of ions in the same medium. The force (F) between two ions A and B in dilute solution is given by **Coulomb's law**:

$$F = \frac{Z_A.Z_B.e^2}{Dr^2}$$

where Z_A is the number of unit charges carried by ion A, Z_B the number carried by ion B, e is one unit of electronic charge, r is the distance between the two ions and D is the dielectric constant of the medium. Ions with like charge repel each other while those of opposite charge attract.

Thus the tertiary and quaternary structures of proteins could involve electrostatic linkages between amino-acids with side chains of opposite charge, between e.g. lysine and glutamic acid. In fact in an aqueous environment it is energetically more favourable for a charged group to form linkages with surrounding water molecules rather than with another charged group in the protein. However such linkages do occur in hydrophobic regions of proteins (see below) and so could play an important role in the stabilisation of the three-dimensional structure.

It often happens that covalent bond formation does not lead to an *equal* sharing of a pair of electrons between two atoms; the electrons may be associated with one of the components more than the other, producing a slight separation of charge between the atoms, called a **dipole effect**. The atom having the greater association with the shared pair of electrons will have a partial negative charge and can thus form weak electrostatic linkages, as can the other atom, which will have a partial positive charge.

The most important example of this phenomenon is the **hydrogen bond**. The oxygen atoms of $-OH$ or $-C=O$ groups have a slight negative charge, while the hydrogen atoms of $>NH$ or $-OH$ groups have a slight positive charge. Hence a weak electrostatic linkage can be formed between the oxygen atom in one group and the hydrogen atom in another, e.g. $-C=O \cdots H-O-$. The bond energy involved is small, but sufficient to add stability to a structure.

All the water molecules present in an aqueous medium link by means of hydrogen bonding to produce a huge three-dimensional network. Hydrogen bonds can be formed between groups in polypeptides and the surrounding water molecules, as well as between different components within polypeptide chains. Such bonds can help to stabilise the secondary, tertiary and quaternary structures of proteins.

Although the arrangement of electrons about an atom may, on average, be symmetrical, the constant fluctuations in electron distribution mean that the arrangement is likely to be asymmetrical at any given instant. Hence a dipole exists, however momentarily, and this induces a corresponding effect in all neighbouring atoms, causing them to attract each other. This is true of all atoms, even those of inert gases. However, when two atoms come into *very* close proximity, the repulsion between their respective clouds of surrounding electrons is greater than the induced attraction. There is an optimal distance between two non-bonding atoms, known as the **van der Waals contact distance**, when the forces of attraction and repulsion are equal. These forces are known as **London dispersion forces** and the weak linkages resulting from dipole effects are sometimes termed **van der Waals bonds**. These play an important part in governing which three-dimensional structure is taken up by a protein.

Non-polar, or **hydrophobic bonds** also have a considerable influence on protein structure. These bonds are not formed as a result of any direct interaction between atoms and may be best considered from the point of view of the complete protein/solvent mixture. The network of hydrogen bonds linking water molecules to each other confers great stability, so the most stable structure for a protein in aqueous solution will be that which gives the greatest possibility of hydrogen bonding between the protein molecule and the surrounding water molecules. Non-polar side chains of amino-acids cannot form hydrogen bonds, so the contact between these and the water molecules must be minimised. Two such side chains in close proximity will tend to come even closer, forcing out all water molecules from between them so that a single non-polar region is formed from the two originally present. Many non-polar side chains may be incorporated into a single non-polar zone, creating a hydrophobic **micro-environment** that is quite different from the micro-environments in other parts of the protein molecule.

In the case of metalloenzymes a further type of bond, the **co-ordinate bond**, needs to be mentioned. Like the covalent bond, this involves the sharing of a pair of electrons between two atoms, but in this case both electrons come originally from the same atom. A metal atom can accept pairs of electrons in this way from donor groups, or **ligands**, until it has the required number of electrons at a particular level. (A ligand is simply something which binds, the word having the same Latin root as the name of the group of enzymes called ligases.) The electrons which may be lost by a metal atom to form an ion are at a different level from those involved in co-ordinate bond formation, so the two processes are quite distinct.

The bonds involved in the maintenance of protein structure will be discussed further, in the light of experimental evidence, in Chapter 2.5.2. In general, the three-dimensional structure taken up by a protein will be that which is energetically most favourable, taking into account all possible interactions involving the types of bond discussed in the present chapter.

2.4 THE DETERMINATION OF PRIMARY STRUCTURE

2.4.1 The Isolation of Each Polypeptide Chain

The first step in the determination of protein structure is to find out how many different types of polypeptide chain are present in the intact protein. Since each polypeptide chain has an N-terminus and a C-terminus, this should be the same as finding out how many different N-terminal or C-terminal amino-acids are present. In view of the possibility that two otherwise dissimilar polypeptide chains might have, for example, the same amino-acid at the N-terminus, it is usually best to determine the number of different N-terminal amino-acids *and* the number of different C-terminal amino-acids; if these are not the same, the larger of the two numbers would be taken to indicate the number of different polypeptide chains present. Another reason for doing this is the possibility that a terminal amino-acid might be buried within the protein molecule and thus not be accessible to the reagents used.

The identity of **N-terminal amino-acids** can be determined by the use of **1–fluoro–2,4–dinitrobenzene (Sanger's reagent)** and of **dansyl chloride**. Both of these reagents form addition compounds with free amino groups. In a polypeptide chain, the only free α-amino group belongs to the N-terminal amino-acid. Hence, after treatment of a protein with one of these reagents, and subsequent complete hydrolysis to the constituent amino-acids (e.g. 6 M HCl at 105° for 24 h), the only α-N-substituted amino-acids present will be those originally at an N-terminus.

With **Sanger's reagent** the reaction sequence is:

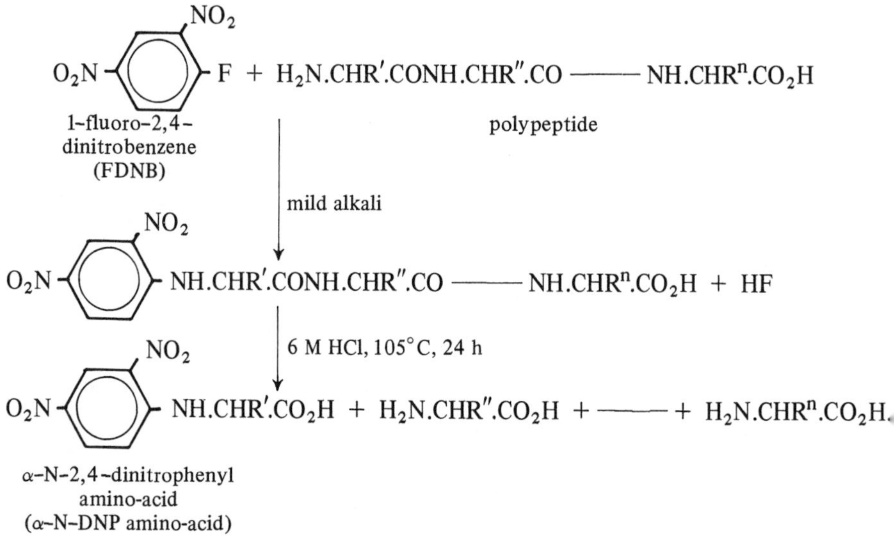

$$O_2N-\langle\!\!\!\langle\,\rangle\!\!\!\rangle-F \; + \; H_2N.CHR'.CONH.CHR''.CO \text{———} NH.CHR^n.CO_2H$$

1-fluoro-2,4-dinitrobenzene (FDNB)   polypeptide

| mild alkali

$$O_2N-\langle\!\!\!\langle\,\rangle\!\!\!\rangle-NH.CHR'.CONH.CHR''.CO \text{———} NH.CHR^n.CO_2H \; + \; HF$$

| 6 M HCl, 105°C, 24 h

$$O_2N-\langle\!\!\!\langle\,\rangle\!\!\!\rangle-NH.CHR'.CO_2H \; + \; H_2N.CHR''.CO_2H \; + \text{———} + \; H_2N.CHR^n.CO_2H.$$

α-N-2,4-dinitrophenyl amino-acid (α-N-DNP amino-acid)

The α-N-DNP amino-acids can be identified by paper or thin layer chromatography, or by high performance liquid chromatography (HPLC), using α-N-DNP amino-acids of known identity as markers. These compounds are yellow, so there is no need to use staining reagents. Amino groups present in the side chains of amino-acids (e.g. of lysine) will also react with Sanger's reagent, but the products can be distinguished chromatographically from α-N-DNP amino-acids, so no confusion will result.

The procedure with **dansyl chloride** is very similar to the above:

$$+ \; H_2N.CHR'.CONH.CHR''.CO \text{———} NH.CHR^n.CO_2H$$

polypeptide

dansyl chloride

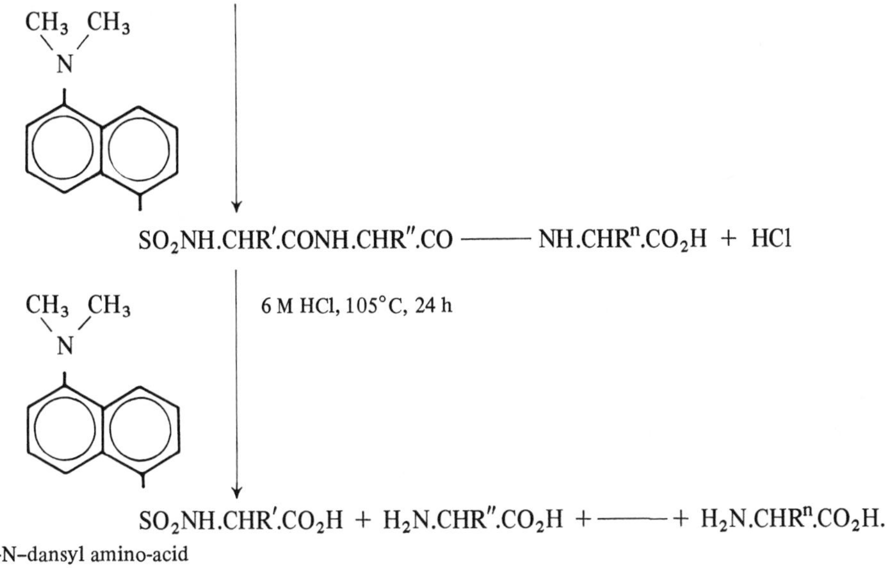

$$CH_3 \quad CH_3$$
$$\diagdown \quad \diagup$$
$$N$$

$$SO_2NH.CHR'.CONH.CHR''.CO \text{———} NH.CHR^n.CO_2H \ + \ HCl$$

$$CH_3 \quad CH_3$$
$$\diagdown \quad \diagup$$
$$N$$

6 M HCl, 105°C, 24 h

$$SO_2NH.CHR'.CO_2H \ + \ H_2N.CHR''.CO_2H \ + \text{———} + \ H_2N.CHR^n.CO_2H.$$

α–N–dansyl amino-acid

Dansyl amino-acids are highly fluorescent and hence can be identified in very small quantities, e.g. by HPLC.

The **C-terminal amino-acids** may similarly be identified by treating the protein with a reagent which attacks free carboxyl groups, e.g. with the reducing agent **sodium borohydride**. This requires preliminary esterification of the carboxyl groups and protection of the free amino groups by acetylation, neither of these processes being shown in the following simplified scheme:

$$H_2N.CHR'.CONH.CHR''.CO \text{———} NH.CHR^n.CO_2H$$
$$\downarrow NaBH_4$$
$$H_2N.CHR'.CONH.CHR''.CO \text{———} NH.CHR^n.CH_2OH$$
$$\downarrow 6 \text{ M HCl, } 105°C, 24 \text{ h}$$
$$H_2N.CHR'.CO_2H \ + \ H_2N.CHR''.CO_2H \ + \text{———} + \ H_2N.CHR^n.CH_2OH$$

α–amino alcohol

Only C-terminal amino acids will be converted to α-amino alcohols. Side chain carboxyl groups (e.g. in glutamic and aspartic acids) will also be reduced, with the subsequent production of amino alcohols, but these will not be α-amino alcohols, so can be distinguished by chromatography.

By means of these techniques, the N-terminal and C-terminal amino-acids in a protein can be identified, enabling the number of different polypeptide chains present to be deduced. The linkages between the various polypeptide chains may then be broken in a variety of ways. **Disulphide bridges** may be cleaved by treatment with **performic acid**, each cystine unit being oxidised to two cysteic acid units without the breakage of any peptide bonds.

$$
\underset{\substack{\text{cystine residue}}}{\overset{\mid}{\underset{\mid}{C}}H.CH_2-S-S-CH_2.\overset{\mid}{\underset{\mid}{C}}H} \xrightarrow{\substack{\text{performic} \\ \text{acid}}} \underset{\substack{\text{cysteic acid} \\ \text{residue}}}{\overset{\mid}{\underset{\mid}{C}}H.CH_2.SO_3H} + \underset{\substack{\text{cysteic acid} \\ \text{residue}}}{HO_3S.CH_2.\overset{\mid}{\underset{\mid}{C}}H}
$$

Performic acid also oxidises methionine and tryptophan residues, if these are present, so an alternative approach is to cleave the disulphide bridge by reduction, e.g. with mercaptoethanol, and then alkylate the sulphydryl groups produced to prevent them reforming $-S-S-$ bonds. **Non-covalent bonds** may be broken at **extremes of pH**, at **high salt concentrations**, or in presence of reagents such as **urea** or **guanidine hydrochloride**.

Each of the polypeptide chains known to be present can then be separated from the others by chromatographic or electrophoretic techniques, as described in Chapter 16.2.2. Hence a pure specimen of each polypeptide chain may be obtained.

A great many enzymes consist of a number of identical polypeptide chains linked only by non-covalent bonds (see Chapter 5.2): this is indicated by the finding of only a single N-terminal and a single C-terminal amino-acid, by the failure to separate any polypeptide chains from any others after the breaking of non-covalent linkages, *and* by demonstrating that a several-fold decrease in molecular weight occurs when these linkages are broken (see Chapter 16.3).

2.4.2 Determination of the Amino-Acid Composition of each Polypeptide Chain

The molecular weight of the polypeptide should first be determined, using such techniques as gel filtration or ultracentrifugation (see Chapter 16.3). The polypeptide is then completely hydrolysed to its component amino-acids and the concentration of each determined.

A common procedure for the **quantitative determination of amino-acids** is **ion exchange chromatography**. The sample is applied to a cation exchange column, e.g. sulphonated $(-SO_2O^-)$ polystyrene, and buffers of increasing pH and salt concentration are pumped through. In general, amino-acids with acidic or neutral polar side chains are eluted from the column before those with basic or non-polar side chains, largely according to their relative attraction for the charges on the ion exchange resin, conditions being chosen so that each amino-acid is eluted separately. Traditionally, glass columns have been used, but recent developments in HPLC technology has led to the use of stainless steel columns. The column eluate is mixed with a reagent such as **ninhydrin**, which reacts with most amino-acids to give a blue–purple colour, the absorbance being measured at 570 nm.

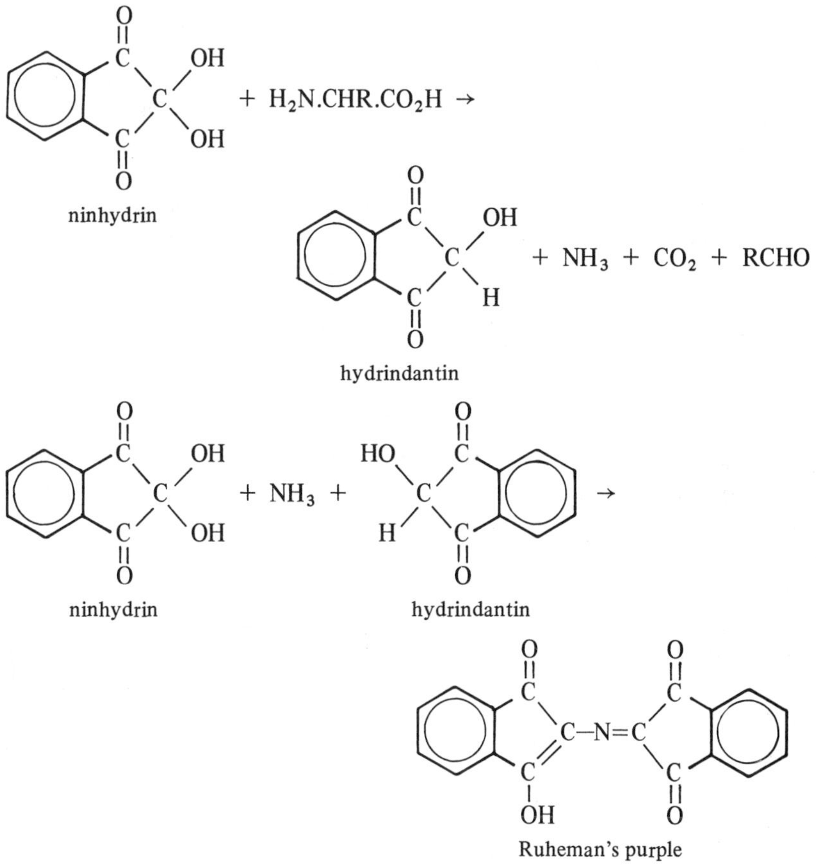

ninhydrin

$+ H_2N.CHR.CO_2H \rightarrow$

hydrindantin

$+ NH_3 + CO_2 + RCHO$

ninhydrin $+ NH_3 +$ hydrindantin $\rightarrow$

Ruheman's purple

The imino acid, proline, reacts differently, giving a yellow colour, the absorbance usually being measured at 440 nm. Hence, if the colour produced is monitored continuously at both 570 nm and 440 nm, and if the instrument has been pre-calibrated with amino-acid standards, it is possible to determine the concentration of each amino-acid present. As an alternative to ninhydrin, more sensitive fluorimetric reagents, such as o–phthalaldehyde (OPA), may be employed.

Reversed-phase HPLC, where the stationary phase is a C_8 or C_{18} hydrocarbon, may be used to analyse pre-derivatised (e.g. DNP or OPA) amino-acids in less than 30 minutes, with mixtures of aqueous and organic solvents as the moving phase.

Before the results of amino-acid analysis are interpreted, the type of hydrolysis used needs to be taken into consideration. Acid hydrolysis (6 M HCl, 105°C, 24 h) leads to complete breakdown of the polypeptide chain to amino acids, but some tryptophan is lost in the process. Also, the amides glutamine and asparagine are hydrolysed to the corresponding acids, liberating ammonia.

Measurement of the ammonia produced gives the total amount of amide initially present.

Hydrolysis of the amides also occurs if alkali is used instead of acid. Such alkaline hydrolysis, e.g. with 5 M NaOH, again leads to total breakdown of the polypeptide. Under these conditions there is complete recovery of tryptophan, but only partial recovery of several other amino acids, including cysteine, cystine, serine and threonine. Thus the results of both acid and alkaline hydrolysis are required to determine the total concentration of all amino acids present, and even then only the combined amide content can be assessed. With this proviso, these results, and those of the molecular weight determinations, can be used to give a reasonable estimate of the number of molecules of each amino acid in the polypeptide.

2.4.3 Determination of the Amino-Acid Sequence of each Polypeptide Chain

The **N-terminal amino-acid** of a polypeptide may be identified by the use of **Sanger's reagent** or **dansyl chloride**, as described in Chapter 2.4.1. However an even more valuable reagent is that of **Edman**, since it allows extra information to be obtained.

Edman's reagent (phenylisothiocyanate) forms an addition compound with a free amino group under alkaline conditions, so will attach itself to the α–amino group of the N-terminal amino-acid of a polypeptide chain, exactly like the other reagents mentioned. Its special property is that if the conditions are then made mildly acidic, the addition compound formed between the reagent and the N-terminal amino-acid will become detached from the polypeptide chain without any other peptide bond being broken.

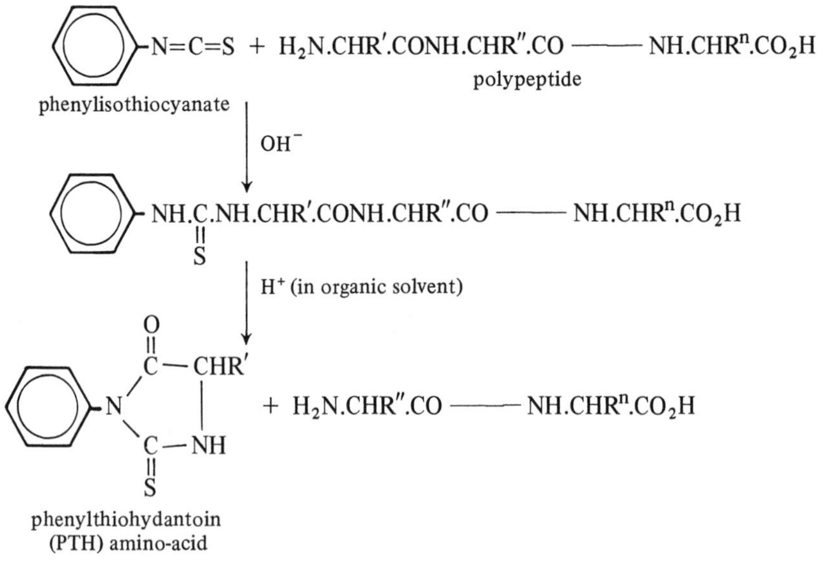

$$\langle\bigcirc\rangle\text{-N=C=S} + H_2N.CHR'.CONH.CHR''.CO \text{------} NH.CHR^n.CO_2H$$

phenylisothiocyanate
$\qquad\qquad$ polypeptide

$\downarrow$ OH$^-$

$$\langle\bigcirc\rangle\text{-NH.C.NH.CHR'.CONH.CHR''.CO} \text{------} NH.CHR^n.CO_2H$$
$$\underset{\text{S}}{\overset{||}{}}$$

$\downarrow$ H$^+$ (in organic solvent)

$$\langle\bigcirc\rangle\text{-N}\underset{\underset{||}{\underset{S}{C}}-NH}{\overset{\overset{O}{\overset{||}{C}}-CHR'}{}} + H_2N.CHR''.CO \text{------} NH.CHR^n.CO_2H$$

phenylthiohydantoin
(PTH) amino-acid

The PTH–amino-acid, and thus the original N-terminus, can be identified by chromatography, while the rest of the polypeptide chain remains intact to be investigated further. The procedure can be repeated to reveal the new N-terminal amino-acid (that with side chain R''),and so on. Protein sequencers involving separation of PTH–amino-acids by reversed-phase HPLC are now commercially available (e.g. the Beckman System 890M).

The **C-terminal amino-acid** may be identified by the use of a **reducing agent** (Chapter 2.4.1), but an alternative approach employs **carboxypeptidase enzymes**, which cleave peptide bonds sequentially, starting from the C-terminus. The order in which particular free amino-acids appear in solution after treatment gives the sequence from this end.

Thus, in theory, it is possible to start at either the N-terminus or the C-terminus of a polypeptide chain and determine the entire amino-acid sequence. However, in practice, the size and complexity of such chains increase the problems of analysis. For this reason, long polypeptide chains are usually split into smaller, more manageable, units by the use of **specific reagents** before sequence analysis is attempted. These reagents include **cyanogen bromide** and a variety of **proteolytic enzymes** (i.e. enzymes which cleave peptide bonds). Consider the following polypeptide chain:

$$H_2N \text{———} NH.CHR'.\overset{\overset{\displaystyle H}{|}}{\underset{\underset{\displaystyle O}{||}}{C}}\text{–}N.CHR''.CO \text{———} CO_2H.$$

The peptide bond shown, which represents any peptide bond in the molecule, may be broken by a specific reagent if R' or R'' has a specific identity. For example, it is hydrolysed by the enzyme **trypsin**, at pH 7–9, where R' is a **lysine** or **arginine** side chain, or by the enzyme **chymotrypsin**, again at pH 7–9, where R' is a **phenylalanine, tryptophan** or **tyrosine** side chain. Other proteolytic enzymes have a less clearly defined specificity (Chapter 5.1.3). **Cyanogen bromide (CNBr)** cleaves the bond where R' is the side chain of methionine, by the following reaction:

$$H_2N \text{———} NH.\overset{\overset{\displaystyle CH_2.CH_2.SCH_3}{|}}{CH}.\overset{\underset{\underset{\displaystyle O}{||}}{C}}{}\text{–}NH.CHR''.CO \text{———} CO_2H$$

$$\downarrow CNBr$$

$$H_2N \text{———} NH.\overset{\overset{\displaystyle CH_2CH_2}{|}}{CH}.\overset{\underset{\underset{\displaystyle O}{||}}{C}}{}\text{–}O + H_2N.CHR''.CO \text{———} CO_2H + CH_3SCN.$$

Thus, action by any one of these reagents on a polypeptide chain produces a number of peptide fragments, which may be separated from each other by chromatography or electrophoresis techniques. Then, since each fragment is relatively small, it is usually possible to determine its complete amino-acid sequence by use of the methods discussed above.

The next task is to find the order in which the peptide fragments join together to form the complete polypeptide. Partial hydrolysis is again performed on a sample of intact polypeptide, this time using a different specific reagent; the resulting peptide fragments are separated as before and the amino acid sequence of each is determined. **Overlapping sequences** between the first and second sets of peptide fragments enable the complete primary structure to be deduced, as in the following example.

A peptide consisting of 11 amino-acid residues is known to have alanine at its N-terminus. Hydrolysis by trypsin gives the following three peptide fragments (each depicted conventionally with the N-terminus to the left):

> Pro–Trp–Gly–Arg   Ala–Glu–Phe–Asp–Lys   Thr–Ser

Hydrolysis by chymotrypsin also gives three fragments:

> Gly–Arg–Thr–Ser   Ala–Glu–Phe   Asp–Lys–Pro–Trp

The complete sequence may be deduced by starting with the known N-terminus and looking for overlaps between the fragments, as follows:

Ala–Glu–Phe	chymotrypsin fragment
Ala–Glu–Phe–Asp–Lys	trypsin fragment
Asp–Lys–Pro–Trp	chymotrypsin fragment
Pro–Trp–Gly–Arg	trypsin fragment
Gly–Arg–Thr–Ser	chymotrypsin fragment
Thr–Ser	trypsin fragment

Ala–Glu–Phe–Asp–Lys–Pro–Trp–Gly–Arg–Thr–Ser	complete sequence

The amides, glutamine and asparagine, are easily hydrolysed to their respective acids and may thus be wrongly identified in the determination of amino-acid sequence. The result may be checked by subjecting intact polypeptide to partial hydrolysis by proteolytic enzyme under the mildest possible conditions, in order to minimise hydrolysis of any amide present, and then quickly investigating the peptide fragments for the presence of amide.

This may be done by electrophoresis, or by performing complete hydrolysis on a fragment and determining the ammonia produced. For example, the fragment Ala–Glun–Phe would not leave the origin on electrophoresis at neutral pH, whereas Ala–Glu–Phe would move towards the anode. The first of these fragments would yield one molecule of ammonia on complete hydrolysis, but no ammonia would be produced by hydrolysis of the other.

Once the complete primary structure of a polypeptide has been determined, the amino-acid residues may be numbered, always starting with the N-terminus as residue number 1 and working sequentially from this.

2.4.4 Determination of the Positions of Disulphide Bridges

Since the amino-acid sequence of a polypeptide chain is most conveniently determined if the peptide bonds of the backbone are the only covalent bonds present which link amino-acid residues, the procedures discussed above usually commence with the splitting of all inter- and intra-chain disulphide bridges (see Chapter 2.4.1); thus they can give no indication as to which particular cysteine units are linked by each disulphide bridge. In order to elucidate this, it is necessary to start again with a sample of intact protein.

Partial hydrolysis is performed to break some of the peptide bonds without disturbing any of the disulphide bridges. However under certain hydrolysis conditions, disulphide bridges may be cleaved and reformed, not necessarily reconnecting the original partners. This is minimised by using $5 M H_2SO_4$ for partial hydrolysis or by using the enzyme pepsin at pH 2: pepsin breaks the peptide bonds between a wide range of amino-acids, but particularly those between two hydrophobic residues.

The peptide fragments formed may be separated from each other as before, and each is then treated with performic acid. Those peptides containing a disulphide bridge will break into two smaller fragments as the —S—S— bond is oxidised, enabling the fragments to be separated and the sequence of each determined. Thus the sequence of amino-acids around each end of a particular disulphide bridge is made known and, by reference to the previously worked out primary structures, the position of the bridges in the intact protein can be deduced.

2.4.5 Some Results of Experimental Investigation of Primary Structure

The general approach described above was first used by Sanger, who in 1953 elucidated the complete primary structure of **insulin**, an achievement which was rewarded with the Nobel Prize. Insulin was found to consist of two polypeptide chains: an A chain, of 21 amino-acid residues, and a B chain, of 30 residues; the chains are linked by two disulphide bridges and there is also an intrachain disulphide bridge in the A chain (see Chapter 3.1.4).

The first enzyme to have its complete amino-acid sequence determined was bovine pancreatic **ribonuclease A**, the result of work by Smyth, Stein and Moore (1963) and others. This enzyme consists of a single polypeptide chain of 124 amino-acid residues, with four intrachain disulphide bridges being present.

In general, work on a variety of **globular proteins** have revealed that these usually incorporate all 20 amino-acids without there being any recurring features in the primary structure. The amino-acid sequence is absolutely specific, so that an error of synthesis resulting in one amino-acid residue being out of place can affect the functioning of the protein.

Let us consider the example of **haemoglobin**, an iron-containing protein which occurs in erythrocytes (red blood cells) and acts as a carrier for oxygen and carbon dioxide. Haemoglobin consists of four polypeptide chains: two α chains (each of 141 amino-acid residues) and two β chains (each of 146 residues). Well over one hundred different abnormal structures, resulting from genetic mutations, have been described in humans: some of these mutations are harmless, the abnormal haemoglobin molecule functioning as well as the normal one, but some affect the patient very severely indeed. In **sickle-cell anaemia**, where the erythrocytes are sickle-shaped and are broken down more easily than normal, the abnormality is at position 6 (from the N-terminus) of the β chain of haemoglobin, where **valine** is present instead of **glutamate**. Normal haemoglobin molecules obtained from different species also show some differences in primary structure, these being fewer the more closely related the species. Haemoglobin has been investigated in more detail than most proteins because it is readily available in large amounts, but the same general conclusions have been found with other globular proteins.

Fibrous proteins, on the other hand, often contain only three or four different amino-acid residues, and recurring sequences of amino-acids are frequently found. For example, **fibroin**, the protein in silk, consists of only glycine, serine and alanine residues, the glycine residues occurring alternately throughout the molecule. Hence it has not yet proved possible to specify the complete primary structure for a fibrous protein.

2.5 THE DETERMINATION OF PROTEIN STRUCTURE BY X-RAY CRYSTALLOGRAPHY

2.5.1 The Principles of X-Ray Crystallography

Crystals, including those of globular proteins, consist of repetitions of a basic structural component called a **unit cell**, which may be a single molecule or a symmetrical arrangement of several molecules. Thus each atom in a crystal must lie in a specific position with regard to all other atoms in the crystal, enabling the structure to be determined by x-ray diffraction analysis. This consists of directing a beam of x-rays of a single wavelength at a crystal and studying the characteristics of the emerging rays. Most rays pass straight through the crystal without being affected, but those which come into contact with an atom in the crystal are scattered by the clouds of electrons surrounding it. More precisely, these electrons act as secondary sources of x-rays, which then radiate out from the atom in all directions. The intensity of the x-rays leaving an atom of high electron density, such as a heavy metal, is much greater than for those leaving an atom of low electron density, such as hydrogen. Thus, areas of high electron density can be said to scatter x-rays more strongly than areas of low electron density, but it should be realised that in each case the radiation emerges from the scattering centre with spherical symmetry.

X-rays, like other forms of electromagnetic radiation, are best regarded as waves of characteristic **length** and **amplitude** (Fig. 2.3); the **intensity** of a ray is proportional to the square of its amplitude. If two rays of identical wavelength are directed along a common path so that they are exactly **in phase**, i.e. the crests and troughs of the waves correspond exactly, then they will combine to give a ray of the same wavelength and phase, but greater amplitude (Fig. 2.3(a)). The amplitude, and hence intensity, obtained under these conditions will be the maximum that can be obtained by combination of these two rays. If the two rays are one quarter of a cycle out of phase (Fig. 2.3(b)), the intensity of the combined ray will be about one quarter of the maximum possible value, and the phase of the combined ray will be a combination of the phases of its component rays. If the two rays are exactly half a cycle of phase (Fig. 2.3(c)), the waves will cancel out and the intensity of the combined ray will be zero.

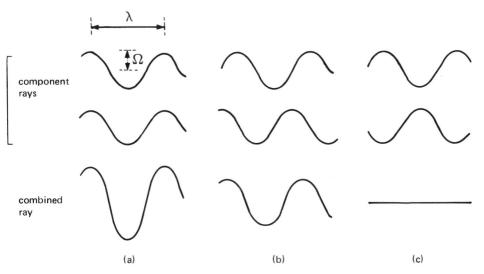

Fig. 2.3   The combination of rays of identical wavelength (λ) and amplitude (Ω) when directed along a common path, where component rays are (a) exactly in phase, (b) one quarter of a cycle out of phase, and (c) half a cycle out of phase.

The scattered x-rays emerging from a crystal can combine in this way: rays emerging at certain angles to the incident ray will combine to give rays of maximum intensity, while those emerging at other angles will cancel each other out. The results can be observed by placing a photographic plate behind the crystal to register the impact of emerging rays. In general, development of the plate will show a spot at the centre, due to the undeflected x-rays, which of course will all be in phase; this is surrounded by a pattern of other spots, corresponding to the angles where emerging rays combine to give intensity maxima. The overall effect is known as a **diffraction pattern**.

Before discussing x-ray crystallography in more detail, let us consider why we cannot observe the atoms in a molecule by the use of optical or electron microscopy. Vision consists of two processes: beams of light (another form of electromagnetic radiation) which strike an object are scattered by the atoms exactly as discussed above, and these scattered rays are brought back together (**focussed**) by the lens in the eye to produce an image of the object on the retina. A magnified image may be produced by the use of further lenses (**optical microscopy**), enabling features to be clearly distinguished (**resolved**) which are too close to be seen separately by the unaided eye. The limit of resolution in microscopy depends on the wavelength of the type of electromagnetic radiation used and the focussing properties of the instrument. With optical microscopy, the limit of resolution is about half the wavelength of the light used. Hence individual atoms, which are separated in a molecule by distances in the order of 1-2 Å (1 Å = 0.1 nm), cannot be resolved by an optical microscope, since the wavelength of visible light is in excess of 4000 Å. **Electron microscopes** give much greater resolving power than optical microscopes, but despite the very low wavelengths of electron beams, individual atoms still cannot be visualised because of the generally poor performance of the electromagnetic lenses used in electron microscopy.

X-rays similarly have wavelengths much smaller than those of light rays; in fact they are of the same order of magnitude as inter-atomic distances. However, no procedure has yet been devised for focussing x-rays, so no image can be produced. Nevertheless, the detailed structure of a crystal scattering x-rays can be deduced from the diffraction patterns obtained.

Each unit cell in a crystal may contain many atoms in a complex arrangement, but let us for the moment consider it simply as a region of high electron density which can act as a scattering centre for x-rays. Thus the crystal consists of a regular arrangement of major scattering centres, each corresponding to a unit cell, as shown in Fig. 2.4.

First of all let us look at the plane of scattering centres containing A, B and C, which is inclined at an angle θ to the incident beam of x-rays. Some rays will be scattered by the electron-dense regions in the plane, while most will pass straight through. Each scattered ray has the same wavelength and phase as the incident ray, so in respect of these properties can be regarded as simply being deflected. Rays will be scattered in all directions, so some will emerge at an angle ϕ to the plane. Those leaving A and B at the same angle to the incident beam will reach a given point having travelled exactly the same distance through space (so that distance QUA = distance RBT) only when $\phi = \theta$. This is known as the **reflection condition** since the phenomenon of reflection at a planar surface is also characterised by these angles being equal. Therefore, all rays **reflected** by the scattering centres on the plane ABC, i.e. all those emerging at an angle such that $\phi = \theta$, will be exactly in phase and will combine to give a ray of maximum intensity.

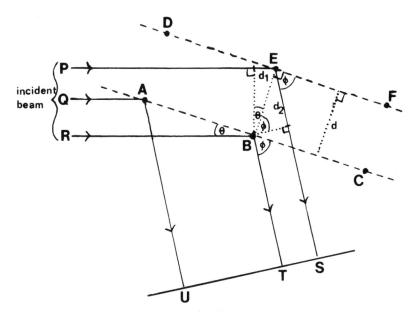

Fig. 2.4 An incident beam of x-rays striking a crystal at an angle θ to planes of scattering centres ABC and DEF. Rays are scattered with spherical symmetry, only those emerging at an angle ϕ to the planes of scattering centres being shown (see text for discussion).

Now let us consider a second plane of scattering centres, containing D, E and F, which is parallel to the plane ABC and separated from it by a distance d. Rays leaving B and E in the same direction can never reach a given point having travelled the same distance through space, since distance PES must be greater than distance RBT. However if the difference in distance $(d_1 + d_2 = d.\sin\theta + d.\sin\phi)$ is exactly a whole number of wavelengths, the emerging waves will still be exactly in phase. Hence, waves which emerge at the same angle to the incident beam from scattering centres on different but parallel planes will combine to give a ray of maximum intensity if $n\lambda = d.\sin\theta + d.\sin\phi$, where n is a whole number, λ is the wavelength of the rays, and the other terms are as defined in Fig. 2.4.

So, to summarise, all the scattering centres in a single plane will combine to give a diffracted ray of maximum intensity at the angle where the reflection condition is met, while centres in different planes will combine to give rays of maximum intensity at angles where $n\lambda = d.\sin\theta + d.\sin\phi$. If these two conditions are put together, rays emerging from all the scattering centres on any number of parallel planes a distance d apart will combine to give an intensity maximum where $n\lambda = 2d.\sin\theta$. This was first stated by the Braggs (father and son) in 1913 and is known as the **Bragg condition**.

Thus it can be seen that regular repeating units are essential for the establishment of clear diffraction patterns, since patterns from different scattering centres may reinforce each other under these conditions. Crystals are rotated in a beam of x-rays, allowing time in each position for the investigation of the diffraction pattern, until the pattern obtained indicates that planes of scattering centres are inclined at a suitable angle to the incident beam for reinforcement to take place. A clear diffraction pattern may be obtained without rotating the specimen if this is not a single crystal but is composed of separate regions of repeating units set at random angles to each other, and thus to the incident beam; this is often the case where the specimen is a powder of fine crystals or a natural fibre.

In general, the greater the repeating distances within a specimen, the closer are the intensity maxima on the photographic plate. If clear diffraction patterns can be obtained with a crystal in three different orientations, then the dimensions of the unit cell and the arrangement of unit cells within the crystal can be deduced from those maxima nearest the centre, which correspond to the largest repeat distances.

If we now turn our attention to the structure of the molecule or molecules making up the unit cell, we must immediately realise that not all the atoms in a complex molecule can lie on the same plane. Hence, although we have hitherto considered a unit cell to be a single scattering centre lying on a specific plane, in fact it may consist of a large number of component scattering centres (atoms), some of which lie at various distances in front of the plane and some at various distances behind the plane. This inevitably affects the diffraction patterns obtained.

If the intensity (and hence amplitude) and the phase is known for each x-ray causing a spot in the diffraction pattern, then a three-dimensional contour map showing the distribution of electron density within the unit cell can be drawn up using a mathematical procedure called **Fourier synthesis**. From this, the structure of the molecule can be deduced. The intensity of an x-ray can be determined from a photograph of a diffraction pattern or measured directly using a geiger counter; however there is no direct way of determining the phase. This has been called the **phase problem**.

The phase problem can be overcome in either of two ways. A **model** may be built of a possible molecular structure and the theoretical diffraction patterns this would give are compared to those actually obtained. This may be useful in explaining certain repeating features, but otherwise the number of possible structures of a complex molecule is too immense to enable this method to be very successful if employed alone. The alternative method is that of **isomorphous replacement**, introduced by Perutz in 1954. A heavy metal atom, such as mercury or uranium, is attached to a specific site of each molecule in the crystal without altering the three-dimensional structure of the molecule. The heavy metal atoms, being regions of very high electron density, will cause appreciable changes to the amplitude and phase of the rays producing a diffraction pattern. If the intensities of the spots are compared to those of the spots in the original diffraction pattern, it is possible to deduce the location of the substituted atoms within the unit cells. The contribution of rays from the heavy metal atoms to each spot in the diffraction pattern may then be calculated, in terms of both phase and amplitude. This enables two possible solutions of the phase problem to be obtained for each spot, one where the phase of the original ray is in advance of that of the ray from the substituted atom, and one where it is an equal distance behind. Which of the alternative solutions is correct may be determined by substituting with a heavy metal atom in a different place, or possibly several different places. Thus it is possible to deduce the complete structure of a molecule.

Low resolution analysis (to about 5 Å), showing the main features of the molecular structure but not the fine detail, may be performed using only the spots near the centre of the diffraction pattern. For **high resolution analysis**, showing the complete structure of the molecule, all the spots must be used. Apart from the extra labour involved, which usually necessitates the use of computers, high resolution analysis is hindered by the fact that the outermost spots in a diffraction pattern are of lower intensity than those nearer the centre. Also, isomorphous replacement almost inevitably introduces some changes, however slight, in the three-dimensional structure and this becomes more significant the higher the resolution attempted. Finally, hydrogen atoms are extremely weak scatterers of x-rays, so are very difficult to pinpoint by these techniques.

It has often been found advantageous to use the **model-building** and **isomorphous replacement** techniques to complement each other. Low resolution studies indicate the general shape of the molecule and, from this data, models can be built to help elucidate the fine structure.

X-rays are commonly obtained by accelerating **electrons** released from an **incandescent tungsten filament** against a **copper** target. This produces rays of approximate wavelength 1.5 Å.

As regards to current developments, x-rays selected from the electromagnetic radiation emitted by highly expensive devices called **synchrotrons** or **electron storage rings** are of higher intensity than those from a conventional source. This is advantageous for the structural determination of proteins of high molecular weight and of those proteins whose crystalline structure is unstable over the relatively long periods required for exposure to weak radiation. Another possibility is the use of **neutron beams**, which are scattered by atomic nuclei rather than electrons. Thus they may be employed for high resolution analysis, since they are scattered strongly by hydrogen atoms. Also, they cause very little radiation damage to macromolecules, enabling irradiation to be carried out for far longer periods than is possible with x-rays.

2.5.2   Some Results of X-ray Crystallography
In 1939, Pauling, Corey and their co-workers began a systematic investigation of the three-dimensional structures of amino-acids, dipeptides and other molecules

to provide data with a view to the eventual elucidation of protein structure. X-ray diffraction analysis soon showed that the C—N bond length on a peptide bond was shorter than would be expected for a single covalent bond. Therefore, some degree of double bond character must be present, the actual structure being between the two extremes shown below:

$$
\begin{array}{cc}
\overset{\displaystyle O}{\underset{\displaystyle \|}{}} & \overset{\displaystyle O^-}{\underset{\displaystyle |}{}} \\
-\overset{|}{\underset{|}{C}}-\overset{}{\underset{|}{N}}- & \overset{+}{\underset{|}{C}}=\overset{+}{\underset{|}{N}}- \\
H & H
\end{array}
$$

$$-\overset{O}{\underset{H}{\overset{\|}{C}-\overset{|}{N}}}- \longleftrightarrow \overset{O^-}{\underset{H}{\overset{|}{C}=\overset{+}{\overset{|}{N}}}}-.$$

A consequence of this partial double bond character is that rotation about the bond is restricted and all of the atoms involved lie in the same plane. Two isomeric arrangements are possible: the *trans* form (as shown above), with the oxygen and hydrogen atoms diametrically opposed, and the *cis* form, with these atoms adjacent. In fact, only the *trans* isomer is found, a more detailed representation of this being given in Fig. 2.5.

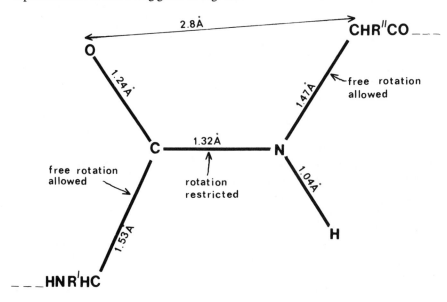

Fig. 2.5 The dimensions of the peptide bond.

The most significant factor accounting for the stability of the *trans* form is the spacing between the α-carbon atom and the oxygen atom (2.8 Å), which is only marginally less than the van der Waals contact distance between these atoms (3.4 Å), so repulsion is slight (see Chapter 2.3.2). In the unstable *cis* form the two α-carbon atoms would be adjacent to each other, separated by a distance (2.8 Å) much less than the van der Waals contact distance between two carbon atoms (4.0 Å), so repulsive forces would be great.

Pauling, Corey and their colleagues also noted that, in crystals, there is a high degree of hydrogen bonding between oxygen atoms in one peptide bond and nitrogen atoms in another. The distance between such atoms is often about 2.9 Å, much less than the van der Waals contact distance between non-bonded oxygen and nitrogen atoms, thus indicating the presence of the hydrogen bond. Furthermore, the N—H⋯O linkage is usually approximately linear.

On the basis of these findings they suggested various theoretical types of secondary structure which might be found in proteins. In particular, both right and left handed α-helices seemed consistent with the available data, but the arrangement of London dispersion forces was more favourable in the former structure.

Astbury had already used x-ray scattering to demonstrate regular features in the structures of several fibrous proteins, and in some cases these were found to be consistent with the postulated right handed α-helix. The main features of the x-ray diffraction pattern of α-keratin (Fig. 2.6(a)) show a periodicity of 5.2 Å along the axis of the fibres, which is the distance between each turn of the α-helix (Fig. 2.6(b)), and a periodicity of 9.7 Å at right angles to this, presumably the distance between adjacent α-helices.

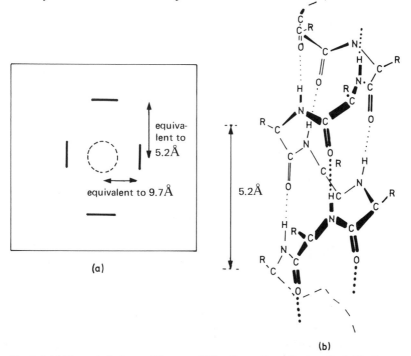

(a)

(b)

Fig. 2.6 (a) The main features of the x-ray diffraction pattern of α-keratin, indicating an α-helix structure, and (b) a polypeptide chain in the form of a right handed α-helix (n.b. the α-hydrogen atom has been omitted from the diagram to minimise congestion).

An α-helix contains approximately 3.6 amino acid residues per turn. This results in each peptide oxygen and nitrogen atom being in a suitable position to form hydrogen bonds with the corresponding atoms in the next turn of the helix (Fig. 2.6(b)); these hydrogen bonds are all approximately parallel to the axis of the helix.

Other types of secondary structure have also been demonstrated. If α-keratin is stretched under humid conditions it is converted to β-keratin, which has characteristic periodicities of 3.3 Å along the axis of the fibre and of 4.7 Å and 9.7 Å perpendicular to this, evidence of a more extended form of secondary structure called a β-**pleated sheet**. Again, hydrogen bonds can be formed between oxygen and nitrogen atoms in different peptide bonds, but in this instance they are perpendicular to the axis of the fibre. The structure is unstable in keratin but not in other fibrous proteins such as silk fibroin. Keratin is said to form **parallel β-pleated sheets**, since the N-termini of adjacent polypeptide chains lie in the same direction, whereas silk fibroin forms **anti-parallel β-pleated sheets**, the N-termini of adjacent chains being in opposite directions (Fig. 2.7). Anti-parallel β-pleated sheets may also be formed by the doubling-back of a single polypeptide chain.

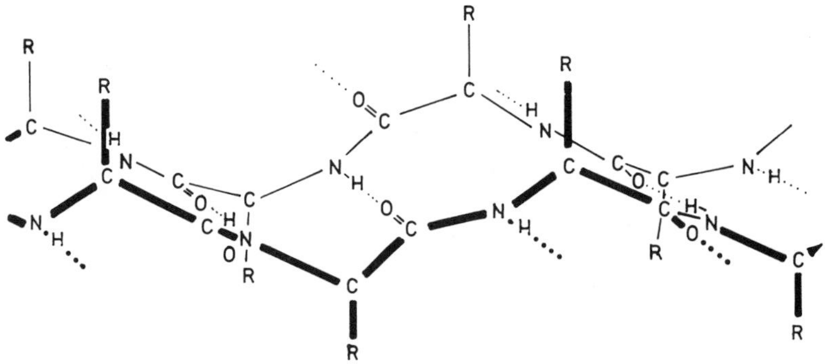

Fig. 2.7 A section of anti-parallel β-pleated sheet (n.b. as with Fig. 2.6, the α-hydrogen atom has been omitted from the diagram).

The stability of these α-helix and β-pleated sheet structures depends on the nature of the amino-acid side chains present, large or charged ones tending to be disruptive. Hence fibrous proteins, which usually consist only of amino-acids with small and uncharged side chains, have well-developed secondary structures. Proline, because of the restricted rotation resulting from its ring structure, is another amino-acid which cannot form part of an α-helix or β-pleated sheet, but it is incorporated into the unique triple-helix structure of the fibrous protein **collagen**.

The first globular protein to have its three-dimensional structure elucidated by x-ray crystallography was sperm-whale **myoglobin**. This close relative of haemoglobin is a single polypeptide chain of 153 amino-acid residues. Despite its relatively small size, over 10,000 diffraction spots had to be accurately analysed to give a resolution to 2 Å. Shortly afterwards, the structural analysis of **haemoglobin** itself was completed: each of the four component polypeptide chains was found to have a tertiary structure almost identical to that of myoglobin. Kendrew, for his work on myoglobin structure, and Perutz, for his studies on haemoglobin, were rewarded with the Nobel Prize in 1962. The structure of **lysozyme**, an enzyme from egg white consisting of 129 amino-acid residues in a single chain, was given by Blake, Phillips and colleagues in 1965.

These and other studies on globular proteins have shown that a limited degree of secondary structure is usually present. Lysozyme, for example, has about 25% of its amino-acids in α-helical zones and some in sections of β-pleated sheet (Fig. 2.8).

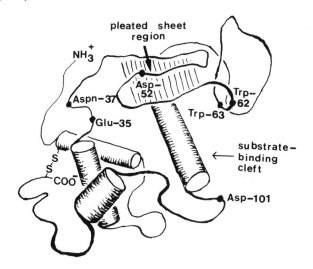

Fig. 2.8 A simplified representation of the three-dimensional structure of egg white lysozyme, as revealed by the x-ray diffraction studies of Phillips and colleagues (1965). Only the backbone of the polypeptide chain is shown, and α-helical regions are represented by cylinders. The amino-acid side chains would fill up most of the available space within a molecule, but a clearly-defined cleft for the binding of substrate is apparent. The positions of certain important amino-acid residues are indicated.

The amino-acids at the positions where the secondary structure is disrupted are, as expected, those with large side chains, e.g. proline and leucine, or those with charged side chains where two or more with like charge are close together. The molecules are very compact, with space for very few water molecules within the interior. Most of the amino-acids with non-polar side chains are found within the interior of the molecule, where they are unlikely to come into contact with water, while those with polar side chains are usually exposed to the solvent.

2.6 THE INVESTIGATION OF PROTEIN STRUCTURE IN SOLUTION

X-ray diffraction analysis is not suitable for the investigation of proteins in solution, since the molecules are not fixed in a regular arrangement. However other techniques may be used to give information as to the structures of proteins in solution, in particular as to the degree of secondary structure present.

There will be differences in both the infra-red and ultra-violet spectra between a polypeptide chain in an α-helix conformation and one existing as a **random coil** (i.e. one without regular, repeating three-dimensional features): these are due to the presence or absence of hydrogen bonding between atoms in different peptide bonds. Also, since a right handed α-helix is an asymmetric structure, there will be differences in optical rotation between a polypeptide in such a conformation and one consisting of the same amino-acid residues in a **random coil**.

Such investigations have helped to demonstrate that polypeptide chains can exist as α-helices in solution. Such a structure is most readily formed if all amino-acid side chains present are small and uncharged, as is the case with polyalanine, a synthetic polypeptide consisting entirely of L-alanine residues. If all the side chains are large, as with polyisoleucine, no α-helix is formed. In the case of synthetic polypeptides with ionisable side chains, e.g. polyglutamic acid, the structure in solution varies with pH. At acid pH, the glutamic acid side chains are uncharged (see Chapter 3.2.2) and an α-helix is formed. However at alkaline pH the side chains all have a negative charge; these repel each other and the α-helix is disrupted, as shown by optical rotation measurements (Fig. 2.9). The reverse effect is found with polylysine, whose side chains are uncharged at alkaline pH but have a positive charge at acid pH.

Spectrophotometry (see Chapter 18.1.3) may give useful information about protein structure since peptide bonds, aromatic and imidazole side chains and disulphides all give absorbance bands in the ultraviolet range which may vary according to the conformation of the protein and the micro-environment of the absorbing group. **Spectrofluorimetry** (see Chapter 18.1.4) too may be of value, for example in investigations of the fluorescence of tryptophan side chains. All data obtained by such investigations and those using techniques such as **nuclear magnetic resonance (nmr)** and **electron spin resonance (esr)** spectrometry is consistent with the assumption that three-dimensional structures found in protein crystals may also occur in solution (see Chapter 10.2). The nmr technique gives currently the best structural information about proteins in solution while esr, which detects unpaired electrons, is particularly useful for the investigation of metal ions in enzymes.

SUMMARY OF CHAPTER 2

Proteins consist of L-amino-acid residues linked by peptide bonds. The sequence of amino-acids in each polypeptide chain constitutes the **primary** structure of

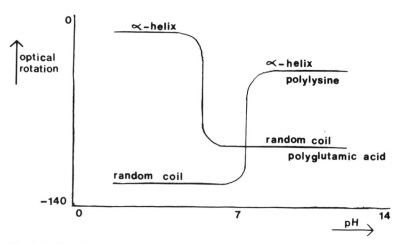

Fig. 2.9   The effect of pH on the optical rotation of polyglutamic acid and polylysine, demonstrating that α-helices are formed when the amino-acid side chains are uncharged.

the protein. This can be determined by a systematic use of chemical procedures. Regular, repeating three-dimensional features constitute the **secondary** structure. This is largely uninterrupted in fibrous, structural proteins but disrupted at many points in globular, functional proteins, including enzymes. The overall three-dimensional structure of each polypeptide chain is termed the **tertiary** structure. Proteins may consist of one or more polypeptide chains, the complete structure being called the **quarternary** structure.

The three-dimension structure of proteins in fibres and crystals may be determined by x-ray diffraction analysis. The structures found in crystals may also occur in solution.

FURTHER READING

Blake, C. C. F. and Johnson, L. N. (1984), Protein structure, *Trends in Bio-chemical Sciences,* **9** (pages 147–151).

Bull, A. T., Lagnado, J. R., Thomas, J. D. and Tipton, K. F. (eds.), *Companion to Biochemistry,* Volume 1 (1975), Chapters 2–4, Volume 2 (1979), Chapter 10, Longman.

Colowick, S. P. and Kaplan, N. O. (eds.), *Methods in Enzymology,* **47** (1977), **48** (1978), **49** (1978), **61** (1979), **91** (1983): Enzyme Structure, Academic Press.

Creighton, T. E. (1984), *Proteins,* Freeman.

Freifelder, D. (1982), *Physical Biochemistry,* 2nd edition (Parts 3 and 5), Freeman.

Hammes, G. G. (1982), *Enzyme Catalysis and Regulation,* Academic Press.

Hunkapiller, M. W., Strickler, J. E. and Wilson, K. J. (1984), Contemporary methodology for protein structure determination, *Science*, 226 (pages 304-311).

Palmer, T. (1985), Amino acid analysis, *Chromatography International*, Issue 6 (pages 5-6).

PROBLEMS

2.1 A peptide consisting of 18 amino-acid residues, with glycine at the N-terminus, gave the following four peptide fragments on digestion with trypsin:

    Ser-Phe-Val-Leu-Lys   Ala-Phe-Ser-Lys
    Gly-Ala-Thr-Arg   Ile-Trp-Glu-Thr-Ser.

Hydrolysis of the intact peptide with chymotrypsin gave the following four fragments:

    Ser-Lys-Ile-Trp   Gly-Ala-Thr-Arg-Ser-Phe
    Glu-Thr-Ser   Val-Leu-Lys-Ala-Phe.

On fresh hydrolysis of the peptide with chymotrypsin, the fragment initially identified as Glu-Thr-Ser remained at the origin on electrophoresis at neutral pH.

Deduce the primary structure of the peptide.

2.2 A series of investigations were carried out on two peptide chains originally linked by one or more disulphide bridges. The N-termini of the peptides were found to be valine and proline.

The disulphide bridges were cleaved by performic acid, two cysteic acid ($Cys.SO_3H$) residues being formed from each cystine residue initially present. Subsequent hydrolysis with trypsin gave the following five fragments:

    Val-Ser-Lys   Gly-Cys.SO$_3$H-Phe-Ile-Ala
    Pro-Arg   Glu-Trp-Cys.SO$_3$H-Gly.
    Cys.SO$_3$H-Leu-Tyr-Cys.SO$_3$H-Arg

Hydrolysis of the original peptides with pepsin, following treatment with performic acid, also gave five fragments:

    Val-Ser-Lys-Cys.SO$_3$H-Leu   Ile-Ala
    Pro-Arg-Gly-Cys.SO$_3$H-Phe   Trp-Cys.SO$_3$H-Gly.
    Tyr-Cys.SO$_3$H-Arg-Glu

A sample of the original specimen was hydrolysed with pepsin without first treating it with performic acid. The fragments produced were then separated from each other. One of these fragments, on subsequent treatment with performic acid, split further to give the following:

    Val-Ser-Lys-Cys.SO$_3$H-Leu
    Trp-Cys.SO$_3$H-Gly.

What can be deduced from this data?

CHAPTER 3

The Biosynthesis and Properties of Proteins

3.1 THE BIOSYNTHESIS OF PROTEINS

3.1.1 The Central Dogma of Molecular Genetics

As we have seen in Chapter 2.4.5, the amino-acid sequence of each globular protein in a given species is absolutely specific. This implies that the biosynthesis of proteins is under genetic control and hence connected with **nucleic acid** structure and function.

Nucleic acids are macromolecules composed of sequences of **mono-nucleotides**. Each mononucleotide consists of a nitrogenous base, a pentose sugar and a phosphate group, the sugar-phosphate component forming the recurring unit in the backbone of a polynuclueotide (or nucleic acid) chain:

$$
\begin{array}{ccccccc}
\text{base} & & \text{base} & & \text{base} & \\
| & & | & & | & \\
-\text{pentose} - \text{phosphate} - \text{pentose} - \text{phosphate} - \text{pentose} - \text{phosphate} - . &
\end{array}
$$

In a molecule of **ribonucleic acid (RNA)**, the pentose is always ribose; in **deoxyribonucleic acid (DNA)** it is $2'$-deoxyribose. (The number given to an atom in the pentose unit of a nucleotide is conventionally followed by an oblique dash, e.g. $2'$, to distinguish it from the number of an atom in the nitrogenous base.) The **purine** bases **adenine** and **guanine**, together with the **pyrimidine** bases **cytosine** and **uracil** (Fig. 3.1), are found in RNA. A small number of methyl-derivatives of these compounds may also be present. In DNA, only the **purine** bases **adenine** and **guanine** and the **pyrimidine** bases **cytosine** and **thymine** (Fig. 3.1) are present.

Covalent glycosidic bonds link the C1$'$ atom of each pentose with either the N9 atom of a purine or the N1 atom of a pyrimidine, the base-pentose unit being termed a **nucleoside**. The common nucleosides forming part of RNA are adenosine, guanosine, cytidine and uridine, and of DNA are deoxyadenosine, deoxyguanosine, deoxycytidine and deoxythymidine, the names being derived from those of the bases present. Thus mononucleotides may be regarded as nucleoside-phosphates and named accordingly, e.g. adenosine $5'$-monophosphate (AMP).

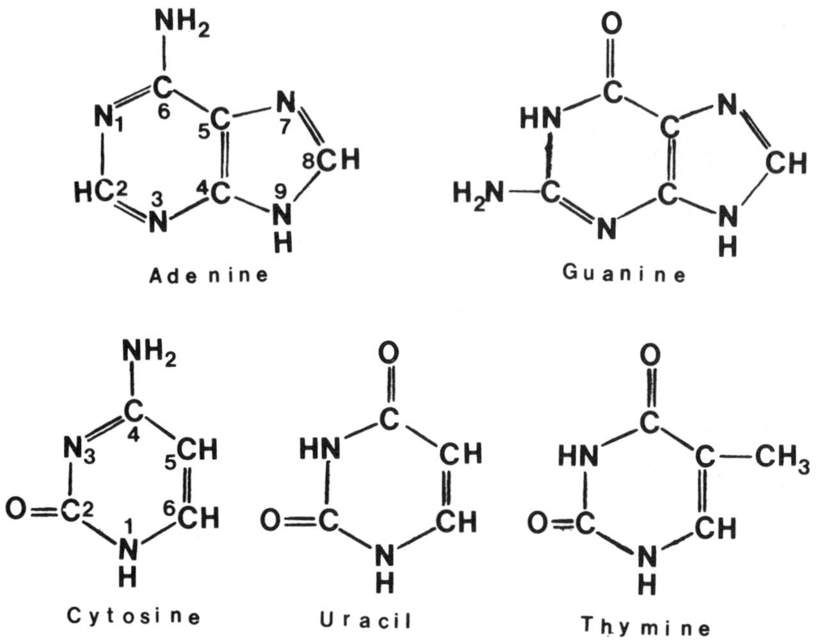

Fig. 3.1 The structures of purine and pyrimidine bases found in nucleic acids.

The structures of the polynucleotide chains in RNA and DNA are indicated in Fig. 3.2. It can be seen that the linkages of the phosphodiester backbone involve both atoms C3′ and C5′. The only C3′ atom not involved in such a linkage is at one end of the chain (termed the 3′ end) and the only free C5′ atom is at the other end (the 5′ end).

It was clearly established by about 1950 that DNA is a store of inherited information, since the total amount and the base composition of DNA in a cell are constant and are not affected by environmental factors. In **eukaryotic cells**, e.g. cells of higher animals and plants, DNA is located in the nucleus where, in association with protein, it forms the chromosomes (see Fig. 14.3). **Prokaryotic cells**, e.g. simple micro-organisms, are much smaller than eukatyotic cells and possess no membranous organelles or nucleus, the single molecule of DNA, largely free from protein, being tightly coiled in a nuclear zone.

The **central dogma of molecular genetics**, formulated by Crick in 1953, is that the genetic information in a cell is the sequence of bases in DNA. The DNA in a cell can **replicate** itself exactly, prior to cell division, so that each daughter cell has the identical genetic information to the parent cell. RNA is synthesised using DNA as a template, so the information contained in the base sequence of DNA is **transcribed** into the base sequence of RNA. Three different types of RNA molecule are known: messenger RNA (mRNA), transfer RNA (tRNA) and

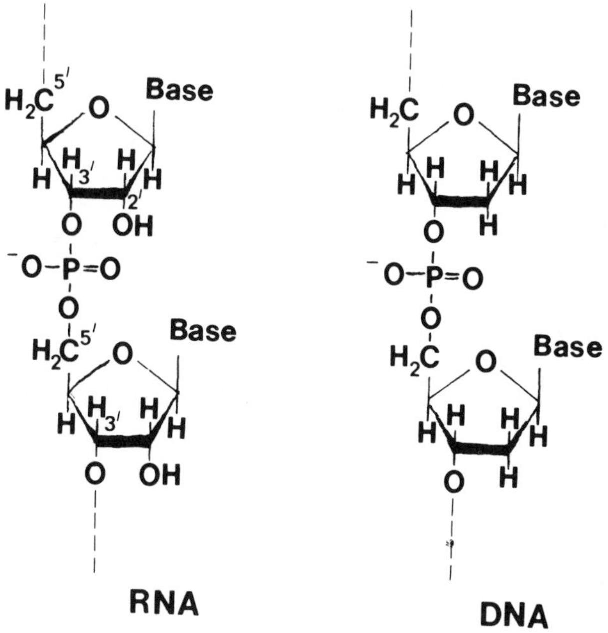

Fig. 3.2 The structures of the polynucleotide chains in RNA and DNA.

ribosomal RNA (rRNA), the last mentioned being associated with protein in the structures called **ribosomes**, which are found in the cytoplasm of cells. Protein biosynthesis involves all three types of RNA, including mRNA, whose base sequence is **translated** into the amino-acid sequence of the polypeptide chain. Thus, the flow of genetic information towards its expression as protein structure, according to the central dogma of molecular genetics, may be summarised as follows:

replication

transcription   translation

DNA ——————→ RNA ——————→ protein

3.1.2 The Double-Helix Structure of DNA

In 1953 Crick and Watson proposed a structure for the DNA molecule consistent with experimental evidence obtained by Chargaff, Franklin and Wilkins: Chargaff (1950) had shown that the amount of adenine (A) present in a DNA molecule was exactly the same as that of thymine (T), whilst the guanine (G) content was exactly equal to the cytosine (C) content; Franklin and Wilkins (1952), in x-ray diffraction studies on pure DNA fibres, had demonstrated a 3.4 Å periodicity along the axis. According to Crick and Watson, these features could be explained if the DNA molecule consisted of two antiparallel polynucleotide chains arranged in a double-helix about a common axis; the hydrophilic sugar–phosphate backbones form the outer surface of the molecule, the hydrophobic bases being directed towards the inside.

The adenine bases in each chain are always paired with thymine bases in the other chain, guanine and cytosine bases being similarly paired (Fig. 3.3(a)). The distance between the backbones is just sufficient to accommodate a purine–pyrimidine unit linked by hydrogen bonds, only A–T and G–C pairs having suitable complementary groups to accomplish this (Fig. 3.3(b)). Pyrimidine molecules are planar and purines approximately so, and the hydrogen bonded base pair arrangement is such that their planes are perpendicular to the axis of the double helix and occuring at a repeat distance of 3.4 Å along it. This model quickly gained general acceptance and Crick, Watson and Wilkins were awarded the Nobel Prize in 1962.

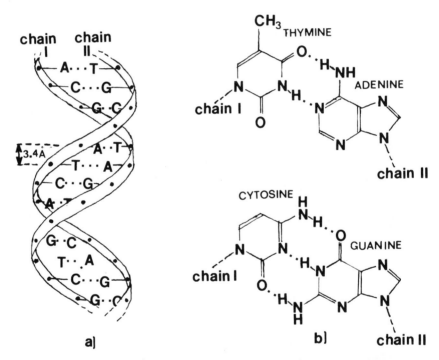

Fig. 3.3 (a) The double-helix structure of DNA, proposed by Crick and Watson. The structure is stabilised by hydrogen bonding between complementary purine and pyrimidine bases in the two helices. Hydrogen bonds can be formed as shown in (b).

The central dogma of molecular genetics (Chapter 3.1.1) can easily be explained in terms of the double-helix structure of DNA. In DNA replication, each strand of double-helix acts as a template, an entire new complementary strand being synthesised alongside; thus, two identical DNA molecules result from the molecule initially present, each possessing in its double-helix one of the original strands and one newly-synthesised strand. In transcription, a section of one of the DNA strands acts as a template for the synthesis of a complementary strand of RNA, which is then released from the template; the synthesis of RNA is sequential, one nucleotide at a time adding to the $5'$ end, the process being catalysed by **RNA polymerase**. Translation, which is of particular relevance to the subject of this book, is discussed in the next section.

3.1.3 The Translation of Genetic Information into Protein Structure

Before amino-acids can be incorporated into a protein, they must first be **activated** by linkage to a molecule of tRNA.

$$\text{amino-acid} + \text{tRNA} + \text{ATP} \rightleftharpoons \text{aminoacyl-tRNA} + \text{AMP} + (\text{PP})_i$$

The reaction is catalysed by an **aminoacyl-tRNA synthetase** and proceeds via an enzyme-bound AMP-amino-acid complex. These enzymes are highly specific, governing which amino-acids can be activated (only L-amino-acids) and to which tRNA molecule they can be linked. As a result of this, each tRNA molecule is specific for a particular amino-acid.

The base sequences of several tRNA molecules have been elucidated by Holley (1965) and others, using a systematic sequencing procedure similar in principle to that described for the analysis of peptides (Chapter 2.4.3) but, of course, using different enzymes and reagents. X-ray crystallography studies have also been carried out; the results show that tRNA molecules have a characteristic structure maintained by hydrogen bonding between complementary bases $(A \cdots U, G \cdots C)$ at certain points (Fig. 3.4). All molecules have a $-C-C-A$

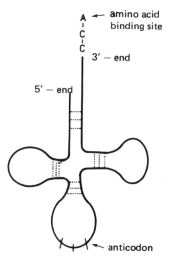

Fig. 3.4 A simplified, two-dimensional representation of tRNA.

sequence at the 3' terminus, added in the cytoplasm after transport of the rest of the molecule from the nucleus, where it is synthesised.

The amino-acid binds to the terminal 3'-OH of the tRNA by an ester linkage. Diametrically opposite this, in an exposed position, is a sequence of three bases termed the **anticodon**. Each molecule of tRNA has a unique anticodon and, as we have seen, will bind only one particular amino-acid. Here then is the clue to the **genetic code**, the link between the base sequence of nucleic acids and the amino-acid sequence of proteins.

The three bases constituting the anticodon are complementary to a triplet of bases, called a **codon**, in mRNA. The structure of mRNA, governed by the structure of a section of DNA, consists of a sequence of such codons and thus specifies the amino-acid sequence of a polypeptide chain. By adding synthetic polyribonucleotides of known composition to cell-free protein-synthesising systems and investigating the structure of the polypeptide chains synthesised, Nirenberg and Ochoa (1961–1966) were able to decipher the entire genetic code (Fig. 3.5). This is apparently the same for all known forms of life.

Protein synthesis is **initiated** by the presence of mRNA, whose initiating codon binds to a **peptidyl binding site (P-site)** of a ribosome sub-unit, stimulating

First position	Second position				Third position
(5' end)	U	C	A	G	(3' end)
U	Phe	Ser	Tyr	Cys	U
	Phe	Ser	Tyr	Cys	C
	Leu	Ser	Terminate	Terminate	A
	Leu	Ser	Terminate	Trp	G
C	Leu	Pro	His	Arg	U
	Leu	Pro	His	Arg	C
	Leu	Pro	Glun	Arg	A
	Leu	Pro	Glun	Arg	G
A	Ile	Thr	Aspn	Ser	U
	Ile	Thr	Aspn	Ser	C
	Ile	Thr	Lys	Arg	A
	fMet/Met	Thr	Lys	Arg	G
G	Val	Ala	Asp	Gly	U
	Val	Ala	Asp	Gly	C
	Val	Ala	Glu	Gly	A
	Val	Ala	Glu	Gly	G

Fig. 3.5 The genetic code.

assembly of the intact ribosome. The initiating codon is AUG or, less commonly, GUG. In prokaryotic cells, where protein synthesis has been investigated in the greatest detail, these code for N-formylmethionine (fMet) at the start of a sequence but for methionine and valine respectively elsewhere, different tRNA molecules being involved. In the cytoplasm of eukaryotic cells the initial amino-acid is methionine (without the formyl group), which again is not carried by the same tRNA molecule as other methionine residues in the chain.

In prokaryotic cells, initiation results in the P-site being occupied by N-formylmethionine-tRNA, whose anticodon pairs with the initiating codon of the mRNA. The overall process is accompanied by the hydrolysis of GTP to GDP and P_i, which provides the energy required. Although the code is always read in the $5' \rightarrow 3'$ direction, the initiating codon is not necessarily at the $5'$-terminus of the mRNA; the intervening bases possibly act as a further component of the initiation process.

When this process is complete, the next codon after the initiating one (reading towards the $3'$-end) lies in close proximity to an **aminoacyl binding site (A-site)** in the ribosome. The **elongation cycle** commences when the appropriate aminoacyl-tRNA enters this site, its anticodon pairing with the codon of the mRNA (Fig. 3.6(a)); this process also is linked to the hydrolysis of a molecule of GTP. A peptide bond is then formed between the α-carboxyl group of the amino-acid in the P-site and the α-amino group of that in the A-site, the dipeptide formed being held by the tRNA in the A-site (Fig. 3.6(b)). This reaction is catalysed by a **peptidyl transferase** enzyme, which is an integral part of the ribosome.

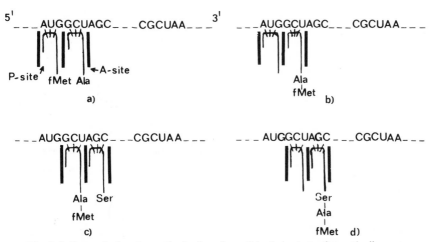

Fig. 3.6 Stages during the synthesis of a polypeptide chain at a prokaryotic ribosome (see text for detail). In the cytoplasm of eukaryotic cells, the initiating codon codes for Met rather than fMet. In either case, removal of the N-terminal residues by deformylase and/or aminopeptidase enzymes after translation would result in a polypeptide chain of outline structure Ala—Ser — — — —Arg.

Translocation then takes place, the uncharged tRNA leaving the A-site and being replaced by the entire contents of the P-site, the mRNA also moving to maintain its pairing with the tRNA anticodon. Once again, energy for this process is provided by the hydrolysis of GTP. Translocation brings a new codon into association with the vacated A-site, which is then filled by the appropriate aminoacyl-tRNA (Fig. 3.6(c)). A peptide bond is formed as before, the growing peptide chain being held by the tRNA in the A-site (Fig. 3.6(d)) prior to a further translocation step.

This cyclic sequence of events continues, an appropriate amino-acid being added to the peptide chain for each codon in the mRNA. During each turn of the cycle, the peptidyl–tRNA waits in the P-site for the next aminoacyl–tRNA to fill the A-site, this being the basis for the naming of these sites.

Eventually a **termination** codon is reached, when release factors cause the mRNA and the complete polypeptide chain to leave the ribosome; the last amino-acid to be added to the peptide chain is the C-terminus. When no longer associated with a mRNA molecule, each ribosome breaks down into its sub-units. However, as long as the mRNA remains intact, it can recombine with a ribosome sub-unit and initiate the process once again.

3.1.4 Modification of Protein Structure after Translation

Each polypeptide chain synthesised in bacteria by translation of the message carried by mRNA has N-formylmethionine as the N-terminus. The formyl group does not appear in the final protein, being removed by the action of a **deformylase**. The resulting methionine residue may also be quickly removed, by an **aminopeptidase**, as may several other amino-acid residues near the N-terminus. Even more modifications, e.g. the attachment of prosthetic groups, may take place before the polypeptide chain folds to take up its correct tertiary structure. It is likely that generally similar processes take place in all organisms, including eukaryotic ones, although the details may differ.

The section of DNA which carries the information for the synthesis of a single protein is called a **gene**; that for the synthesis of a single polypeptide chain may be termed a **cistron**. Eukaryotic cistrons include untranslated sequences (**introns**) interspersed with expressed sequences (**exons**), the RNA complementary to introns being discarded at the transcription stage, when that complementary to the exons is linked together (**spliced**). If a protein consists of more than one polypeptide chain, it might be assumed that more than one cistron is responsible for its synthesis. However this is not necessarily so, particularly where the poly-peptide chains are all identical. Even if they are not, it is possible that the protein could be synthesised originally as a single polypeptide chain and sub-sequently cleaved in one or more places. For example, the hormone **insulin** is synthesised as the inactive polypeptide **proinsulin** and subsequently activated by enzymic cleavage. The process may be represented diagramatically as shown at top of opposite page.

Some proteolytic enzymes, e.g. **chymotrypsin** and **trypsin**, may similarly be synthesised as inactive polypeptides, the active enzyme being produced by cleavage of peptide bonds (see Chapter 5.1.2).

3.1.5 Control of Protein Synthesis

Various mechanisms are present within the living cells to prevent the unnecessary synthesis of protein. Molecules of mRNA have a relatively short life, being broken down by ribonucleases; therefore, a continuing requirement for the protein in question necessitates further synthesis of mRNA. On the other hand, if there is no further requirement for the protein, **transcriptional control** may prevent mRNA synthesis.

The first control system to be investigated in detail was that of the **lactose operon** of *Escherichia coli*. This micro-organism can utilise lactose as its sole source of glucose, the enzymes **β-galactosidase, galactoside permease** and **galactoside transacetylase** being produced in large amounts. Galactose permease is required for the transport of lactose across the bacterial cell membrane, while β–galactosidase catalyses its subsequent hydrolysis to glucose and galactose; the precise physiological function of the third enzyme, galactoside transacetylase, is not known. If the organism is grown in the presence of the source material, lactose, synthesis of these enzymes may be **induced**; if grown in presence of the catabolic end-product, glucose, synthesis is **repressed**.

In 1961 Jacob and Monod, after a study of mutants unable to produce one or more of these enzymes, came to the conclusion that all three are normally synthesised together under the control of a **regulator gene** (i) and an **operator gene** (o); these are both involved in the induction process. Catabolite repression is achieved at a **promotor site** (p), which normally binds RNA polymerase to the DNA template in readiness for transcription to take place. The i, p and o sites are adjacent to the actual **structural genes** for the synthesis of β-galactosidase (z), permease (y) and transacetylase (a), the whole unit being termed the **lactose operon** (Fig. 3.7). Later work, especially that of Gilbert and Müller-Hill, has confirmed this model and enabled many of the details of the mechanism to be elucidated. The i gene is responsible for the synthesis of a **repressor protein** which binds tightly to the o site, overlapping onto the p site and preventing binding of RNA polymerase, and hence transcription of the z, y and a genes. This repressor protein has four sub-units, each with a binding site for allolactose,

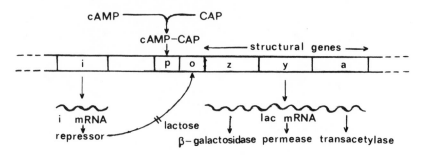

Fig. 3.7 The lactose operon of *E. coli.*

a metabolite of lactose. The presence of lactose results in the formation of some allolactose, which becomes attached to the repressor protein and prevents it from binding to the operator.

The promotor, on the other hand, is activated by the binding of a complex formed between $3',5'$-cyclic AMP (cAMP) and a **catabolite activator protein** (*CAP*). The synthesis of the *CAP* is the responsibility of yet another gene. Glucose inhibits the synthesis of cAMP so, in the presence of glucose, the cAMP-*CAP* complex is not formed, the promotor remains unactivated and RNA polymerase cannot be bound. Therefore, for transcription to take place, the cAMP-*CAP* complex must bind to the *p* site, but the repressor protein must *not* bind to the *o* site.

In the presence of both lactose and glucose, repressor protein binding does not take place but neither does promotor activation. In the absence of both lactose and glucose, the promotor is activated but the operator is repressed. In the absence of lactose and the presence of glucose, both of the negative influences are at work. In none of these situations does transcription of the lactose operon occur: it takes place only when the organism really requires the enzymes in question, i.e. when lactose is present and glucose is not otherwise available. Jacob and Monod were awarded the Nobel Prize in 1965, partly for their work in the elucidation of the lactose operon.

A similar control mechanism is found with the **tryptophan operon** of *E. coli*, the enzymes for tryptophan synthesis not being produced in the presence of this amino-acid. In this example the metabolic end-product, tryptophan, acts as a **co-repressor**: only a tryptophan-repressor complex can bind to the operator to prevent transcription.

In general, transcriptional control of protein synthesis is very important in micro-organisms, but some degree of **translational control**, i.e. control of the rate of the translation process, may also be present. In eukaryotic cells too there is evidence for both types of control, hormones usually being involved. For example, there is evidence that **glucocorticoids**, e.g. cortisol, bind tightly to the acidic proteins associated with DNA in mammalian liver, increasing the rate of synthesis of mRNA. Glucocorticoids have long been known to mobilise amino acids for glucose synthesis in the liver, so it is possible that the mechanism of action involves increasing the production of the enzymes involved, e.g. **aminotransferases**, by increasing the rate of transcription. The hormones involved in transcriptional control are usually those which act on a relatively narrow range of target organs. Glucocorticoids act only on cells which contain a specific receptor protein in the cytoplasm (i.e. on liver cells and those of peripheral tissue such as muscle, adipose and lymphoid tissue): this receptor protein is essential for binding the glucocorticoid and assisting its transport from the cytoplasm into the nucleus of the cell. Another example of a hormone involved in transcriptional control is the mineralocorticoid aldosterone, which acts on kidney cells to increase the synthesis of a protein required for the transport of Na^+ across membranes, thus increasing the renal reabsorption of this cation.

All cells in the same organism are the carriers of identical genetic information and therefore have the potential to synthesise the same range of proteins. In general, it seems likely that transcriptional control in mammals is important in **differentiation**, i.e. in determining which types of protein may or may not be synthesised by cells of particular organs. In contrast, translational control is apparently associated with hormones such as thyroxine, growth hormone and insulin, which influence important metabolic processes in cells of a wide range of target tissues. The ribosomes in eukaryotic cells, unlike those in prokaryotic cells, are often bound to the **endoplasmic reticulum** and the ratio of bound to unbound ribosomes can apparently be a factor in translational control.

3.2 THE PROPERTIES OF PROTEINS

3.2.1 Chemical Properties of Proteins

The chemical properties of proteins are largely those of the side chains of the constituent amino-acids. Thus **arginine** side chains, each containing a **guanidine** group, can react with α-naphthol in the presence of an oxidising agent such as sodium hypochlorite to produce a red colour: this is the **Sakaguchi** reaction. Similarly, **tryptophan** side chains, being **indoles**, can react with glyoxylic acid in the presence of concentrated sulphuric acid to produce a purple colour: this is the **Hopkins–Cole** reaction.

Tyrosine side chains, each possessing a **phenolic** group, can undergo a vareity of reactions. If treated with mercuric sulphate and sodium nitrate and then heated, a red complex is produced by the **Millon** reaction. They also undergo the **Folin–Ciocalteu** reaction if treated with tungstate and molybdate, a blue colour being formed.

All of these procedures, particularly the last mentioned, can be used for the **quantitative estimation of proteins**, the intensity of the colour produced being dependent on the number of reacting groups present. However, it is usually necessary to assume that the protein being estimated has an average distribution of amino-acid residues, and no reacting groups other than those in the protein must be present.

An alternative method for the quantitative estimation of protein is the **biuret** reaction: treatment with cupric sulphate in alkali gives a purple complex; the reacting unit in this case is the peptide bond, so free amino-acids do not react. The widely-used **Lowry** method for protein determination combines the **biuret** and **Folin–Ciocalteu** procedures. A further method for protein analysis makes use of the fact that **tyrosine** and **tryptophan** side chains absorb light at 280 nm.

As will be seen later (Chapters 10 and 11), functional groups in amino-acid side chains play an important role in the catalytic activity of enzymes. Many agents can inactivate enzymes by binding to these functional groups: for example, **heavy metal ions** (e.g. Ag^+) bind strongly to the **sulphydryl** groups of **cysteine** residues and thus may act as poisons to a great many enzymes.

$$E - SH + Ag^+ \rightarrow E - S - Ag + H^+$$
enzyme

3.2.2 Acid-Base Properties of Proteins

According to the definitions given by Brönsted and Lowry in 1923, which are particularly applicable to acid-base reactions in dilute aqueous solution, an **acid** is a **proton donor** and a **base** is a **proton acceptor**. The equation for the dissociation of an acid, e.g. acetic acid, is as follows:

$$CH_3COOH \rightleftharpoons H^+ + CH_3COO^-.$$

The equation for the protonation of a base, e.g. ammonia (NH_3), can be written in the same form, if looked at from the point of view of the dissociation of the corresponding conjugate acid (NH_4^+):

$$NH_4^+ \rightleftharpoons H^+ + NH_3.$$

Thus, *any* acid-base reaction can be written in the form:

$$HA \rightleftharpoons H^+ + A^-$$

where HA is an acid, or proton donor, and A^- is a base, or proton acceptor. A **weak acid** has a high affinity for its proton and dissociates only partially at most pH values, whereas a **strong acid** has a low affinity for its protein and dissociates readily. Thus, the **strength** of an acid (which has nothing to do with its overall **concentration**) is indicated by its **dissociation constant K_a**. For the acid HA, the dissociation constant at a particular temperature is given by:

$$K_a = \frac{[H^+][A^-]}{[HA]}.$$

K_a should really be called the *apparent* dissociation constant since the terms in square brackets are usually taken to be the experimentally determined concentrations. To obtain the true thermodynamic dissociation constant, allowance would have to be made for departures from ideal behaviour. Also, to be strictly accurate, the expression K_a should involve $[H_2O]$ and $[H_3O^+]$. However, if investigations are performed in dilute aqueous solution, water will always be present in great excess, its concentration not being measurably affected by the reactions taking place. Therefore, these terms may be ignored, provided the aim is simply to compare systems which are all in dilute aqueous solution.

The expression for K_a may be rearranged to give:

$$[H^+] = \frac{K_a.[HA]}{[A^-]}.$$

Taking the negative logarithm of both sides:

$$-\log_{10}[H^+] = -\log_{10}\left(\frac{K_a.[HA]}{[A^-]}\right)$$

$$= -\log_{10}K_a - \log_{10}\left(\frac{[HA]}{[A^-]}\right)$$

$$= -\log_{10}K_a + \log_{10}\left(\frac{[A^-]}{[HA]}\right).$$

By definition, $-\log_{10}[H^+] = pH$, and $-\log_{10}K_a = pK_a$.

Hence, $\quad pH = pK_a + \log_{10}\left(\frac{[A^-]}{[HA]}\right).$

This is known as the **Henderson–Hasselbalch equation**; it gives the relationship between the pH and the degree of ionisation of any ionisable species *HA* (Fig. 3.8).

When the pH is equal to the pK_a value, exactly half of the molecules present are in the dissociated form, i.e. $[HA] = [A^-]$. If strong acid (e.g. HCl) is added to this solution, one mole of A^- is converted to HA for every mole of acid added. Similarly one mole of HA is converted to A^- for every mole of strong base (e.g. NaOH) added. Therefore, if a known amount of strong acid or base is added to a known amount of HA, the new value of $[HA]$ and $[A^-]$ can easily be calculated, and the resulting pH obtained from the Henderson–Hasselbalch equation.

If the solution where $pH = pK_a$ is treated with 0.5 moles of HCl for every mole of HA present (including that present in the dissociated form), then the HA will be fully protonated. Addition of 0.5 moles of NaOH per mole of HA will return the system to its initial position, and addition of a further 0.5 moles of NaOH per mole of HA will result in complete deprotonation. To return once more to the initial position, 0.5 moles of HCl are required for every mole of HA present. The pH changes taking place throughout this process are indicated in Fig. 3.8.

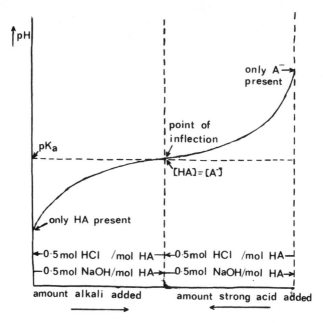

Fig. 3.8 Titration curve for the ionisable species HA, as predicted by the Henderson–Hasselbalch equation.

It can be seen that the pK_a value is characterised by a point of inflection on the curve. Therefore, any ionisable group can be titrated against strong acid or alkali and, from the graph of pH against titre obtained, the pK_a value of the group can be determined. If two groups are present which ionise over completely different pH ranges (e.g. the α-carboxyl and α-amino groups of free amino-acids), a graph of pH against titre will show two separate ionisation curves, enabling the pK_a value of each group to be determined with ease. However, if two groups are present which ionise over approximately the same pH range, then the graph of pH against titre obtained will be a composite of two ionisation curves and accurate determination of the pK_a values will be difficult. If several ionisable groups are present, the situation will be even more complicated.

The groups contributing to the acid-base properties of **proteins** are the ionisable side chains of aspartate, glutamate, histidine, cysteine, tyrosine, lysine and arginine residues, together with N-terminal α-amino groups and C-terminal α-carboxyl groups; all other α-amino and α-carboxyl groups present are, of course, involved in peptide bonds and thus not free to ionise. The pK_a values of all these groups are given in Fig. 3.9; however, it must be pointed out that the values given are only approximate ones, since the actual values depend to a considerable extent upon the environment within the protein (see Chapter 10.1.5).

Ionisable group	Dissociation reaction	Approximate pK_a
α-carboxyl	$-COOH \rightleftharpoons H^+ + -COO^-$	3.0
aspartyl carboxyl	$-CH_2COOH \rightleftharpoons H^+ + -CH_2COO^-$	3.9
glutamyl carboxyl	$-CH_2CH_2COOH \rightleftharpoons H^+ + -CH_2CH_2COO^-$	4.1
histidine imidazole	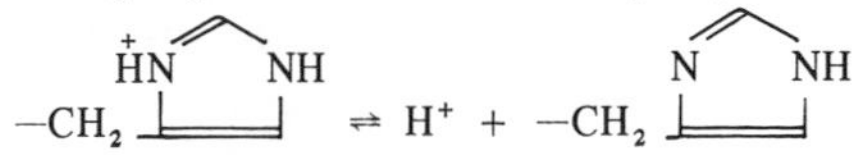 $-CH_2 \overset{\overset{+}{HN}\frown NH}{\underline{}} \rightleftharpoons H^+ + -CH_2 \overset{N \frown NH}{\underline{}}$	6.0
α-amino	$-NH_3^+ \rightleftharpoons H^+ + -NH_2$	8.0
cyteine sulphydryl	$-CH_2SH \rightleftharpoons H^+ + -CH_2S^-$	8.4
tyrosyl hydroxyl	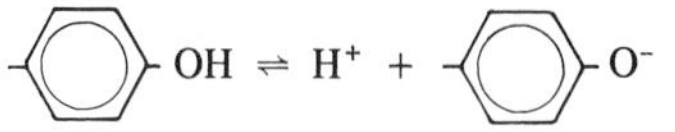 $-\!\langle\!\bigcirc\!\rangle\!- OH \rightleftharpoons H^+ + -\!\langle\!\bigcirc\!\rangle\!- O^-$	10.1
lysyl amino	$-(CH_2)_4NH_3^+ \rightleftharpoons H^+ + -(CH_2)_4NH_2$	10.8
arginine guanidine	$-(CH_2)_3NH.\underset{\overset{\|}{+NH_2}}{C}.NH_2 \rightleftharpoons H^+ + -(CH_2)_3NH.\underset{\overset{\|}{NH}}{C}.NH_2$	12.5

Fig. 3.9 The ionisable groups which contribute to the acid-base properties of proteins, shown with their approximate pK_a values. These can vary by several pH units according to their environment within the protein.

The overall titration curve for a protein will show the superimposed effects of all the ionisations present. For example, if a solution of **ribonuclease** is taken to acid pH and then titrated against NaOH, the observed graph of pH against titre can be analysed to show that all the ionisable groups known to be present, i.e. one α-carboxyl, ten side chain carboxyl, four histidine imidazole, six tyrosyl hydroxyl, one α-amino, ten side chain amino and four arginine guanidine groups, have contributed to the overall effect.

If we now return to Fig. 3.8 we can see that the change in pH brought about by the addition of a given amount of strong acid or base reaches a minimum value where $pH = pK_a$: in other words, the **buffering capacity** of an ionisable group is greatest where the pH is near its pK_a value. Proteins (excluding fibrous proteins which are largely insoluble) can act as **buffers**, their buffering capacity being greatest near the pK_a value of their most abundant ionisable side chain. Thus, haemoglobin is able to play an important buffering role in erythrocytes (red blood cells), since it contains a relatively large amount of an amino-acid (histidine) whose side chain has a pK_a value near the intracellular pH.

Fig. 3.8 also shows that the total electrical charge present depends on the degree of dissociation of ionisable groups and hence on pH. Therefore, most proteins will have a net positive charge at low pH values and will travel towards the cathode on electrophoresis. Conversely, at high pH values most proteins will have a net negative charge and travel towards the anode on electrophoresis. The pH at which there is no net charge on the molecule (i.e. at which there is an equal balance between positive and negative charges) is termed the **isoelectric point**: a protein having a relatively high content of basic amino-acids will have a relatively high isoelectric point, one having a high content of acidic amino-acids will have a low isoelectric point.

Globular proteins often function correctly only when certain ionisable side chains are in a specified form, making their usefulness pH-dependent. Each enzyme, therefore, has a characteristic pH optimum and is active over a relatively small pH range; in many cases a bell-shaped plot of activity against pH is obtained (Fig. 3.10). This is discussed in more detail in Chapter 10.1.5.

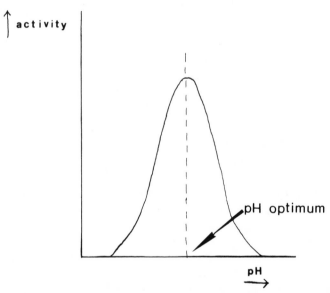

Fig. 3.10 A bell-shaped plot of activity against pH, which is obtained for many enzymes.

3.2.3 The Solubility of Globular Proteins

Solubility in aqueous solvents is enhanced by the formation of weak ionic interactions, including hydrogen bonds, between solute molecules and water. Therefore, any factor which interferes with this process must influence solubility. Electrical interactions between solute molecules will also affect solubility, since repulsive forces will hinder the formation of insoluble aggregates. In general, the solubility of a globular protein in an aqueous solvent is influenced by four main factors: salt concentration, pH, the organic content of the solvent and the temperature.

Addition of a small amount of **neutral salt** to a solution can increase the solubility of a protein. The added ions can cause small changes in ionisation of amino-acid side chains and can also interfere with interactions between protein molecules, the overall effect being to increase interactions between solute and solvent. This phenomenon, known as **salting-in**, depends solely on the **ionic strength** of the salt solution. (The **ionic strength** is the value obtained by multiplying the concentration of each ion by the square of its charge, adding together the results obtained for the different ions present and dividing by two: i.e. ionic strength = ½ Σ [A] Z_A^2.) Thus, divalent ions such as Mg^{2+} and SO_4^{2-} are relatively more effective than monovalent ions such as Na^+ and Cl^-. At very high salt concentrations, the abundance of interactions between the added ions and water decreases the possibilities for protein-water interactions, often resulting in the protein being precipitated from solution (Fig. 3.11). This is termed **salting-out**.

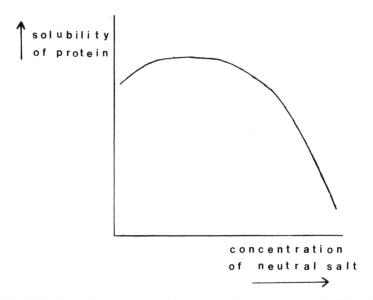

Fig. 3.11 The effect of neutral salt concentrations on protein solubility. The phenomena of salting-in, at low salt concentration, and salting-out, at high salt concentration, may be observed.

Some cations, e.g. Zn^{2+} and Pb^{2+}, have a more direct action on protein solubility, linking with protein anions to form **insoluble complexes**. Proteins may also be precipitated by treatment with various acids, e.g. trichloroacetic, perchloric, picric or sulphosalicylic acids, which form **acid-insoluble salts** with protein cations. Such techniques are often used to remove proteins from solutions prior to the analysis of other substances.

At **extremes of pH**, the pattern of charges carried by the ionisable side chains will be very different from that under normal conditions, so the compact tertiary structure will usually be disrupted, a more random structure being formed: this process is termed **denaturation**. Since the tertiary structure of an active globular protein is characterised by the majority of hydrophobic groups being hidden inside the molecule, disruption will bring these into contact with the aqueous solvent and the solubility of the protein will decrease considerably. Reversion to the original conditions will sometimes, but not always, result in the protein refolding into the tertiary structure required for functional activity: although this structure is likely to be the most favourable one from the point of view of energy, its re-establishment may be extremely slow because of the tangles produced during denaturation.

The solubility of each globular protein will also decrease markedly over a very narrow pH range around its **isoelectric point** (Fig. 3.12). Molecules of the protein are electrically neutral at this particular pH, so there is no net repulsion between them to prevent the formation of insoluble aggregates.

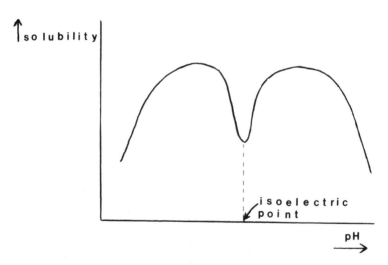

Fig. 3.12 The variation of the solubility of a typical globular protein with pH.

The introduction of a **water-miscible organic component**, e.g. ethanol, into the solvent lowers its **dielectric constant**. This increases the attractive forces between groups of opposite charge within a protein molecule and thus diminishes their linkages with surrounding water molecules (Chapter 2.3.2). The solubility of the protein will decrease in consequence.

Techniques involving varying pH, salt concentration and organic solvent concentration are widely used to separate mixtures of proteins by differential precipitation (see Chapter 16.1.4).

The solubility of globular proteins increases with **temperature** up to about 40-50°C. Above this temperature, thermal agitation tends to disrupt tertiary structure, leading to denaturation and a sharp decrease in solubility. In enzymes, this effect is paralleled to some extent by changes in activity: rates of enzyme-catalysed reactions increase with increasing temperature, since collisions between molecules become more frequent, until the enzyme is denatured and catalytic activity is lost (Fig. 3.13). A few proteins are also denatured if the temperature is reduced to below 10°C.

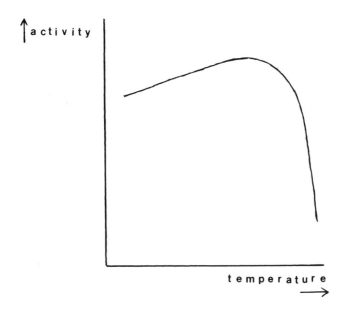

Fig. 3.13 The effect of temperature on enzyme activity.

Other conditions which lead to protein denaturation include the presence of high concentrations (e.g. 6 M) of urea or guanidine HCl, which disrupt the non-covalent interactions stabilising tertiary structure.

SUMMARY OF CHAPTER 3

Genetic information is stored in cells as the base sequence of DNA. This can be **replicated** prior to cell division, so that each cell produced contains the same genetic information as the original cell. The information may be **transcribed** into RNA structure and then **translated** into protein structure in such a way that the amino-acid sequence of each protein synthesised is determined by the base sequence of a section of DNA known as a **gene**.

Various control mechanisms exist to prevent unnecessary protein synthesis taking place. In simple micro-organisms, control of protein synthesis by **transcriptional control** is of particular importance, but some degree of **translational control** may also occur. In higher organisms, it seems likely that transcriptional control can determine which proteins are produced by a particular type of cell, the subsequent rates of synthesis being affected by translational control.

The chemical and acid-base properties of proteins are largely those of the side chains of the amino-acid residues present. The solubility of a protein in an aqueous solvent is influenced by salt concentration, pH, organic solvent content and temperature.

FURTHER READING

Austin, S. A. and Kay, J. E. (1982), Translational regulation of protein synthesis in eukaryotes, in Campbell, P. N. and Marshall, R. D. (eds.), *Essays in Biochemistry,* **18**, Academic Press.

Stryer, L. (1981), *Biochemistry,* 2nd edition, (Chapters 2, 24–29), Freeman.

Walker, R. (1983), *The Molecular Biology of Enzyme Synthesis,* Wiley.

PROBLEMS

3.1 (a) What are the pH values of the solutions obtained by mixing:

(i) 0.2 mol. dm^{-3} sodium acetate with an equal volume of 0.1 mol. dm^{-3} HCl, and

(ii) 0.2 mol. dm^{-3} sodium acetate with an equal volume of 0.1 mol. dm^{-3} acetic acid?

(b) Serial dilutions of an acetic acid solution were prepared so that tube number 1 contained 5 cm^3 of 0.32 mol. dm^{-3} acetic acid, tube number 2 contained 5 cm^3 of 0.16 mol. dm^{-3} acetic acid, tube number 3 contained 5 cm^3 of 0.08 mol. dm^{-3} acetic acid, and so on. To each was added 1 cm^3 of a solution of the protein casein in 0.1 mol. dm^{-3} sodium acetate. The greatest degree of protein precipitation was observed to occur in tube number 4. What does this indicate about the isoelectric point of casein?

(Assume the pK_a of acetic acid is 4.74.)

3.2 The imidazole side chain of histidine is ionisable. Calculate the percentage present in the protonated form at (a) pH 3, (b) pH 5, (c) pH 7 and (d) pH 9. (Assume the pK_a of the imidazole group is 6.0.)

CHAPTER 4

Specificity of Enzyme Action

4.1 TYPES OF SPECIFICITY

A characteristic feature of enzymes is that they are specific in action. Some enzymes exhibit **group specificity**, i.e. they may act on several different, though closely related, substrates to catalyse a reaction involving a particular chemical group. An example of this kind of enzyme is alcohol dehydrogenase, which will catalyse the oxidation of a variety of alcohols. Another is hexokinase, which will assist the transfer of phosphate from ATP to several different hexose sugars. Other enzymes will act only on one particular substrate, when they are said to exhibit **absolute specificity**. For example, glucokinase catalyses the transfer of phosphate from ATP to glucose and to no other sugar.

Uncatalysed reactions often give rise to a wide range of products, but enzyme-catalysed reactions are **product-specific** as well as being **substrate-specific**. Also, in addition to showing chemical specificity, enzymes exhibit **stereochemical specificity**: if a substrate can exist in two stereochemical forms, chemically identical but with a different arrangement of atoms in three-dimensional space (Chapter 2.2.2), then only one of the isomers will undergo reaction as a result of catalysis by a particular enzyme. For example, L-amino-acid oxidase mediates the oxidation of L-amino-acids to oxo acids; a separate enzyme, D-amino-acid oxidase, is required for the corresponding oxidation of D-amino-acids.

Even greater specificity is shown by the fungal enzyme glucose oxidase, which catalyses the reaction:

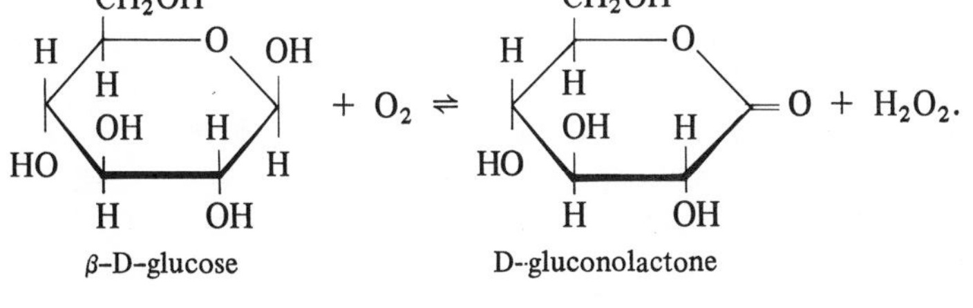

β–D-glucose   D-gluconolactone

No other naturally occurring sugar, including α-D-glucose and β-D-galactose, can be acted upon to any appreciable extent.

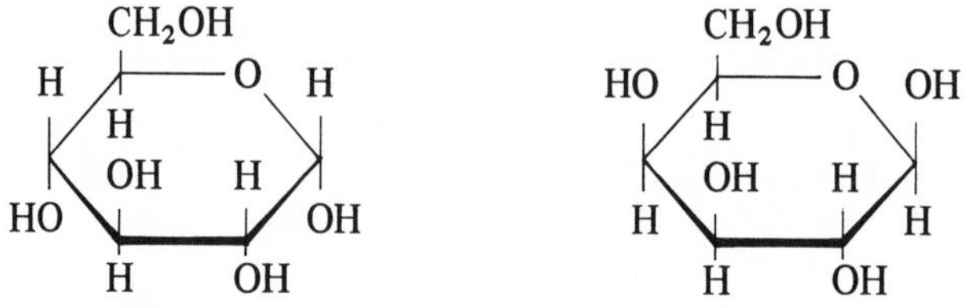

α-D-glucose   β-D-galactose

The only enzymes which act on both stereoisomeric forms of a substrate are those whose function is to interconvert L- and D-isomers. An example is alanine racemase, which catalyses the reaction:

L-alanine ⇌ D-alanine.

Enzyme-catalysed reactions may yield stereo-specific products even when the substrate possesses no asymmetric carbon atom. For example, the action of glycerol kinase on glycerol always results in the production of L-glycerol-3-phosphate:

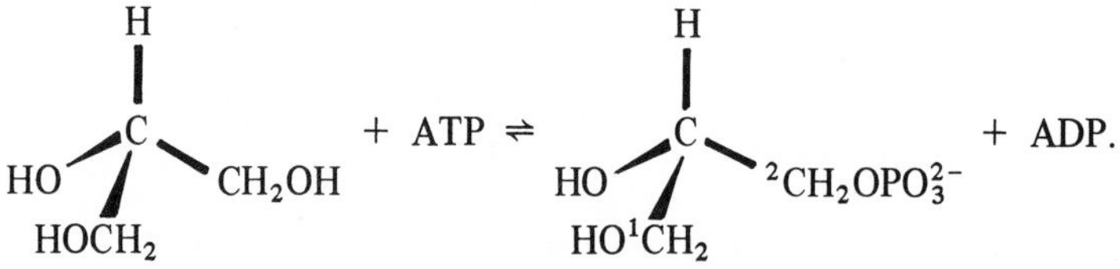

glycerol   L-glycerol-3-phosphate

No L-glycerol-1-phosphate is formed, even though the two $-CH_2OH$ groups of glycerol are chemically identical.

4.2 THE ACTIVE SITE

In order to explain the stereochemical specificity of enzymes (Chapter 4.1), Ogston (1948) pointed out that there must be *at least three different points of interaction* between enzyme and substrate (Fig. 4.1).

These interactions can have either a binding or a catalytic function: **binding sites** link to specific groups in the substrate, ensuring that the enzyme and substrate molecules are held in a fixed orientation with respect to each other, with the reacting group or groups in the vicinity of **catalytic sites**. For example, sites A'' and A''' (Fig. 4.1) might represent binding sites for R'' and R''' respectively, and A' a catalytic site for a reaction involving R'. Thus, even if R' and R'' are chemically identical (as with glycerol in the glycerol kinase reaction mentioned in Chapter 4.1), the asymmetry of the enzyme–substrate complex means that only R' can react, providing binding site A''' is specific for R''': R'' can never undergo reaction under these conditions, since it is not brought into the vicinity of site A' even when R' binds to site A''.

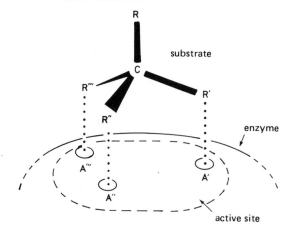

Fig. 4.1 Diagrammatic representation of three-point interaction between enzyme and substrate. A', A'', and A''' are sites on the enzyme which interact with groups R', R'' and R''', respectively, of the substrate. Each point of interaction may have a binding or a catalytic function.

Generally similar considerations apply to enzymes catalysing reactions involving more than one substrate. In this case, the reacting groups of each substrate are brought together in the vicinity of one or more catalytic sites.

The region which contains the binding and catalytic sites is termed the **active site**, or **active centre**, of the enzyme. This comprises only a small proportion of the total volume of the enzyme and is usually at or near the surface, since it must be accessible to substrate molecules. In some cases, x-ray diffraction studies have revealed a clearly-defined pocket or cleft in the enzyme molecule into which the whole or part of each substrate can fit (see Fig. 2.8).

Although the active site is given a planar representation in Fig. 4.1, it should be realised that it has, in fact, a three-dimensional structure since it consists of portions of a polypeptide chain. The amino-acid residues involved may be widely separated in the primary structure, being brought together in space because of the twists and turns within the molecule. The binding and catalytic sites must be either amino-acid residues or cofactors, the latter being themselves bound to amino-acid side chains. Substrate-binding may involve a variety of linkages (see Chapter 2.3.2), but the bonds formed are usually relatively weak (i.e. non-covalent).

Those amino-acid residues in the active site which do not have a binding or catalytic function may nevertheless contribute to the specificity of the enzyme. Their side chains must be of suitable size, shape and character not to interfere with the binding of the substrate, but they might interfere with the binding of other, chemically similar, substances.

The active site often includes both polar and non-polar amino-acid residues, creating an arrangement of hydrophilic and hydrophobic **microenvironments** not found elsewhere on an enzyme molecule. Thus, the function of an enzyme may depend not only on the spatial arrangement of binding and catalytic sites, but also on the **environment** in which these sites occur.

4.3 THE FISCHER "LOCK-AND-KEY" HYPOTHESIS

As early as 1890, Fischer suggested that enzyme specificity implied the presence of complementary structural features between enzyme and substrate: a substrate might fit into its complementary site on the enzyme as a key fits into a lock. This is entirely consistent with the more detailed aspects of active site structure discussed in Chapter 4.2. According to the **lock-and-key** model, all structures remain fixed throughout the binding process (Fig. 4.2).

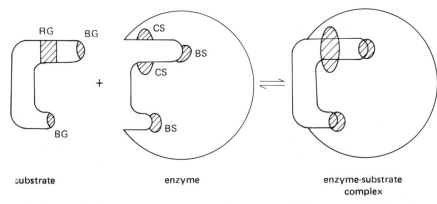

Fig. 4.2 Diagrammatic representation of the interaction between an enzyme and its substrate, according to the lock-and-key model. In the example illustrated, the single substrate is bound at two points, bringing the reacting group into the vicinity of two different catalytic sites. (BS = a binding site on the enzyme, CS = a catalytic site, BG = a binding group on the substrate and RG = a reacting group, i.e. a group undergoing enzyme-catalysed reaction.)

4.4 THE KOSHLAND "INDUCED-FIT" HYPOTHESIS

The lock-and-key hypotheses explains many features of enzyme specificity, but takes no account of the known flexibility of proteins. X-ray diffraction analysis and data from several forms of spectroscopy, including nuclear magnetic resonance (nmr), have revealed differences in structure between free and substrate-bound enzymes. Thus, the binding of a substrate to an enzyme may bring about a **conformational change**, i.e. a change in three-dimensional structure but not in primary structure. This is not necessarily surprising, for the bonds formed between a substrate and its binding sites may have replaced previously existing linkages between each binding site and neighbouring groups on the enzyme. Also, the presence of a substrate at the active site may exclude water molecules and thus make the region more non-polar. Both of these factors could be responsible for some degree of change in tertiary structure taking place.

Koshland, in his **induced-fit** hypothesis of 1958, suggested that the structure of a substrate may be complementary to that of the active site in the enzyme-substrate complex, but not in the free enzyme: a conformational change takes place in the enzyme during the binding of substrate which results in the required matching of structures (Fig. 4.3). The induced-fit hypothesis essentially requires the active site to be floppy and the substrate to be rigid, allowing the enzyme to wrap itself around the substrate, in this way bringing together the corresponding catalytic sites and reacting groups. In some respects, the relationship between a substrate and an active site is similar to that between a hand and a woollen glove: in each interaction the structure of one component (substrate or hand) remains fixed and the shape of the second component (active site or glove) changes to become complementary to that of the first.

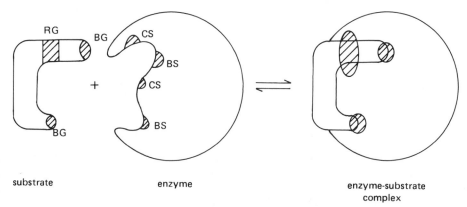

substrate   enzyme   enzyme-substrate
                                                           complex

Fig. 4.3 Diagrammatic representation of the interaction between an enzyme and its substrate, according to the induced-fit model (abbreviations as for Fig. 4.2).

Such a mechanism could help to achieve a high degree of specificity for the enzyme. In the lock-and-key mechanism, the active site is always structurally intact, with the catalytic sites aligned and freely accessible. Thus a suitable reacting group, whether part of an appropriately bound substrate or not, can come into contact with the region of catalytic activity and some degree of reaction take place. In the induced-fit mechanism, on the other hand, different catalytic components might be separated by a considerable margin in the free enzyme, minimising the risk of a chance collision of a reactive group with both of them (see Fig. 4.3); it is also possible that access to the catalytic groups of the free enzyme might be blocked. Only when a binding group of the substrate is recognised by the corresponding site of the enzyme and the binding process proceeds does the conformational change take place which results in all the relevant groups in substrate and enzyme coming together. Of course, a similar binding group in a substance other than the substrate might trigger off a conformational change but, in general, this would not result in catalytic groups being brought together in the vicinity of an appropriate reacting group, so no reaction would take place (Fig. 4.4). This would be termed **non-productive binding**.

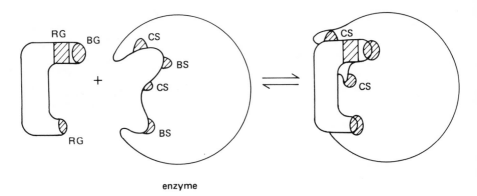

enzyme

Fig. 4.4 Diagrammatic representation of the non-productive interaction between an enzyme and a compound resembling its substrate, according to the induced-fit model (abbreviations as for Fig. 4.2).

An example of a reaction which appears to proceed via an induced-fit mechanism is that catalysed by **yeast hexokinase**:

$$\text{D-hexose} + \text{ATP} \rightleftharpoons \text{D-hexose-6-P} + \text{ADP}.$$

In the absence of hexose, bound ATP is hydrolysed extremely slowly, even though, in chemical terms, this hydrolysis could be brought about by the action of a water molecule in the solvent just as well as by an $-OH$ group in the hexose; this, together with x-ray diffraction evidence, suggests that the binding of the hexose causes a conformational change in the enzyme which activates the ATP.

Conformational changes of the type discussed in this section have been shown to play a part in the mechanism of action of several other enzymes, e.g. **carboxypeptidase A** (Chapter 11.4.4). They have also been useful in explaining the behaviour of **allosteric enzymes** (Chapter 13.3.1).

4.5   HYPOTHESES INVOLVING STRAIN OR TRANSITION-STATE STABILISATION

Although the lock-and-key and induced-fit models can explain enzyme specificity, neither suggests any direct mechanism by which the catalysed reaction may be driven forward. Substrate-binding often involves the expenditure of a considerable amount of energy and, although it serves a very useful purpose in bringing reacting and catalytic groups together, further energy must be supplied before the reaction can proceed. Haldane, in 1930, pointed out that if the binding energy was used to distort the substrate in such a way as to facilitate the subsequent reaction, then less energy would be required for the reaction to take place; this concept was developed further by Pauling, in 1948.

Let us assume, for example, that the structure of the active site is almost complementary to that of a substrate, but not exactly so. If the structure of the active site is rigid, the substrate must be distorted slightly in order to bind to the enzyme. This distortion might result in the stretching, and thus weakening, of a bond which is subsequently to be cleaved, thus assisting the forward reaction (Fig. 4.5).

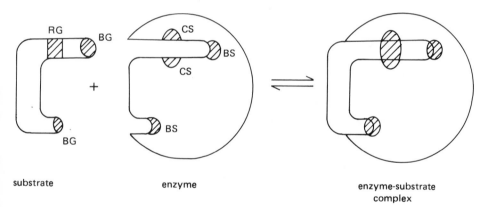

Fig. 4.5   Diagrammatic representation of the interaction between an enzyme and its substrate, incorporating a 'strain' effect (abbreviations as for Fig. 4.2).

In fact, little clear-cut evidence has been obtained for the occurrence of distorted binding. An alternative, and possibly more likely, mechanism for driving the reaction forward is **transition-state stabilisation**. This assumes that the substrate is bound in an undistorted form, but the enzyme–substrate complex possesses various unfavourable interactions. These tend to distort the substrate in such a way as to favour the following reaction sequence: enzyme–substrate complex → transition-state → products (see Chapters 6.2 and 6.6). As the reaction proceeds, the unfavourable interactions diminish, and are absent from the transition-state.

Thus, the overall effects of **strain** and **transition-state stabilisation** are very similar, but the sequence of events is slightly different in the two cases. An example of an enzyme-catalysed reaction proceeding via a transition-state stabilisation mechanism is the hydrolysis of peptides by **chymotrypsin. Lysozyme** (Chapter 11.3.5) is often cited as an example of an enzyme which operates by a strain mechanism, but even in this case the true mechanism may be transition-state stabilisation.

4.6   FURTHER COMMENTS ON SPECIFICITY

Specificity of enzyme action is determined by two separate factors: the relative ability of a potential substrate to bind to the enzyme and, once bound, its relative ability to undergo a reaction to form products. Only the overall rate of product formation indicates whether the enzyme can utilise a particular potential substrate.

No single one of the hypotheses discussed in Chapters 4.3–4.5 is able to account for the features of catalysis and specificity observed in all enzyme-catalysed reactions. Moreover, in some cases at least, a contribution from more than one of these factors appears to be present.

The induced-fit and strain/transition-state stabilisation mechanisms are not necessarily mutually exclusive. If the active site of an enzyme has a floppy structure and moulds itself around the substrate during the binding process, then distortion of the substrate would be unlikely to occur. However, a more precise conformational change taking place in the protein during the binding of substrate could result in some degree of strain being present in the latter. In either case the enzyme–substrate complex formed could posses internal stress and thus assist the subsequent reaction by transition-state stabilisation. The mechanism for the hydrolysis of peptides by **papain** appears to involve a conformational change in the enzyme in addition to either a strain or transition-state stabilisation factor.

In general, irrespective of the mechanism of an enzyme-catalysed reaction, the major factor governing specificity is the stability of the enzyme-bound transition-state which exists during the conversion of enzyme-bound substrate to products. A potential substrate which can form a relatively stable transition-state when bound to the enzyme will be converted to products at an appreciable rate. This was first pointed out by Pauling in 1948 and has since been confirmed by experiments with **transition-state analogues.** These are stable compounds which resemble the transition-state compounds thought to be formed as part of a reaction sequence. Such analogues have been shown to bind very tightly to the active sites of the appropriate enzymes, more tightly in fact than the corresponding substrate or products.

The investigation of transition-state structure is difficult because it occurs only transiently under normal conditions. However, if a reaction is carried out at low temperatures (e.g. at $-21°C$ in aqueous dimethyl sulphoxide), the lifetimes of intermediates are extended and their structures may be studied by techniques such as nmr. This is termed **cryoenzymology**.

SUMMARY OF CHAPTER 4

Enzymes exhibit chemical and stereochemical specificity with respect both to substrates and products. Such specificity requires at least three different points of interaction between enzyme and substrate. Each substrate is bound to the enzyme at specific sites to form an enzyme–substrate complex in which reacting groups are held in close proximity to each other and to catalytic sites. That region of the enzyme's three-dimensional structure which contains the substrate-binding sites and the catalytic sites is termed the active site.

According to the Fischer **lock-and-key** hypothesis, the active site has rigid structural features which are complementary to those of each substrate. In contrast, the Koshland **induced-fit** hypothesis suggests that at least some active sites are flexible, possessing a structure complementary to that of a substrate only when the latter is bound to the enzyme. These models can explain some aspects of enzyme specificity, but do not suggest any mechanism for driving forward the enzyme-catalysed reaction. The Haldane and Pauling concept of **strain**, and the **transition-state stabilisation** modification of this, explains how distortion of the substrate during or after binding can facilitate the subsequent reaction. This is not necessarily inconsistent with an induced-fit mechanism.

The most important factor in determining whether an enzyme will act on a particular substrate to produce a product appears to be the stability of the enzyme-bound transition-state which would have to be formed.

FURTHER READING

Ferdinand, W. (1976), *The Enzyme Molecule* (Chapter 4), Wiley.

Fersht, A. (1985), *Enzyme Structure and Mechanism,* 2nd edition (Chapters 2, 12 and 13), Freeman.

Hammes, G. G. (1982), *Enzyme Catalysis and Regulation* (Chapter 6), Academic Press.

Mackenzie, N. E., Malthouse, J. P. G. and Scott, A. I. (1984), Studying enzyme mechanism by ^{13}C nmr, *Science,* **225** (pages 883–889).

Stryer, L. (1981), *Biochemistry,* 2nd edition (Chapter 6), Freeman.

CHAPTER 5

Monomeric and Oligomeric Enzymes

5.1 MONOMERIC ENZYMES

5.1.1 Introduction

Monomeric proteins are those which consist of only a single polypeptide chain, so they cannot be dissociated into smaller units. Very few monomeric enzymes are known, and all of these catalyse hydrolytic reactions. In general they contain between 100 and 300 amino-acid residues and have molecular weights in the range 13,000–35,000 daltons. Some, e.g. **carboxypeptidase A**, are associated with a metal ion, but most act without the help of any cofactor.

A number of monomeric enzymes are **proteases** (or **proteolytic enzymes**), i.e. they catalyse the hydrolysis of peptide bonds in other proteins. In order to prevent them doing generalised damage to all cellular proteins, they are often synthesised in an inactive form known as a **proenzyme** or **zymogen**, and activated as required. Such enzymes include the **serine proteases**, so called because of the presence in the active site of an essential serine residue, i.e. a serine residue whose presence is essential for enzymic activity (see Chapter 10.1.2).

5.1.2 The Serine Proteases

The serine proteases **chymotrypsin**, **trypsin** and **elastase**, which are produced in an inactive form by the mammalian pancreas, form a closely related group of enzymes. Although only about 40% of the primary structure is common to all three enzymes, most of the catalytically important amino-acid residues correspond exactly. X-ray crystallography studies have also shown that their tertiary structures are very similar.

They are believed to function by an identical mechanism (see Chapter 11.3.2) and show a similar pH optimum of about pH 8. All are **endopeptidases**, hydrolysing peptide bonds in the middle of polypeptide chains, but their specificities are different. **Chymotrypsin** has a large hydrophobic binding pocket which will bind **phenylanine**, **tryptophan** and **tyrosine** side chains, enabling cleavage of the peptide bond at the carbonyl side of one of these residues. In **trypsin**, aspartate replaces serine at the bottom of the binding pocket, giving this enzyme a specificity for cleaving bonds adjacent to amino-acid residues with basic side chains, i.e. **lysine** or **arginine**. In the case of **elastase**, two glycine residues at the mouth of the binding pocket in chymotrypsin or trypsin are replaced by valine and threonine, whose bulky side chains block the pocket and result in the enzyme specifically cleaving bonds adjacent to residues with small non-polar side chains, e.g. **alanine**.

Chymotrypsin is synthesised in the pancreas as the zymogen chymo-trypsinogen (or pre-chymotrypsin). This is a single polypeptide chain of 245 residues containing five intrachain disulphide bridges. On passing into the intestine, where proleolytic enzymes are required to digest dietary proteins, chymotrypsinogen is attacked by trypsin. This breaks the peptide bond between arginine-15 and isoleucine-16, producing π-chymotrypsin. Already the molecule has full enzymatic activity, but further changes then take place: a dipeptide is removed from positions 14 and 15 by the action of another molecule of π-chymotrypsin, producing δ-chymotrypsin, and further chymotrypsin digestion removes a dipeptide from positions 147 and 148 to give the final product, α-chymotrypsin (Fig. 5.1). This contains three polypeptide chains linked by disulphide bridges, so it is not strictly a monomeric enzyme, but the sequential numbering system of the original chymotrypsinogen molecule is usually maintained.

In contrast, **trypsin** *is* a genuine monomeric enzyme. Trypsinogen lacks nine amino-acid residues at the N-terminus, by comparison with chymotrypsinogen, so cannot form the equivalent of the 1-122 disulphide bridge. The action of enteropeptidase (or trypsin itself) in the intestine removes a hexapeptide from the N-terminus of trypsinogen to produce the active trypsin, which is equivalent to the main chain of π-chymotrypsin and has the same N- and C- termini. **Elastase** is similarly produced from its corresponding zymogen by the action of trypsin.

The similar primary structures and almost identical tertiary structures of these three enzymes suggest that they evolved from a common ancestor by **divergent evolution**. The gene for the ancestral enzyme may have been duplicated several times, enabling different **mutations** (accidental changes of base sequence) in these duplicated genes to result in the production of slightly different enzymes.

Serine proteases involved in **blood clotting**, e.g. thrombin, and some **bacterial serine proteases** (e.g. from *S. griseus*) may also have evolved from this common ancestor, since their structures are similar to those of the enzymes from mammalian pancreas. In contrast, other bacterial serine proteases, e.g. **subtilisin** from *Bacillus amyloliquifaciens*, have very different primary and tertiary structures from those of the mammalian serine proteases. However, the active site structures and mechanism of action of all these enzymes are almost identical. This may suggest **convergent evolution**, i.e. the acquisition of similar characteristics from different starting materials by independent evolutionary pathways.

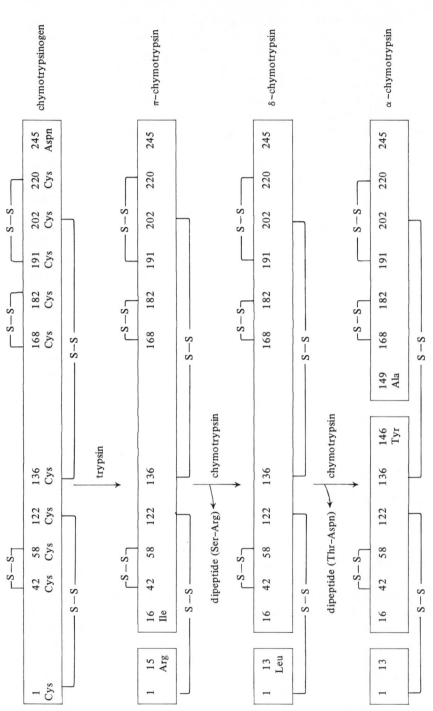

Fig. 5.1 Diagrammatic representation of the formation of bovine α–chymotrypsin from the inactive zymogen chymotrypsinogen, as elucidated by the work of Blow and colleagues (1968).

5.1.3 Some Other Monomeric Enzymes

Pepsin, like the pancreatic serine proteases, plays a role in the digestion of proteins eaten by mammals. It is called an **acid protease** because it functions at the low pH values found in the stomach. Peptide fragments are removed from the inactive form, pepsinogen, by the action of acid or other pepsin molecules to produce the active enzyme. This has a preference for cleaving bonds with a non-polar amino-acid residue on either side. Another acid protease found in the stomach is **chymosin (rennin)**. Others are found in micro-organisms.

A group of **thiol proteases**, similar in structure to each other, are found in plants. These include **papain**, from the pawpaw fruit, and **ficin**, from figs. Other thiol proteases, of different structure, are found in bacteria and mammalian lysosomes. The essential cysteine residue in each of these enzymes plays a similar role to that of serine in the serine proteases.

Several **exopeptidases**, which remove terminal amino-acid residues from polypeptide chains, are well known. Bovine pancreatic **carboxypeptidase A**, a monomeric enzyme containing one zinc ion per molecule, will break the peptide bonds linking C-terminal non-polar amino-acids to the rest of a chain. It is produced when trypsin removes peptide fragments from the zymogen, procarboxypeptidase A. A very similar enzyme, **carboxypeptidase B**, which has a specificity for C-terminal amino-acids with basic side chains, is also secreted as a zymogen by bovine pancreas.

Not all monomeric enzymes act on proteins. Well-known ones which hydrolyse other substrates include **ribonuclease** (Chapter 11.3.4) and **lysozyme** (Chapter 11.3.5).

5.2 OLIGOMERIC ENZYMES

5.2.1 Introduction

Oligomeric proteins consist of two or more polypeptide chains, which are usually linked to each other by non-covalent interactions and never by peptide bonds. The component polypeptide chains are termed **sub-units** and may be identical to or different from each other; if they are identical, they are sometimes called protomers. **Dimeric** proteins consist of **two**, **trimeric** proteins of **three** and **tetrameric** proteins of **four** sub-units. The molecular weight is usually in excess of 35,000 daltons.

The vast majority of known enzymes are oligomeric: for example, all of the enzymes involved in glycolysis possess either two or four sub-units. It is, therefore, reasonable to assume that the sub-units of oligomeric proteins gain properties in association that they do not have in isolation. Such enzymes are not synthesised as inactive zymogens, but their activities may be regulated in a far more precise way by **feed-back inhibition**: this is possible because many oligomeric proteins exhibit **allostery**, i.e. their different binding sites interact (see Chapters 12 and 13). Some examples of oligomeric enzymes are considered below, in order to see what other advantages may result from the association of sub-units.

5.2.2 Lactate Dehydrogenase

Vertebrate **lactate dehydrogenase (LDH)** is an example of an oligomeric enzyme where each sub-unit has the same function, in this case to catalyse the reaction:

$$CH_3.\underset{\underset{\text{OH}}{|}}{C}H.CO_2^- + NAD^+ \rightleftharpoons CH_3.\underset{\underset{\text{O}}{||}}{C}.CO_2^- + NADH + H^+.$$

$$\text{L-lactate} \qquad\qquad\qquad \text{pyruvate}$$

The enzyme, as found in many species, is a tetramer of molecular weight 140,000 daltons. However, although each sub-unit has a molecular weight of about 35,000 daltons, two types, of different amino-acid composition, are found within each species: they are the **M-form**, which predominates in skeletal muscle and other largely anaerobic tissues, and the **H-form**, the predominant sub-unit in the **heart**. The two types of sub-unit are produced by separate genes. Each monomer is catalytically inactive, but it can combine with others of the same or different type to produce the active tetrameric enzyme. All combinations of H and M sub-units are equally possible, so five **isoenzymes** of LDH can exist:

$$H_4, H_3M, H_2M_2, HM_3 \text{ and } M_4 \text{ (LDH}_1\text{-LDH}_5 \text{ respectively).}$$

Although these catalyse the same reaction, they do so with different characteristics, (the properties of H_3M, H_2M_2 and HM_3 being intermediate between those of H_4 and M_4) which enables the different isoenzymes to play different physiological roles. The precise mechanisms involved are far from clear, but Kaplan and his colleagues (1964) have proposed an **aerobic-anaerobic hypothesis** whose major features may be seen if we consider LDH in its metabolic context (Fig. 5.2).

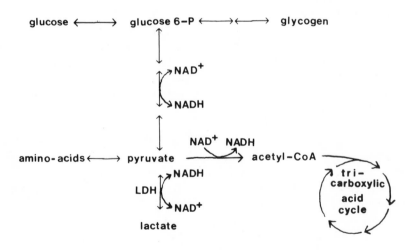

Fig. 5.2 The main features of pyruvate metabolism in vertebrates.

Pyruvate may be produced from carbohydrates, by glycolysis, or from amino-acids; under anaerobic conditions it may undergo LDH-mediated conversion to lactate, but when oxygen is freely available, pyruvate is metabolised to enter the tricarboxylic acid cycle. The tricarboxylic acid cycle and the pathway of glycolysis are important from the point of view of making energy available in a suitable form within the cell: both lead to the synthesis of ATP, an important intermediate in energy metabolism (Chapter 11.5.4). Lactate can only be produced from pyruvate and can only be metabolised back to pyruvate.

Under aerobic conditions, most of the pyruvate formed is channelled into the tricarboxylic acid cycle to ensure maximum ATP production. Thus, tissues which have a plentiful and constant oxygen supply do not normally require much lactate production to take place and tend to be rich in the H_4 isoenzyme: this converts pyruvate to lactate at a relatively low rate, the process possibly even being inhibited by pyruvate. However such tissues may require to utilise lactate (produced elsewhere) as substrate; when lactate levels are high, forcing the reaction in favour of net pyruvate formation, the inhibiting effect of pyruvate on H_4 is removed.

Under anaerobic conditions, the tricarboxylic acid cycle cannot operate, leaving the cell dependent upon glycolysis for ATP production. Without a constant supply of NAD^+, this too would break down, but the LDH-mediated conversion of pyruvate to lactate can ensure that NAD^+ levels are maintained. Therefore, in tissues which can become oxygen-starved, an LDH isoenzyme (M_4) with a high capacity for converting pyruvate to lactate is required. The lactate produced eventually finds its way to the heart or liver, via the blood stream, and pyruvate is reformed.

This hypothesis is not completely supported by the available experimental evidence: for example, although the M_4 isoenzyme has a higher turnover number (see Chapter 7.1.3) than the H_4 for pyruvate to lactate conversion, it apparently has a *lower* affinity than H_4 for the substrate; also, the predominant isoenzyme in human liver is M_4, whereas a predominance of H_4 might have been expected from the aerobic–anaerobic hypothesis (see also Chapter 14.2.1). Moreover, this hypothesis does not explain why five (and not just two) different isoenzymes are required, presumably, enabling each tissue to have the type (or types) appropriate to its needs.

Nevertheless, whatever the precise reasons for the presence of five isoenzymes of LDH within an organism, it is apparent that an arrangement enabling this to be achieved with only two different sub-units (and two genes) is more efficient than the possible alternative of having five different active monomeric isoenzymes (requiring five genes).

5.2.3 Lactose Synthase

Mammary gland **lactose synthase** is an example of an oligomeric enzyme where a non-functional sub-unit modifies the behaviour of a functional sub-unit.

This enzyme, as isolated from milk, consists of two sub-units: one of these is a catalytically inactive protein, **α-lactalbumin**, found only in mammary gland; the other is **galactosyl transferase**, an enzyme present in most tissues.

Galactosyl transferase, in the absence of α-lactalbumin, catalyses the reaction:

UDP-galactose + N-acetylglucosamine ⇌ UDP + N-acetyllactosamine.

This is important in the synthesis of the carbohydrate components of glycoproteins; the enzyme is also produced and stored in the mammary gland during pregnancy, when levels of α-lactalbumin are low. After the birth of the baby, reduced synthesis of the hormone progesterone in the mother leads to increased synthesis of the luteotrophic hormone (prolactin), stimulating the production of α-lactalbumin in the mammary gland. This combines with the stored galactosyl transferase to form lactose synthase, an enzyme which facilitates production of the lactose component of the milk required for the new-born baby. Lactose synthase catalyses the reaction:

UDP-galactose + glucose ⇌ UDP + lactose.

Thus it can be seen that the presence of the α-lactalbumin sub-unit changes the specificity of the enzyme, causing it to transfer galactose to glucose rather than to N-acetylglucosamine. Presumably the active site of the enzyme is modified as a result of the association between the two sub-units (see Chapter 14.2.6).

5.2.4 Tryptophan Synthase

Tryptophan synthase of *E. coli* is an example of an oligomeric enzyme which contains two different functional sub-units. The enzyme catalyses the reaction:

$$\text{indole-3-glycerol phosphate + L-serine} \xrightarrow{\substack{\text{pyridoxal} \\ \text{phosphate}}} \text{L-tryptophan +} \\ \text{glyceraldehyde 3-phosphate.}$$

It can be dissociated into two α sub-units, each of molecular weight 29,000 daltons, and a $β_2$ sub-unit, of molecular weight 90,000 daltons. The $β_2$ sub-unit further dissociates in the presence of 4 M urea to give two β sub-units, each of which has a binding site for the coenzyme pyrodoxal phosphate.

The isolated α sub-units will catalyse the reaction:

indole-3-glycerol phosphate ⇌ indole + glyceraldehyde-3-phosphate.

The isolated $β_2$ sub-unit also has catalytic activity, but for the reaction:

$$\text{indole + L-serine} \xrightarrow{\text{pyridoxal phosphate}} \text{L-tryptophan.}$$

So, the different sub-units of tryptophan synthase can be seen to catalyse separate halves of the overall reaction. However, the rates of these partial reactions are less than 5% the rate of the reaction catalysed by the intact $\alpha_2\beta_2$ enzyme. Also, significantly, indole is not released from the intact enzyme. Therefore, it is apparent that the oligomer has a degree of organisation not possessed by the isolated sub-units: the intermediate compound, indole, is passed directly from the active site of an α sub-unit to that of a β sub-unit, which is presumably in close proximity, to increase the efficiency of the overall process.

Another example of an enzyme where different sub-units catalyse different, though linked, reactions is discussed in Chapter 11.5.7.

5.2.5 Pyruvate Dehydrogenase

Pyruvate dehydrogenase of bacteria and animal cells is an example of a **multienzyme complex**. It shows the same type of organisation as tryptophan synthase (Chapter 5.2.4), but on an even larger scale. The Enzyme Commission (Chapter 1.3) recommended that such a complex should be regarded as a system of separate enzymes rather than as a single enzyme.

Pyruvate dehydrogenase enables pyruvate to enter the tricarboxylic acid cycle, by catalysing its overall conversion to acetyl–CoA:

$$\text{pyruvate} + \text{CoASH} + \text{NAD}^+ \rightarrow \text{acetyl-CoA} + \text{CO}_2 + \text{NADH}.$$

Reed and colleagues (1968) have shown that the *E. coli* enzyme consists of some 60 polypeptide chains and has a molecular weight of about 4,600,000 daltons. Three separate catalytic activities are present: a **pyruvate decarboxylase-dehydrogenase (E_1)** known as **pyruvate dehydrogenase (lipoamide), dihydrolipoamide transacetylase (E_2)** and **dihydrolipoamide reductase (E_3)**. The reaction sequence is shown in Fig. 5.3.

The whole process takes place with the substrate bound to the enzyme, either directly or via the cofactors **thiamine pyrophosphate (TPP)** and **lipoate**. TPP is associated with E_1, while the side chain of lipoate is covalently bound, by an amide linkage, to a lysyl residue of E_2. Hence the cofactor is actually **lipoamide** rather than **lipoate**. Protein E_3 also contains a prosthetic group, FAD. The reaction mechanism is discussed in Chapter 11.5.5.

The enzyme complex is about 300 Å in diameter and its features have been observed by electron microscopy. It has a polyhedral structure, with each of the sub-units appearing approximately spherical. The complex is held together by non-covalent forces and may easily undergo dissociation: at alkaline pH, the sub-units of the E_1 protein can be separated from those of the E_2 and E_3 proteins; at neutral pH and high urea concentration, the E_2 and E_3 proteins can be separated from each other. If the various sub-units are mixed together at neutral pH in the absence of urea, the multienzyme complex will spontaneously reform, but E_1 and E_3 sub-units will not re-associate unless E_2 is present. It appears that 24 sub-units of E_2 form a core to the complex, with a symmetrical

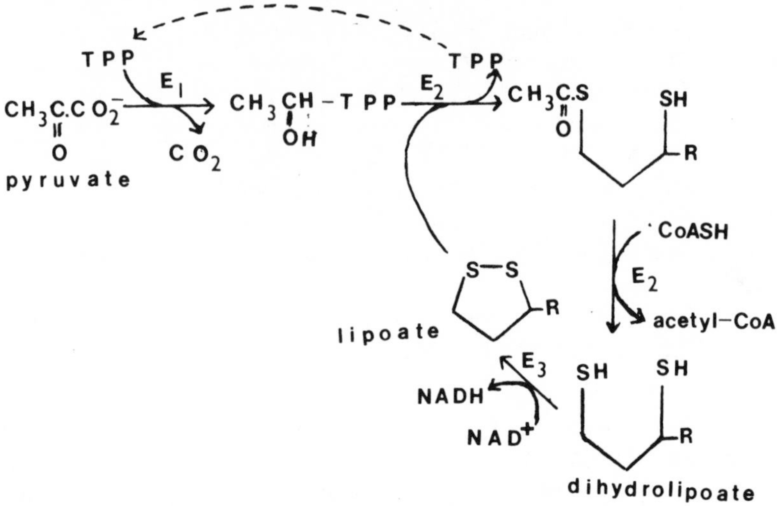

Fig. 5.3   The reactions catalysed by the pyruvate dehydrogenase complex of *E. coli*. E_1, E_2 and E_3 are the separate enzymes making up the complex (see text for details).

arrangement of E_1 and E_3 sub-units around this core; along each of the 12 edges of a cube is a molecule of E_1, probably consisting of two sub-units, and on each of the six faces of the cube is a molecule of E_3, again probably a dimer. It is possible that the flexible side chain of each lipoamide cofactor enables the lipoyl head to make contact with the active groups on adjacent enzymes and thus link the various processes taking place.

SUMMARY OF CHAPTER 5

Monomeric proteins consist of a single polypeptide chain; oligomeric enzymes have two or more such chains. Only a few enzymes, mainly hydrolases, are monomeric. These are often synthesised as inactive zymogens and activated by the removal of peptide fragments.

Oligomeric enzymes are often allosteric, enabling their activities to be regulated by feed-back inhibition. Varying combinations of different sub-units making up an oligomeric enzyme can enable a wide range of expression to be obtained. The association of different sub-units can also increase the efficiency of an enzyme, since a sequence of reactions can take place without the intermediate products being released. Such organisation is most notable in multienzyme complexes.

FURTHER READING

Fersht, A. (1985), *Enzyme Structure and Mechanism*, 2nd edition (Chapters 1 and 15), Freeman.

Stryer, L. (1981), *Biochemistry*, 2nd edition (Chapters 7, 8, 13, 16 and 21), Freeman.

Zubay, G. (1983), *Biochemistry* (Chapters 4, 9, 12 and 22), Addison-Wesley.

Part 2

Kinetic and Chemical Mechanisms of Enzyme-Catalysed Reactions

CHAPTER 6

An Introduction to Bioenergetics, Catalysis and Kinetics

6.1 SOME CONCEPTS OF BIOENERGETICS

6.1.1 The First and Second Laws of Thermodynamics

Bioenergetics, a branch of thermodynamics, is concerned with the changes in energy and similar factors as a biochemical process takes place, and not with the mechanism or speed of the process.

The **first law of thermodynamics** states that energy can neither be created nor destroyed, but can be converted into other forms of energy or used to perform work.

The **second law of thermodynamics** states that the entropy, or degree of disorder, of the universe is always increasing. However, it says nothing about a particular system under study, be it a mechanical engine, a living cell or a chemical reaction. Life, a state of high organisation or low entropy, can be maintained for a while by the consumption of a highly organised form of chemical energy (food) and, in the case of photosynthetic organisms, light energy; this energy is either converted to a less organised form of energy (heat) or utilised to perform work. However, the approach towards ultimate thermodynamic equilibrium is certain for every organism – death and decay.

Heat cannot be used to perform work by systems operating at constant temperature and constant pressure, such as the living cell. Thus, we have the concept of two forms of energy: that which can be used to perform work (called **free energy**) and that which cannot. For any system under study, processes can only take place **spontaneously** which result in a decrease in the free energy of the system, i.e. a transfer of free energy to the surroundings, unless there is interference from outside the system.

6.1.2 Enthalpy, Entropy and Free Energy

In the terminology of thermodynamics, a **closed system** is one which can exchange energy but not matter with its surroundings. The exchange of energy must involve thermal transfer and/or the performance of work. If, in a closed system at constant temperature and pressure, a process takes place which involves a transfer of heat to or from the surroundings and a change in volume of the system, then from the first law of thermodynamics

$$\Delta E = \Delta H - P\Delta V$$

where ΔE is the increase in **intrinsic energy** of the system, H is the increase in **enthalpy** and $P\Delta V$ is the **work done** on the surroundings by increasing the volume of the system by ΔV at constant pressure P and temperature T. The enthalpy change is defined simply as the quantity of heat absorbed by the system under these conditions and can be determined by calorimetric experiments.

Under these conditions, the increase in **entropy** of the surroundings is $-\dfrac{\Delta H}{T}$. If the process took place under conditions of thermodynamic reversibility, i.e. infinitely slowly, the increase in entropy of the system, ΔS, would be $\dfrac{\Delta H}{T}$. For the process to take place spontaneously under thermodynamically *irreversible* conditions, ΔS must be greater than $\dfrac{\Delta H}{T}$, giving an overall increase in the entropy of the system plus surroundings, as required by the second law of thermodynamics.

Hence $\Delta S - \dfrac{\Delta H}{T} > 0$

and $\Delta H - T\Delta S < 0$

Gibbs (1878) defined the increase in **Gibbs free energy** of the system, ΔG, such that

$$\Delta G = \Delta H - T\Delta S.$$

So, for any spontaneous process at constant temperature and pressure, $\Delta G < 0$.

6.1.3 Free Energy and Chemical Reactions

For a chemical reaction, the change in **Gibbs free energy** (ΔG) is the energy which is available to do **work** (for example, osmotic work, muscular work or biosynthetic work) as the reaction proceeds from given concentrations of reactants and products (the **"initial conditions"**) to **chemical equilibrium**. Consider a reaction:

$$A \rightleftharpoons B.$$

If, at given concentrations of A and B, the sign of ΔG is negative, then the system would lose free energy to its surroundings as the reaction proceeded to equilibrium in the direction written, i.e. it would be energetically favourable for the reaction to proceed in that direction. Assuming no interference with the system, the concentration of B would increase and that of A would decrease until chemical equilibrium was reached.

If, on the other hand, at given concentrations of A and B the sign of ΔG is positive, then the system could only make free energy available to its surroundings if the reaction proceeded to equilibrium in the *opposite* direction to that written, and this is the only process that could happen spontaneously. Again, assuming no interference with the system from outside, the concentration of B would decrease and that of A would increase until chemical equilibrium was reached.

If, under given conditions, the value of ΔG is zero, then chemical equilibrium has been attained.

It should be realised that at each point ΔG for the forward reaction is of equal value but opposite sign to ΔG for the reverse reaction.

The relationship between ΔG and the concentrations of reactants and products may be seen by considering the reaction:

$$A + B \rightleftharpoons C + D.$$

At a particular temperature ($T°K$)

$$\Delta G = -RT \log_e \left(\frac{[C_{eq}][D_{eq}]}{[A_{eq}][B_{eq}]} \right) + RT \log_e \left(\frac{[C_o][D_o]}{[A_o][B_o]} \right)$$

where R is the gas constant

[C_{eq}] is the equilibrium concentration of C (and similarly for [D_{eq}] etc.)

[C_o] is the initial concentration of C (and similarly for [D_o] etc.)

and $\log_e ( ) = 2.303 \log_{10} ( ).$

However, by definition, the **equilibrium constant**, $K_{eq} = \left(\frac{[C_{eq}][D_{eq}]}{[A_{eq}][B_{eq}]} \right)$

$$\therefore \Delta G = -RT \log_e K_{eq} + RT \log_e \left(\frac{[C_o][D_o]}{[A_o][B_o]} \right)$$

Under the special conditions where the initial concentrations of all reactants and products are 1.0 mol.l^{-1},

$$\Delta G = -RT \log_e K_{eq} = \Delta G^{\ominus} \text{(called the } \textbf{standard free energy change}\text{)}.$$

Thus, in general,

$$\Delta G = \Delta G^{\ominus} + RT \log_e \left(\frac{\text{product of initial concentrations of products}}{\text{product of initial concentrations of reactants}} \right).$$

6.1.4 Standard Free Energy

The **standard free energy change**, $\Delta G^{\ominus}$, is a useful concept. As we have seen above, it can be calculated if the equilibrium constant of a reaction is known:

$$\Delta G^{\ominus} = -RT \log_e K_{eq} = -2.303 \, RT \log_{10} K_{eq}.$$

It is also the difference between the standard free energy of formation of the reactants and the standard free energy of formation of the products, each term being adjusted to the stoichiometry of the reaction equation. For a reaction where a molecules of A react with b molecules of B:

$$aA + bB \rightleftharpoons cC + dD,$$

$$\Delta G^{\ominus} = (c.G_C^{\ominus} + d.G_D^{\ominus}) - (a.G_A^{\ominus} + b.G_B^{\ominus})$$

where $G_C^{\ominus}$ is the **standard free energy of formation** of C, etc.

The **standard free energy of formation** is a measure of the total amount of free energy a compound can yield on complete decomposition. Values for many compounds and groups have been calculated and are available in the literature. Standard free energies of formation are additive, and so are standard free energy changes. Therefore, if the standard free energy change for $A \rightleftharpoons B$ is $\Delta G_1^{\ominus}$ and for $B \rightleftharpoons C$ is $\Delta G_2^{\ominus}$, then the standard free energy change for $A \rightleftharpoons C$ can be determined by addition and is $\Delta G_1^{\ominus} + \Delta G_2^{\ominus}$:

reaction	standard free energy change
$A \rightleftharpoons B$	$\Delta G_1^{\ominus}$
$B \rightleftharpoons C$	$\Delta G_2^{\ominus}$
$A + \cancel{B} \rightleftharpoons \cancel{B} + C$	$\Delta G_1^{\ominus} + \Delta G_2^{\ominus}$.

Standard free energies for redox actions can be calculated from the **standard electode potential**, if this is known:

$$\Delta G^{\ominus} = -nF\Delta E_{\ominus}$$

where   n is the number of electrons transfered,
F is the Faraday
$\Delta E_{\ominus}$ is the standard electrode potential = $E_{\ominus}$ (oxidising couple) −
$E_{\ominus}$ (reducing couple).

So, $\Delta G^{\ominus}$ can be calculated by a variety of methods, of which that from K_{eq} is particularly useful. Strictly speaking, $\Delta G^{\ominus}$ refers to the standard free energy change at pH 0. Biochemists, therefore, prefer to use instead the term $\Delta G^{\ominus\prime}$, calculated at some defined physiological pH, usually pH 7. Apart from the pH difference, the meanings of $\Delta G^{\ominus}$ and $\Delta G^{\ominus\prime}$ are the same. If ΔG at pH 7 is being calculated for a reaction involving H^+, the $[H^+]$ term has already been taken into account if $\Delta G^{\ominus\prime}$ is used in the calculation, but needs to be included if $\Delta G^{\ominus}$ is used instead. For reactions where water is a reactant or product, its concentration is set by convention at 1.0 mol.l^{-1} to remove it from the calculations, since water is always present in excess.

6.1.5 Bioenergetics and the Living Cell

Although it is ΔG and not $\Delta G^{\ominus\prime}$ which actually determines how a reaction will proceed, the relationship between the two terms means that $\Delta G^{\ominus\prime}$ will give some idea of whether the reaction is likely to go largely in the forward or the reverse direction over most concentrations of reactant and product, or whether it will be a finely balanced reaction. This, of course, assumes that the reaction is allowed to proceed without interference from outside. In the living cell this is unlikely to be the case, for free energy made available from one reaction can be used to drive another in an energetically unfavourable direction, provided the two reactions have a **common intermediate** (this is termed the **principle of common intermediates**). Such coupled reactions often involve ATP, for this can be readily synthesised within the cell by either **substrate-level phosphorylation** or **oxidative phosphorylation**, yet is readily hydrolysed to ADP or AMP with the release of much free energy. For example, for the reaction:

$$\text{ribulose 5-phosphate} + P_i \rightleftharpoons \text{ribulose 1,5-bisphosphate}$$

$\Delta G^{\ominus\prime}$ is $+13.4$ kJ.mol^{-1}. This is a large enough positive value to suggest that the reaction should go in the direction of ribulose-5-phosphate formation at most concentrations of reactant and product, but in practice it goes almost exclusively in the opposite direction, due to coupling with ATP breakdown.

reaction	*standard free energy change*
ribulose-5-P + P$_i$ $\rightleftharpoons$ ribulose 1,5-bisP	$+ 13.4$ kJ.mol^{-1}
ATP $\rightleftharpoons$ ADP + P$_i$	-31.0 kJ.mol^{-1}
ribulose 5-P + ATP $\rightleftharpoons$ ribulose 1,5-bisP + ADP + P$_i$	-17.6 kJ.mol^{-1}

Another consideration in the living cell is that concentrations of reactants and products are constantly changing due to the effects of other reactions and to the intake of food. All this has the effect of keeping most reactions away from chemical equilibrium. If they were all allowed to proceed to equilibrium there would be no free energy available to perform work and life would be impossible.

6.2 FACTORS AFFECTING THE RATES OF CHEMICAL REACTIONS

6.2.1 The Collision Theory

Molecules can react only if they come into contact with each other. Therefore, any factor which increases the rate of collisions, e.g. increased concentration of the reactants or increased temperature, will increase the reaction rate (according to the principles of Arrhenius and van't Hoff).

However, not all colliding molecules react. This can be partly explained by steric reasons, for not all collisions will result in the appropriate groups of molecules coming into contact, particularly if the reactants are complex. A further and more important reason is that not all colliding molecules possess between them sufficient energy to undergo a reaction.

6.2.2 Energy of Activation and the Transition-State Theory

Not all molecules of the same type will possess the same amount of energy, taking all forms of energy into account. The energy of an individual molecule will depend, for example, on what collisions that molecule has recently been involved in. In order for a reaction to take place, colliding molecules must have sufficient energy to overcome a **potential-barrier** known as the **energy of activation**. This is true even of energetically favourable reactions, i.e. those which can proceed spontaneously with liberation of free energy.

The fact that a reaction *can*, in general, proceed spontaneously does not mean it *will* necessarily do so *in all circumstances*. By way of analogy, consider a ball on a hillside. The ball would tend to roll downhill; it would not spontaneously roll uphill. Nevertheless, if a stone was in its path, the ball would stay where it was. The stone would be the **potential-barrier** which would have to be overcome before the ball could roll down the hill. It is a similar potential-barrier which stops human beings bursting spontaneously into flame, for we burn with a great liberation of energy, if heated to a high-enough temperature.

The best explanation for the requirement for activation energy in a chemical reaction is the **transition-state theory**, developed by Eyring. This postulates that every chemical reaction proceeds via the formation of an unstable intermediate between reactants and products. Consider the hydrolysis of an ester:

$$R'C\overset{\displaystyle O}{\underset{OR''}{\diagup\diagdown}} + H_2O \rightarrow R'C\overset{\displaystyle O}{\underset{OH}{\diagup\diagdown}} + R''OH.$$

It is believed that the reaction mechanism involves the addition of water to the ester to form a transition-state compound having regions of positive and negative charge which cannot be stabilised:

$$O=\underset{OR''}{\overset{R'}{C}} \ + \ O\overset{H}{\underset{H}{<}} \ \rightarrow \ \overset{\delta-}{O}\!\!=\!\!\!\cdots\underset{OR''}{\overset{R'}{C}}\cdots\overset{\delta+}{O}\overset{H}{\underset{H}{<}} \ \rightarrow \ O=\underset{OH}{\overset{R'}{C}} \ + \ R''OH.$$

An unstable compound, by definition, will break down to give a more stable one and so must possess more free energy than the stable compound. A free energy profile of a reaction involving an unstable transition-state, such as the one given above, is given in Fig. 6.1.

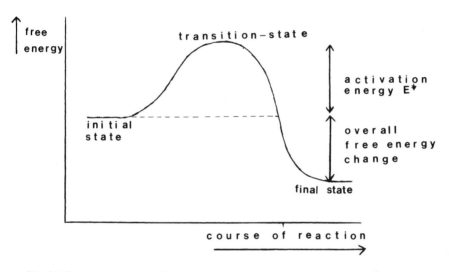

Fig. 6.1 Free energy changes for an energetically favourable reaction proceeding via the formation of a transition-state.

The activation energy is, therefore, the energy needed to form the transition-state from the reactants. The transition-state is unstable and will very quickly break down to form the products (or back to reactants) but no products can be formed from reactants unless the transition-state has been formed. The free energy of activation thus acts as a potential-barrier to the reaction taking place.

An estimate of the activation energy may be obtained from the reaction rate at different temperature, as will now be discussed.

The probability that a particular molecule possesses energy in excess of a value E is $e^{-E/RT}$ (where R is the gas constant and T the absolute temperature).

For a molecule of energy E to react it must collide with a molecule of energy at least $E^{\ddagger} - E$, where $E^{\ddagger}$ is the **activation energy.** (To be more precise, the free energy of activation, $\Delta G^{\ddagger}$, should be used instead of $E^{\ddagger}$.) The probability for a colliding pair to have sufficient energy to react is

$$e^{-E/RT} \cdot e^{-(E^{\ddagger}-E)/RT} = e^{-E^{\ddagger}/RT}.$$

This is consistent with the **Arrhenius equation**, first derived experimentally in 1889:

$$k = \text{constant} \times e^{-E^{\ddagger}/RT}$$

(where k is the **rate constant**, a characteristic of the reaction at temperature T; the other constant has been shown to be equal to PZ, where P is a **steric factor** and Z the **collision frequency**).

Taking logs,
$$\log_e k = \log_e PZ - \left(\frac{E^{\ddagger}}{RT}\right).$$

If the rate constant $= k_{T_1}$ at absolute temperature T_1
and $= k_{T_2}$ at absolute temperature T_2,

$$\log_e \left(\frac{k_{T_1}}{k_{T_2}}\right) = \log_e k_{T_1} - \log_e k_{T_2} = \left(\log_e PZ - \frac{E^{\ddagger}}{RT_1}\right) - \left(\log_e PZ - \frac{E^{\ddagger}}{RT_2}\right)$$

$$= \frac{E^{\ddagger}}{R}\left(\frac{1}{T_2} - \frac{1}{T_1}\right) = \frac{E^{\ddagger}}{R}\left(\frac{T_1 - T_2}{T_1 T_2}\right)$$

$$\therefore E^{\ddagger} = R\left(\frac{T_1 T_2}{T_1 - T_2}\right)\log_e\left(\frac{k_{T_1}}{k_{T_2}}\right) = 2.303\,R\left(\frac{T_1 T_2}{T_1 - T_2}\right)\log_{10}\left(\frac{k_{T_1}}{k_{T_2}}\right)$$

Assuming that the rate of reaction is proportional to the rate constant,

$$E^{\ddagger} = 2.303\,R\left(\frac{T_1 T_2}{T_1 - T_2}\right)\log_{10}\left(\frac{v_{T_1}}{v_{T_2}}\right)$$

(where v_{T_1} is the rate of reaction at temperature T_1, and v_{T_2} is the rate of reaction at temperature T_2).

If the reaction rate (or the rate constant) has been determined at more than two temperatures, $E^{\ddagger}$ can be calculated by plotting $\log_{10} k$ (or $\log_{10} v$) against $\frac{1}{T}$. The slope of the graph is $-\dfrac{E^{\ddagger}}{2.303\,R}$.

It is important to realise that the activation energy calculated depends on the assumptions made and cannot be used to decide between possible reaction mechanisms.

6.2.3 Catalysis

A catalyst accelerates a chemical reaction without changing its extent and can be removed unchanged from amongst the end-products of the reaction. It has no overall thermodynamic effect: the amount of free energy liberated or taken up when a reaction has been completed will be the same whether a catalyst is present or not.

In most cases, a catalyst acts by reducing the energy of activation. The catalyst, or part of it, combines with the reactants to form a different transition-state from that involved in the uncatalysed reaction; one which is more stable and, therefore, of lower energy (Fig. 6.2).

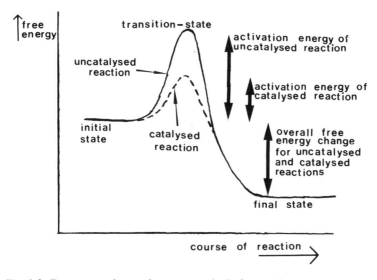

Fig. 6.2 Free energy changes for an energetically favourable reaction, showing the effects of a catalyst.

Note that the initial and final states are at the same free energy levels for the catalysed and uncatalysed reactions, and the overall free energy change as the reaction proceeds is also the same.

As a rough generalisation, therefore, an uncatalysed reaction might proceed as follows:

$$A + B \xrightarrow{\text{very slow}} A\cdots B \xrightarrow{\text{very fast}} \text{products.}$$

The same reaction, but in the presence of a catalyst C, might take place like this:

$$A + B + C \xrightarrow{\text{fast}} A\cdots B\cdots C \xrightarrow{\text{fast}} \text{products} + C.$$

Because the catalyst stabilises the transition-state, the breakdown of the intermediate to the products might be slower than in the uncatalysed reaction. Nevertheless, the overall rate is determined by the slowest step in the sequence, so the overall rate of the catalysed reaction will be faster than that for the uncatalysed reaction.

Catalysts are often **acids** or **bases**: acids stabilise the transition-state by donating a proton, bases by accepting a proton. The hydrolysis of an ester, which we considered earlier, may be catalysed by a base $(X-O^-)$ as follows:

$$O=\underset{\underset{OR''}{|}}{\overset{\overset{R'}{|}}{C}} \; + \; O\overset{H}{\underset{H}{\diagdown}} \; + \; X-O^- \rightarrow O=\underset{\underset{OR''}{|}}{\overset{\overset{R'\;H}{|\;\;|}}{C}}\cdots O\cdots H\cdots\overset{\delta-}{O}-X \rightarrow O=\underset{\underset{OH}{|}}{\overset{\overset{R'}{|}}{C}} \; + \; R''.OH \; + \; X-O^-.$$

Other common forms of catalysis include **covalent catalysis**, where the transition-state is stabilised by changes involving covalent bonds, and **metal ion catalysis**, where the transition-state is stabilised by electrostatic interactions with a metal ion.

In this section we have looked at the free energy profiles of energetically favourable reactions, considering the changes taking place when the molecules of the reactant are converted to molecules of the product. Of course, most reactions are reversible, but note that the activation energy required for the reverse, energetically unfavourable, reaction in Figs. 6.1 and 6.2 is the energy required to form the transition-state from the products and is thus much greater than the activation energy in the energetically favourable direction. The proportion of colliding product molecules possessing between them sufficient energy to overcome the potential-barrier will be even less than for the reactants. However, the greater the product concentration, the greater the number of molecules present capable of undergoing the reverse reaction. As we have seen, the overall trend, indicated by ΔG, towards chemical equilibrium depends on the actual concentrations of reactants and products present. The position of chemical equilibrium is not affected by the presence of a catalyst: the catalyst merely accelerates its attainment.

6.3 KINETICS OF UNCATALYSED CHEMICAL REACTIONS

6.3.1 The Law of Mass Action and the Order of Reaction

Kinetics is the study of reaction rates and the factors influencing them. It is not concerned with the chemical nature of the changes taking place.

All kinetic work is based on the **Law of Mass Action** proposed by Guldberg and Waage in 1867. This states that the rate of a reaction is proportional to the product of the **activities** of each reactant, each **activity** being raised to the power of the number of molecules of that reactant taking part, as indicated by the reaction equation. For a reaction:

aA + bB → products

the rate is proportional to (activity of A)^a × (activity of B)^b.

For practical purposes the term **activity** is usually replaced by **concentration**, although strictly speaking it is only equal to the concentration in ideal gases and in very dilute solutions.

From the Law of Mass Action developed the concept of the **order of reaction**.

A **first-order reaction** is one which proceeds at a rate proportional to the concentration of *one* reactant. Thus, for a first-order reaction A → P taking place at a constant temperature and pressure in a dilute solution, reaction rate at any time *t* is given by:

$$v = -\frac{d[A]}{dt} = +\frac{d[P]}{dt} = k[A]$$

where v = reaction rate at time *t*

      k = rate constant

     [A] = concentration of reactant *A* at time *t*

     [P] = product concentration at time *t*

$-\dfrac{d[A]}{dt}$ = rate of decrease of [*A*]

$+\dfrac{d[P]}{dt}$ = rate of increase of [*P*].

A **second-order reaction** is one which proceeds at a rate proportional to the concentrations of *two* reactants or to the *second power* of a single reactant. So, for a second-order reaction A + B → P, taking place at a constant temperature and pressure in a dilute solution,

$$v = -\frac{d[A]}{dt} = -\frac{d[B]}{dt} = +\frac{d[P]}{dt} = k[A][B]$$

the terms having the same general meanings as above. Similarly, for a second-order reaction of the form 2A → P,

$$v = -\frac{d[A]}{dt} = +\frac{d[P]}{dt} = k[A]^2.$$

A two-reactant reaction can, under certain circumstances, be regarded as a pseudo single-reactant one. For instance, reactions taking place in an aqueous medium and involving water and one other reactant are usually considered pseudo single-reactant since water will be in vast excess and its concentration will not change significantly during the course of the reaction.

It should also be noted that a **zero-order reaction** is possible. This is one whose rate is *independent* of the concentration of any of the reactants.

6.3.2 The Use of Initial Velocity

For any reaction, starting with reactants only and measuring the appearance of product with time, a graph of the form represented in Fig. 6.3 is obtained.

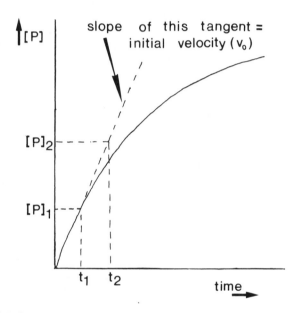

Fig. 6.3 Graph of product concentration against time for a chemical reaction.

The rate of the reaction at any time t is the slope of the curve at that point. This may be constant for a little time at the start of the reaction and then decreases with the decreasing concentration of the reactant(s) as the reaction proceeds, finally falling to zero. At this point either all the reactants have been converted into products or, more commonly, a chemical equilibrium has been set up, with the rate of the forward reaction now being equal to the rate of the back reaction.

The **initial velocity** (v_0) of the reaction is the reaction rate at $t = 0$ and may be determined by drawing a tangent to the graph, as shown in Fig. 6.1. From this tangent,

$$v_0 = \frac{[P]_2 - [P]_1}{t_2 - t_1}.$$

The units for v_0 are those used for product concentration, divided by those used for time.

The importance of initial velocity is that it is a kinetic parameter determined for the reaction in a situation which can be easily specified. The concentrations of each reactant are known from the amounts actually added; this would not be true at any other point during the reaction. Also, since there are no products present at $t = 0$, no back reaction will be taking place.

It will be apparent that the **initial velocity** will depend on the **initial concentrations** of the reactants. For a single-reactant reaction, it has been seen in the previous section that, for the reaction to be first-order, $v = k[A]$. Thus, at time $t = 0$, $v_0 = k[A_0]$, where $[A_0]$ is the initial concentration of A.

Similarly, for the single-reactant second-order reaction, $v_0 = k[A_0]^2$ and for a single-reactant zero-order reaction, $v_0 = k[A_0]^0 = k$. The graphs corresponding to these equations are shown in Fig. 6.4. This approach can also be applied to reactions involving more than one reactant. If all the reactants but one are maintained at fixed concentrations, the order of reaction with respect to the variable reactant can be determined. The overall order of reaction will be the sum of the orders with respect to each of the individual reactants.

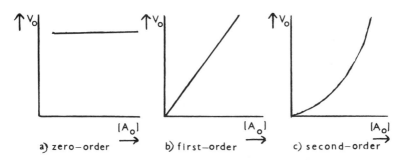

Fig. 6.4 Graph of initial velocity against initial reactant concentration for a single-reactant reaction.

6.4 KINETICS OF ENZYME-CATALYSED REACTIONS: AN HISTORICAL INTRODUCTION

Wilhelmy, in 1850, showed that the rate of acid-hydrolysis of sucrose was proportional to the sucrose concentration, at constant acid concentration. From the terms defined in the previous section, therefore, we would say that this reaction was first-order with respect to sucrose.

When the identical reaction, but catalysed by the enzyme **invertase**, was similarly investigated, different results were obtained. Brown (1902) showed that at low sucrose concentrations the reaction was again first-order with respect to the substrate, but at higher concentrations it became zero-order.

This was found to be generally true for all single-substrate enzyme-catalysed reactions and for multi-substrate reactions where the concentrations of all the substrates but one were kept constant. A graph of initial velocity (v_0) against initial substrate concentration ($[S_0]$) at constant total enzyme concentration ($[E_0]$) was found to be a rectangular hyperbola, as shown in Fig. 6.5.

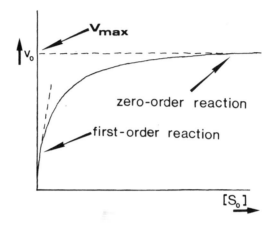

Fig. 6.5 Graph of initial velocity against initial substrate concentration at constant total enzyme concentration for a single-substrate enzyme-catalysed reaction.

Such a graph has the general equation

$$v_0 = \frac{a[S_0]}{[S_0] + b}$$

where a and b are constants. The constant a represents the maximum value of v_0 (called V_{max}) and b is the value of $[S_0]$ where $v_0 = \frac{1}{2} V_{max}$.

Although some enzymes do not give hyperbolic graphs (see Chapters 12 and 13), the attainment of a maximum initial velocity with increasing substrate concentration at constant total enzyme concentration is characteristic of all enzymes.

The explanation for this feature was first given by Brown, with particular reference to the hydrolysis of sucrose, but subsequently found to be of general application. Enzyme and substrate combine to form an **enzyme–substrate complex**, which undergoes a further reaction to breakdown to enzyme and products:

$$E + S \xrightarrow{\text{rate constant } k_1} ES \xrightarrow{\text{rate constant } k_2} E + P.$$

The overall rate of reaction (the rate of formation of P) must be limited by the amount of enzyme available and by the rate of breakdown of the enzyme-substrate complex. If the substrate concentration is sufficiently high it will 'saturate' the enzyme, i.e. force an immediate reaction with each available enzyme molecule to form an enzyme–substrate complex. Under these conditions, therefore, there will be no free enzyme present and the concentration of enzyme–substrate complex ([ES]) will be the total enzyme concentration present ($[E_0]$), making the overall rate of reaction $k_2[E_0]$ (from the **Law of Mass Action**). This is independent of substrate concentration and so cannot be increased by using still higher substrate concentrations. It is, therefore, the maximum initial velocity possible at this enzyme concentration.

Hence $V_{max} = k_2[E_0]$.

In contrast, at very low substrate enzyme concentrations the enzyme will be far from saturated and the overall rate of reaction will be limited by the rate at which enzyme and substrate molecules react to form an enzyme–substrate complex. At constant enzyme concentration this will be proportional to the substrate concentration and, therefore, a first-order reaction will result.

At intermediate substrate concentrations, the enzyme will be partially saturated with substrate and the order of reaction will be somewhere between zero-order and first-order.

Note that the terms 'saturated' and 'partially saturated' refer to the population of enzyme molecules, and not to each individual molecule: an enzyme molecule which binds a single substrate molecule cannot in itself be partially saturated at any given moment, but a population of such molecules can be, some being substrate-bound and some free.

The existence of an enzyme–substrate complex was first demonstrated experimentally in 1936 by spectroscopic studies on the enzymes **peroxidase** and **catalase**. Since then, the presence of such complexes has been amply documented by x-ray crystallography, esr, nmr, fluorimetry and other techniques.

6.5 METHODS USED FOR INVESTIGATING THE KINETICS OF ENZYME-CATALYSED REACTIONS

6.5.1 Initial Velocity Studies

Initial velocity studies have been found particularly useful for investigating the kinetics of enzyme-catalysed reactions. In addition to the general advantages of such studies, mentioned earlier, there is the extra reason here that enzymes are often unstable in solution, so the restriction of investigations to v_0 determinations give the best chance of avoiding errors caused by loss of enzyme activity with time.

The initial velocity of a reaction is usually determined from a graph of product concentration against time, as shown in Fig. 6.3; alternatively, a direct instrumental reading known to be proportional to product concentration (e.g.

absorbance units) may be plotted against time. The rate of reaction could also be determined from the disappearance of substrate as the reaction proceeds, but in general it is considered a better technique to measure the appearance of something from an initial value of nothing rather than the disappearance of something from an initial large value, particularly when it is the initial rate which is all-important.

The experiments are performed at constant temperature and pH using a method which enables the course of the reaction to be monitored continuously, and, preferably, automatically. The actual technique used will depend on the reaction being investigated.

If there is a difference between substrate and product in the absorbance of light of a particular wavelength, **spectrophotometric** or **colorimetric techniques** (preferably the former) may be used. The **molar extinction coefficient** of a substance, if known, will enable the actual concentration of that substance to be calculated from an absorbance reading:

$$\text{absorbance} = \text{molar extinction coefficient} \times \text{concentration (mol.l}^{-1})$$
$$\times \text{ cell light path (cm)}.$$

Of particular suitability for investigation by spectrophotometric techniques are the wide range of oxidation/reduction reactions involving $NAD^+/NADH$ or $NADP^+/NADPH$ interconversion. The reduced forms of these dinucleotides have an absorbance peak at 340 nm whereas the oxidised forms do not (Fig. 6.6), thus making it easy to follow the course of their interconversion by monitoring absorbance at this wavelength.

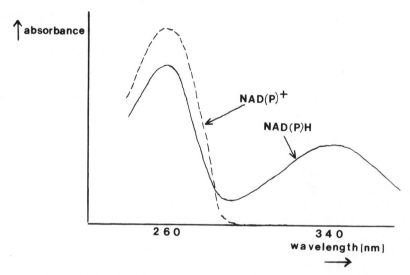

Fig. 6.6 The ultraviolet absorption spectra of $NAD(P)^+$ and $NAD(P)H$.

The interconversion of these substances may also be followed by **fluori-metric techniques**, which give greater sensitivity than spectrophotometric ones.

Reactions where gases are evolved or taken up as the reaction proceeds can be investigated by **manometric** techniques: the reaction is performed in an airtight vessel coupled to a monometer, which records changes in volume or pressure.

Ion-selective electrodes (e.g. the oxygen electrode) can be used where relevant to a particular reaction. More generally, other **electrochemical methods** (e.g. conductometry) can be applied wherever there are differences in electrical properties between substrate and product. These and other techniques are discussed in more detail in Chapter 18.1.

Methods which are not immediately applicable to the investigation of a particular reaction may nevertheless be used if the reaction is coupled to a suitable **indicator reaction**. The principle here is to add excess of the indicator enzyme, so the rate of the indicator reaction will be a measure of the rate of the reaction under investigation. For example, **glucose oxidase** catalyses:

$$\beta\text{-D-glucose} + O_2 \rightarrow \text{gluconolactone} + H_2O_2.$$

This reaction can be followed by manometric or oxygen electrode techniques, but is not suitable for investigation by spectrophotometric methods. However, if excess **peroxidase** is added to break down the hydrogen peroxide as it is formed, this reaction may be linked to the production of a coloured dye from a suitable chromogen, e.g. guaiacum:

$$H_2O_2 + \text{chromogen} \rightarrow H_2O + \tfrac{1}{2}O_2 + \text{dye}.$$

The rate of appearance of dye, determined spectrophotometrically, will be an indicator of the rate of reaction catalysed by glucose oxidase.

An alternative to the use of coupled reactions is to introduce an artificial substrate which, when reacting, results in a measurable change not provided by the natural substrate. An example is p-nitrophenyl phosphate, a colourless compound, which is hydrolysed to the yellow p-nitrophenol by the action of **alkaline phosphatase**.

6.5.2 Rapid-Reaction Techniques

In order to test some of the hypotheses made above, e.g. to investigate the formation and breakdown of the enzyme-substrate complex or complexes, it is necessary to use techniques capable of detecting changes taking place over time scales of the order of magnitude of 1 millisecond. Of particular importance is a detailed kinetic study of the changes taking place over the first fraction of a second of the reaction (termed **transient kinetics**). This is usually performed using rapid mixing techniques of the **continuous-flow** or **stopped-flow** variety, the reaction being monitored by some suitable detector coupled to an oscilloscope and usually to a computer (Fig. 6.7).

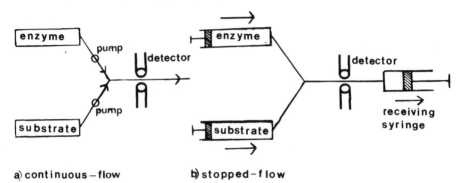

a) continuous – flow   b) stopped – flow

Fig. 6.7 Simplified diagrammatic representation of continuous-flow and stopped-flow techniques.

With **continuous-flow** systems, introduced by Hartridge and Roughton, streams of enzyme and substrate converge and are pumped together at a fixed speed down capillary-bore tubing to ensure rapid mixing and elimination of dead-space. From a knowledge of the dimensions of the tube and the flow rate, the time taken from mixing to the arrival at the detector can be calculated. By altering the flow rate, this time of reaction can be changed. Also, several detectors can be placed at different points along the flow tube. The main disadvantage of continuous-flow methods is that they tend to be wasteful of reagents.

With **stopped-flow** techniques, developed by Chance and by Gibson, solutions of enzymes and substrate are rapidly injected together into an observation chamber and the course of reaction monitored continuously.

The main limitation of both these techniques is the time taken to mix enzyme and substrate(s). This problem does not apply to the alternative approach, the investigation of **relaxation kinetics**. Here the enzyme and substrate(s) are mixed and the system allowed to come to equilibrium; then the position of the equilibrium is rapidly altered by a sudden temperature or pressure change, and the approach to the new position of equilibrium constantly monitored. This technique, pioneered by Eigen (1963), is particularly suitable for the investigation of reactions which are readily reversible, e.g. those catalysed by **aminotransferases**.

6.6 THE NATURE OF ENZYME CATALYSIS

Enzymes behave like any other catalysts in forming with the reactants a **transition-state** of lower **free energy** than that which would be found in the uncatalysed reaction. As we have seen, there is kinetic and spectroscopic evidence for the existence of **enzyme–substrate complexes**; however, these are not synonymous with transition-states. For a single-substrate reaction, the enzyme initially binds the substrate at a specific binding-site to form a relatively stable enzyme–substrate complex, this process taking place via the formation of an unstable transition-state. In the enzyme–substrate complex, the reacting groups are held in close proximity to each other and to the catalytic-site of the enzyme. The catalysed reaction can now take place, via the formation of another unstable transition-state, to give the product. However, at this point the product may well still be bound to the enzyme, in which case another relatively stable reaction intermediate, the **enzyme–product complex**, would exist before the free product was liberated. The free energy profile of a reaction of this type is depicted in Fig. 6.8. For reactions involving several substrates and products, the process would be even more complicated, but the principles remain the same.

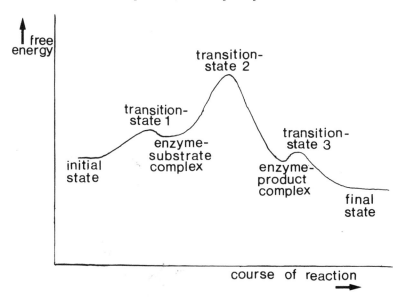

Fig. 6.8 The free energy profile of an enzyme-catalysed reaction involving the formation of an enzyme–substrate and an enzyme–product complex.

The activation energy for the rate-limiting step of enzyme-catalysed reactions may be estimated as for uncatalysed reactions by investigating the rate of reaction at different temperatures.

For a reaction of the form:

$$E + S \xrightarrow{k_1} ES \xrightarrow{k_2} E + P$$

$$\text{or} \quad E + S \xrightarrow{k_1} ES \xrightarrow{k_2} EP \xrightarrow{k_3} E + P$$

(where ES → EP is the rate-limiting-step)

the overall rate of reaction is given by $k_2[ES]$ and the best estimate of k_2 comes from V_{max}, since $V_{max} = k_2[E_0]$.

For such a reaction investigated at absolute temperatures T_1 and T_2, at the same initial enzyme concentration,

$$\text{Activation energy} = 2.303 \, R \left(\frac{T_1 T_2}{T_1 - T_2} \right) \log_{10} \left(\frac{k_2 \text{ at } T_1}{k_2 \text{ at } T_2} \right)$$

$$= 2.303 \, R \left(\frac{T_1 T_2}{T_1 - T_2} \right) \log_{10} \left(\frac{V_{max} \text{ at } T_1}{V_{max} \text{ at } T_2} \right)$$

(the terms being as defined in Chapter 6.2.2).

Relatively small changes in activation energy can greatly alter the rate of reaction: an enzyme which reduces the activation energy from $100 \, kJ.mol^{-1}$ to $60 \, kJ.mol^{-1}$, perfectly reasonable figures, increases the reaction rate by about 10 million.

SUMMARY OF CHAPTER 6

Biochemical processes can only proceed spontaneously in such a direction that the **free energy** of the system, i.e. the energy that can be used to perform work, decreases. However, even energetically-favourable chemical reactions have to overcome a **potential-barrier**, known as the **activation energy**, before the reaction can take place. This is explained by the need to form unstable **transition-states**. **Catalysts**, including enzymes, act by allowing the formation of different, more stable, transition-states and, thus, reduce the activation energy. The position of **chemical equilibrium** is unchanged but is reached much faster than in the equivalent uncatalysed reaction.

The **initial velocity** of enzyme-catalysed reactions has a limiting value at each total enzyme concentration; this occurs when the enzyme is saturated with substrate. Enzymes react with substrates to form **enzyme–substrate complexes**; these are quite distinct from the transition-states which also occur as part of the process of enzyme catalysis.

FURTHER READING

Dawes, E. A. (1980), *Quantitative Problems in Biochemistry* (Chapters 3–5 including problems), Sixth Edition, Longman.

Morris, J. G. (1974), *A Biologist's Physical Chemistry* (Chapters 7–10 including problems), Second Edition, Arnold.

PROBLEMS

(Assume $R = 8.314$ J.K^{-1}.mol^{-1}, $F = 96487$ J.V^{-1}, activity = concentration.)

6.1 (a) The following reaction characteristics have been demonstrated at pH 7.5 and 310°K:

aspartate + citrulline $\rightleftharpoons$ argininosuccinate + H_2O $K'_{eq} = 1.6 \times 10^{-6}$

arginine + fumarate $\rightleftharpoons$ argininosuccinate $K'_{eq} = 93$

arginine + H_2O $\rightleftharpoons$ citrulline + NH_4^+ $K'_{eq} = 1.4 \times 10^5$

aspartate + H_2O $\rightleftharpoons$ malate + NH_4^+ $K'_{eq} = 7.5 \times 10^{-3}$.

Calculate $\Delta G^{\oplus'}$ at 310°K and pH 7.5 for the following reaction:

fumarate + H_2O $\rightleftharpoons$ malate.

If $\Delta H^{\oplus'}$ for this reaction is -16.6 kJ.mol^{-1}, what will be the value of $\Delta S^{\oplus'}$ under these conditions?

(b) If the concentration of fumarate is 10^{-4} mol.l^{-1} and of malate 9×10^{-4} mol.l^{-1}, calculate ΔG for the formation of malate from fumarate at 310°K and pH 7.5. If a system with these initial concentrations was allowed to proceed to equilibrium, deduce the final concentrations of fumarate and malate.

6.2 For the reaction:

malate + NAD^+ $\rightleftharpoons$ oxaloacetate + NADH + H^+,

$\Delta G^{\oplus'} = 27.96$ kJ.mol^{-1} at pH 7.0 and 25°C.

(a) Calculate the difference in $E'_{\oplus}$ between the oxaloacetate|malate couple and the NAD^+|NADH + H^+ couple at pH 7.0 and 25°C.

(b) Calculate the value of ΔG for the reaction at pH 7.0 and 25°C when malate and NAD^+ are both present at 0.01 mol.l^{-1} and oxaloacetate and NADH are both present at 0.02 mol.l^{-1}.

(c) Calculate the concentration of oxaloacetate equal to concentration of NADH present in equilibrium with 0.05 mol.l^{-1} each of malate and NAD^+ at pH 7.0 and 25°C.

6.3 A reaction, in the presence of a certain amount of enzyme, was found to have a V_{max} of 61.7 nmol.s^{-1} at 25°C. In the presence of the same amount of enzyme at 37°C, the V_{max} was 120.8 nmol.s^{-1}. Using sensitive detection techniques it was possible to follow the course of the uncatalysed reaction, which had a rate of 30.2×10^{-3} nmol.s^{-1} at 25°C and 88.4×10^{-3} nmol.s^{-1} at 37°C. Calculate the activation energy for the catalysed and uncatalysed reactions.

6.4   For an uncatalysed reaction $A + B \rightleftharpoons C + D$, the progress of the reaction was followed spectrophotometrically for different initial concentrations of A, the initial concentration of B always being 10 nmol.1^{-1}. No products were present initially in any experiment. The following results were obtained:

Initial concentration of A (mmol.1^{-1})	Absorbance at time t (min.)					
	t = 0.5	1.0	1.5	2.0	2.5	3.0
10.0	0.073	0.127	0.180	0.234	0.272	0.292
8.0	0.062	0.105	0.148	0.191	0.224	0.247
6.0	0.051	0.085	0.116	0.149	0.175	0.191
4.0	0.040	0.061	0.183	0.104	0.123	0.135
2.0	0.030	0.041	0.051	0.062	0.070	0.075

Determine the order of reaction with respect to A.

6.5(a)   The following data were obtained at pH 7.5 and 37°C:

for the reaction glucose 6-P $\rightleftharpoons$ glucose 1-P, $K'_{eq} = 6.12 \times 10^{-2}$;
for the reaction fructose 6-P + UTP $\rightleftharpoons$ UDP-glucose + $2P_i$, $K'_{eq} = 4.52 \times 10^3$;
for the reaction glucose 1-P + UTP $\rightleftharpoons$ UDP-glucose + $(PP)_i$, $K'_{eq} = 1.00$;
for the reaction $2P_i \rightleftharpoons (PP)_i$, $K'_{eq} = 3.06 \times 10^{-5}$.

From this, determine the K_{eq} at pH 7.5 and 37°C for the reaction:

glucose 6-P $\rightleftharpoons$ fructose 6-P.

If the initial concentration of glucose 6-P is 9.0 mmol.1^{-1} and of fructose 6-P is 2.0 mmol.1^{-1}, what is the value of ΔG under these conditions?

(b)   The reaction in the presence of hexose phosphate isomerase was found to have a V_{max} of 3.6 μmol.min^{-1} at 25°C and of 7.0 μmol.min^{-1} at 37°C, the conditions being identical in all other respects. The reaction in the absence of enzyme proceded at a rate of 5.4 nmol.min^{-1} at 37°C, and of 15.5 nmol.min^{-1} at 50°C, the conditions being identical in all other respects. Determine the activation energies of the catalysed and uncatalysed reactions. What assumptions are being made in the above calculations?

CHAPTER 7

Kinetics of Single-Substrate Enzyme-Catalysed Reactions

7.1 THE RELATIONSHIP BETWEEN INITIAL VELOCITY AND SUBSTRATE CONCENTRATION

7.1.1 The Henri and Michaelis–Menten Equations

In Chapter 6.4 we discussed experimental evidence which showed that for many single-substrate and pseudo single-substrate enzyme-catalysed reactions there was a hyperbolic relationship between initial velocity v_o and initial substrate concentration $[S_o]$ (see Fig. 6.5) so that:

$$v_o = \frac{V_{max}[S_o]}{[S_o] + b}$$

where V_{max} is the maximum v_o at a particular total enzyme concentration and b is another constant.

Kinetic models to explain these findings were proposed by Henri (1903) and Michaelis and Menten (1913). These were essentially similar, but Michaelis and Menten did a great deal of experimental work to give their treatment a sounder basis, e.g. unlike Henri they recognised the importance of using initial velocity v_o rather than any velocity v.

Let us consider a single-substrate enzyme-catalysed reaction where there is just one substrate-binding site per enzyme. The simplest general equation for such a reaction would be:

$$E + S \underset{k_{-1}}{\overset{k_1}{\rightleftharpoons}} ES \underset{k_{-2}}{\overset{k_2}{\rightleftharpoons}} E + P$$

If investigations are restricted to the initial period of the reaction, the product concentration is negligible and the formation of ES from product can be ignored. Under these conditions, therefore, the reaction simplifies still further to:

$$E + S \underset{k_{-1}}{\overset{k_1}{\rightleftharpoons}} ES \overset{k_2}{\rightarrow} E + P.$$

The rate of formation of ES at any time t (within the initial period when the product concentration is negligible) $= k_1[E][S]$, where [E] is the concentration of free enzyme and [S] the concentration of free substrate at time t.

Also at time t, the rate of breakdown of ES back to E and $S = k_{-1}[ES]$, where [ES] is the concentration of enzyme-substrate complex at this time.

The **Michaelis-Menton assumption** was that an **equilibrium** between enzyme, substrate and enzyme-substrate complex was almost instantly set up and maintained, the breakdown of enzyme-substrate complex to products being too slow to disturb this equilibrium. Using this assumption, therefore:

$$k_1[E][S] = k_{-1}[ES].$$

The constants may then be separated from the variables, giving:

$$\frac{[E][S]}{[ES]} = \frac{k_{-1}}{k_1} = K_s \quad \text{(where } K_s \text{ is the dissociation constant of ES)}.$$

The total concentration of enzyme present $[E_o]$ must be the sum of the concentration of free enzyme [E] and the concentration of bound enzyme [ES]. Therefore, in order to involve total enzyme concentration (a quantity which can easily be determined and specified) into the above equation, the following substitution is made:

$$[E] = [E_o] - [ES].$$

$$\therefore \frac{([E_o] - [ES])[S]}{[ES]} = K_s$$

$$\therefore K_s[ES] = ([E_o] - [ES])[S]$$

$$= [E_o][S] - [ES][S]$$

$$\therefore [ES][S] + K_s[ES] = [E_o][S]$$

$$\therefore [ES]([S] + K_s) = [E_o][S]$$

$$\therefore [ES] = \frac{[E_o][S]}{[S] + K_s}.$$

[ES] has been isolated in this way because, as discussed in Chapter 6.4, this term governs the rate of formation of products (the overall rate of reaction) according to the relationship:

$$v_o = k_2[ES].$$

If we substitute the expression for [ES] derived above, we obtain:

$$v_0 = \frac{k_2[E_0][S]}{[S] + K_s}.$$

Moreover, we know that when the substrate concentration is very high, all the enzyme is present as the enzyme-substrate complex and the limiting initial velocity V_{max} is reached. Under these conditions:

$$V_{max} = k_2[E_0].$$

Therefore, we can substitute V_{max} for $k_2[E_0]$ in the expression for v_0 and get:

$$v_0 = \frac{V_{max}[S]}{[S] + K_s}.$$

It was further assumed by Michaelis and Menten that the substrate was usually present in much greater concentrations than the enzyme. This is generally true, for enzymes, like all catalysts, are often present at very low concentrations. It is important to realise that we can talk about the substrate concentration being "low" and giving first-order kinetics with a very small degree of **saturation** of the enzyme while at the same time the substrate concentration may be a thousand times that of the enzyme: this is valid because the formation of the enzyme-substrate complex from the enzyme and substrate is a reversible process.

If we make the assumption that the initial substrate concentration $[S_0]$ is very much greater than the initial enzyme concentration $[E_0]$, then the formation of the enzyme-substrate complex will result in an insignificant change in free substrate concentration. Hence, in the expression for v_0 derived above, we can substitute $[S_0]$ (a quantity easily specified) for [S], giving:

$$v_0 = \frac{V_{max}[S_0]}{[S_0] + K_s}.$$

This is an equation of the form required to explain the experimental findings. However it is unsatisfactory in that the Michaelis-Menten equilibrium-assumption cannot be generally applicable: some, possibly many, enzyme-catalysed reactions are likely to proceed at rates fast enough to disturb such an equilibrium.

7.1.2 The Briggs-Haldane Modification of the Michaelis-Menten Equation

The equation derived by Michaelis and Menten (Chapter 7.1.1) was modified by Briggs and Haldane (1925) who introduced a more generally valid assumption, that of the **steady-state**. They argued that since the concentration of enzyme, and thus enzyme-substrate complex, was usually very small compared with the substrate concentration, then the rate of change of [ES] would be negligible compared to the rate of change of [P] over the initial period of the reaction, except during the very brief period at the beginning while *ES* was first being formed. In the absence of product, the contration of *ES* would be determined by the total enzyme concentration, which remains constant throughout, and by the substrate concentration, which changes by a negligible amount, as a percentage of its initial value, over the period of interest. So, once this complex had been produced it would be maintained in a steady-state, i.e. it would be broken down as fast as it was being formed, [ES] remaining constant.

Rapid reaction studies have shown that this is a reasonable assumption for most enzyme-catalysed reactions under these conditions: the steady-state is usually established within a few milliseconds of the start of the reaction and is maintained for a few minutes until the product concentration, and hence the rate of the reverse reaction, becomes significant.

If we again consider a single-substrate single-binding-site reaction:

$$E + S \underset{k_{-1}}{\overset{k_1}{\rightleftharpoons}} ES \overset{k_2}{\rightarrow} E + P$$

the rate of formation of *ES* at any time t (within the initial period when the product concentration is negligible) $= k_1[E][S]$.

The rate of breakdown of *ES* at this time $= k_{-1}[ES] + k_2[ES]$, since *ES* can break down to form products or reform reactants.

Using the steady-state assumption:

$$k_1[E][S] = k_{-1}[ES] + k_2[ES] = [ES](k_{-1} + k_2).$$

Separating the constants from the variables:

$$\frac{[E][S]}{[ES]} = \frac{k_{-1} + k_2}{k_1} = K_m \qquad \text{(where } K_m \text{ is another constant).}$$

Substituting $[E] = [E_o] - [ES]$ as before:

$$\frac{([E_o] - [ES])[S]}{[ES]} = K_m$$

from which

$$[ES] = \frac{[E_o][S]}{[S] + K_m}.$$

Again, since $v_0 = k_2 [ES]$,

$$v_0 = \frac{k_2 [E_0][S]}{[S] + K_m}$$

and since $V_{max} = k_2 [E_0]$,

$$v_0 = \frac{V_{max}[S]}{[S] + K_m}.$$

Finally, since the substrate concentration is usually much greater than the enzyme concentration, $[S] \doteq [S_0]$, so

$$v_0 = \frac{V_{max}[S_0]}{[S_0] + K_m} \text{ at constant } [E_0].$$

This has the same form as the equation derived by Michaelis and Menten: only the definition of the constant in the denomenator has changed. Hence this equation has retained the name **Michaelis–Menten equation** and K_m is called the **Michaelis constant**. The **equilibrium-assumption** is regarded as being a special case of the more general **steady-state-assumption**, occurring where $k_{-1} \gg k_2$.

A graph of v_0 against $[S_0]$ will have the form of a rectangular hyperbola (Fig. 7.1), consistent with experimental findings for many enzyme-catalysed reactions.

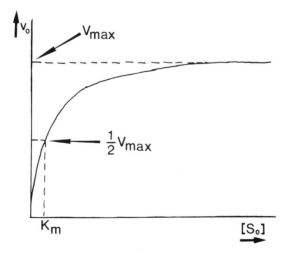

Fig. 7.1 Graph of v_0 against $[S_0]$ at constant $[E_0]$ for a single-substrate enzyme-catalysed reaction, from the Michaelis–Menten equation.

V_{max}, the maximum initial velocity at a particular $[E_0]$, can be obtained from the graph as shown in Fig. 7.1. It will have the same units as v_0.

K_m can also be obtained from the graph, from the following reasoning. When $v_o = \frac{1}{2} V_{max}$,

$$\frac{V_{max}}{2} = \frac{V_{max}[S_o]}{[S_o] + K_m}$$

$$\therefore (V_{max})([S_o] + K_m) = 2(V_{max})[S_o]$$

$$\therefore K_m = [S_o].$$

Therefore, K_m is the value of $[S_o]$ which gives an initial velocity equal to $\frac{1}{2} V_{max}$. Also, K_m will have the same units as $[S_o]$. (However, note that the K_m measured in this way may be different from the true K_m, which is defined in terms of rate constants, if there are any inhibitors or other complicating factors present, as discussed in Chapters 8, 14.2 and 20.2.2: in these circumstances, the measured value may be termed the **apparent K_m**.)

7.1.3 The Significance of the Michaelis–Menten Equation

The Michaelis–Menten equation, both in its original form and as modified by Briggs and Haldane, is derived with respect to a single-substrate enzyme-catalysed reaction with one substrate-binding site per enzyme and involving the formation of a single intermediate complex. Such reactions are very rare indeed. However the requirement for a single-substrate reaction can be taken to include pseudo single-substrate reactions and also variations with respect to one substrate in a multi-substrate reaction, provided the other substrates are maintained at a constant concentration. Similarly the requirement for an enzyme with a single substrate-binding site can include enzymes with more than one binding-site for the substrate in question, provided there is no interaction between the binding sites (see Chapters 12 and 13). Finally, the requirement for the formation of a single intermediate complex can include reactions of the type:

$$E + S \underset{k_{-1}}{\overset{k_1}{\rightleftharpoons}} ES \overset{k_2}{\rightarrow} EP \overset{k_3}{\rightarrow} E + P$$

provided ES → EP is the rate-limiting-step of the overall reaction (see Chapter 6.6).

The constant k_{cat}, called the **turnover number**, is often applied to enzyme-catalysed reactions: this is obtained from the general expression $V_{max} = k_{cat}[E_o]$. It represents the maximum number of substrate molecules which can be converted to products per molecule enzyme per unit time. For reactions of the simple type discussed above, $k_{cat} = k_2$; for more complex reactions, k_{cat} will be a function involving several individual rate constants. The turnover number for most enzymes lies in the range $1-10^4$ per second.

In practice, it is found that the Michaelis–Menten equation is applicable to a great many enzyme-catalysed reactions and the constants V_{max} and K_m can be determined. V_{max} varies with the total concentration of enzyme present, but K_m is independent of enzyme concentration and is characteristic of the system being investigated. It can thus be used to identify a particular enzyme; in most cases, K_m values lie in the range 10^{-2}-10^{-6} mol.l^{-1}. Also, since the catalytic step (ES → E + P, or ES → EP) is often the rate-limiting step, k_1 and k_{-1} are frequently much larger than k_2; where this is the case, the Michaelis-Menten equilibrium is valid and $K_m \simeq K_s$. In general, K_s gives an indication of the **affinity** of the enzyme for the substrate: a low K_s value indicates a high affinity of enzyme for substrate, whereas a high K_s value indicates a low affinity.

If we turn our attention to processes taking place within the living cell, one of the factors which determines which of the several alternative metabolic pathways a substance enters is the K_m value of the first enzyme in each pathway. Consider, for example, the fate of the hexose-phosphates, glucose 6-phosphate, glucose 1-phosphate and fructose 6-phosphate, which are readily interconvertible in the cell and can be regarded as a single unit. The next step in the direction of glycolysis is mediated by **phosphofructokinase**, which has a much lower K_m for its substrate fructose 6-phosphate than the first enzyme in the direction of glycogen synthesis has for its substrate glucose 1-phosphate. At most concentrations of hexose-phosphate, therefore, phosphofructokinase will be more saturated with substrate than the enzymes of glycogen synthesis and so most of the hexose-phosphate will be metabolised via glycolysis. Only at high hexose-phosphate concentrations when phosphofructokinase is fully saturated with substrate will glycogen synthesis become significant. Of course in general the fate of a metabolite depends on other factors besides K_m, including the concentration of each enzyme, the value of V_{max} at this enzyme concentration and the effects of activators and inhibitors (see Chapter 14.2).

It should also be realised that steady-state kinetic constants are determined in highly-purified solutions by *in vitro* laboratory experiments, usually at the pH optimum of the enzyme under investigation. This is not necessarily the same pH as that *in vivo* in the living organism where the enzyme functions, since many enzymes are found in the same environment and not all will have an identical pH optimum, even though of course they must all have some activity at the physiological pH. Similarly the temperature at which the *in vitro* experiment is conducted may not be the *in vivo* temperature. Also, steady-state experiments are usually performed at enzyme concentrations much lower than those found *in vivo*. (This is not necessarily the case with rapid-reaction methods, discussed in Chapter 7.2). Even more important, factors absent from *in vitro* experiments, e.g. membranes, may contribute to the action of an enzyme *in vivo*. Thus despite the usefulness of *in vitro* kinetic studies, it cannot be assumed that they determine exactly how the enzyme behaves in the living organism.

From a practical viewpoint, a knowledge of the K_m of an enzyme is invaluable when assaying that enzyme (Chapter 15.2): enzyme assays should be performed with the enzyme fully saturated with substrate. From the Michaelis-

Menten equation, saturation is approached tangentially and only actually achieved at a substrate concentration of infinity. Nevertheless, if the K_m is known, the equation enables an initial substrate concentration to be calculated which, for all practical purposes, can be regarded as saturating the enzyme. For example, if $[S_o] = 100 \, K_m$, then $v_o = 0.99 \, V_{max}$, irrespective of the actual enzyme concentration but provided it is much less than the substrate concentration.

7.1.4 The Lineweaver-Burk Plot

The graph of the Michaelis–Menten equation, v_o against $[S_o]$ (Fig. 7.1), is not entirely satisfactory for the determination of V_{max} and K_m. Unless, after a series of experiments, there are at least three consistent points on the plateau of the curve at different $[S_o]$ values, then an accurate value of V_{max}, and hence of K_m, cannot be obtained: the graph, being a curve, cannot be accurately extrapolated upwards from non-saturating values of $[S_o]$.

Lineweaver and Burk (1934) overcame this problem without making any fresh assumptions. They simply took the Michaelis–Menten equation:

$$v_o = \frac{V_{max}[S_o]}{[S_o] + K_m}$$

and inverted it:

$$\frac{1}{v_o} = \frac{[S_o] + K_m}{V_{max}[S_o]} = \frac{[S_o]}{V_{max}[S_o]} + \frac{K_m}{V_{max}[S_o]}.$$

$$\therefore \frac{1}{v_o} = \frac{K_m}{V_{max}} \cdot \frac{1}{[S_o]} + \frac{1}{V_{max}} \quad \text{(The \textbf{Lineweaver-Burk equation}).}$$

This is of the form $y = mx + c$, which is the equation of a straight line graph; a plot of y against x has a slope m and intercept c on the y-axis.

A plot of $\dfrac{1}{v_o}$ against $\dfrac{1}{[S_o]}$ (The **Lineweaver-Burk plot**) for systems obeying the Michaelis-Menten equation is shown in Fig. 7.2.

The graph, being linear, can be extrapolated even if no experiment has been performed at a saturating substrate concentration, and from the extrapolated graph the values of K_m and V_{max} can be determined as shown in Fig. 7.2. Departure from linearity for a particular enzyme-catalysed reaction indicates that the assumptions inherent in the Michaelis–Menten equation are not valid in this instance.

7.1.5 The Eadie–Hofstee and Hanes Plots

The Lineweaver-Burk plot has been criticised on several grounds. Firstly, and of least importance, the extrapolation across the $\dfrac{1}{v_o}$ axis to determine the value

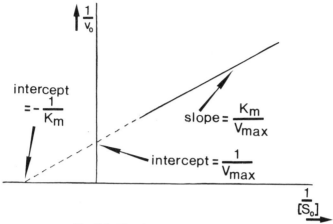

Fig. 7.2 The Lineweaver–Burk plot.

of $-\dfrac{1}{K_m}$ sometimes reaches the edge of the graph paper before reaching the $\dfrac{1}{[S_o]}$ axis, possibly resulting in the graph having to be redrawn with altered axes. Secondly, it is said to give undue weight to measurements made at low substrate concentrations, when results are likely to be most inaccurate (this criticism should be borne in mind when a Lineweaver-Burk plot is being drawn). Thirdly, departures from linearity are less obvious than in some other plots, particularly the Eadie-Hofstee and Hanes plots. This could be very important if a reaction mechanism was being investigated.

The **Eadie-Hofstee plot** takes as its starting point the Lineweaver-Burk equation, based in turn on the Michaelis-Menten equation. Both sides of the equation are multiplied by the factor $v_o.V_{max}$:

$$\frac{1}{(\not{v_o})}.(\not{v_o}).V_{max} = \frac{K_m}{(V_{\not{max}})}.\frac{1}{[S_o]}.v_o.(V_{\not{max}}) + \frac{1}{(V_{\not{max}})}.v_o.(V_{\not{max}}).$$

$$\therefore v_o = -K_m.\frac{v_o}{[S_o]} + V_{max}.$$

Again this is the equation of a straight line graph, from which V_{max} and K_m can be determined as shown in Fig. 7.3(a).

The **Hanes plot** similarly starts with the Lineweaver-Burk equation, which in this instance is multiplied throughout by $[S_o]$:

$$\frac{1}{v_o}.[S_o] = \frac{K_m}{V_{max}}.\frac{1}{[\not{S_o}]}[\not{S_o}] + \frac{1}{V_{max}}.[S_o].$$

$$\therefore \frac{[S_o]}{v_o} = \frac{1}{V_{max}}.[S_o] + \frac{K_m}{V_{max}}.$$

Once more this gives a linear plot, from which V_{max} and K_m can be obtained as in Fig. 7.3 (b).

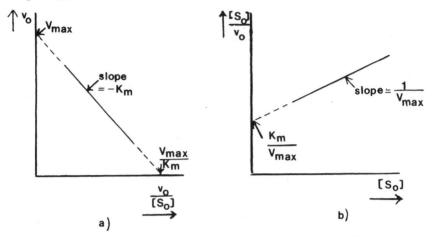

Fig. 7.3 (a) The Eadie–Hoffstee plot; (b) the Hanes plot.

The Eadie–Hofstee and Hanes plots are favoured by enzyme kineticists, but the Lineweaver-Burk plot continues to be widely used by enzymologists in general. The important thing is to obtain good experimental data covering a wide range of substrate concentrations which are chosen so that the points will be evenly spread over the plot being used.

Computerised data processing, usually based on the least-squares approach to curve fitting and sometimes incorporating automatic rejection of points outside an arbitrarily-set limit, gives a quick and convenient method of obtaining an estimate of K_m and V_{max}. However the results are not entirely reliable, and should never be accepted blindly: it is important to consider the actual graphs obtained, particularly to see if there is any evidence of a departure from linearity.

7.1.6 The Eisenthal and Cornish–Bowden Plot
In view of the difficulties in obtaining completely reliable estimates of K_m and V_{max}, even with the aid of statistical analysis, from the plots discussed above, Eisenthal and Cornish-Bowden (1974) suggested a different approach, though one still based on the Michaelis-Menten equation. The reciprocal form of this equation, at constant $[E_o]$, gives

$$\frac{1}{v_o} = \frac{K_m + [S_o]}{V_{max}[S_o]}.$$

$$\therefore \frac{V_{max}}{v_o} = \frac{K_m + [S_o]}{[S_o]} = \frac{K_m}{[S_o]} + 1$$

Therefore, at constant v_o and $[S_o]$, a plot of V_{max} against K_m is linear. The reader can be forgiven for feeling uneasy about this suggestion, for, at first sight, it seems nonsense to plot a constant (V_{max}) against another constant (K_m); nevertheless, the relationship is mathematically sound.

When $K_m = 0$, $V_{max} = v_o$, and when $V_{max} = 0$, $K_m = -[S_o]$. Therefore, each $v_o, [S_o]$ pair can be used to generate a line by marking v_o on the V_{max} axis and $-[S_o]$ on the K_m axis; these two points are then joined up and the line extrapolated. Lines for all $v_o, [S_o]$ pairs, at constant $[E_o]$, must pass through the true values of K_m and V_{max}, so these values must be given by the point of intersection of all the lines. However, because of experimental error, the lines are likely to intersect over a range of values (Fig. 7.3(c)). Nevertheless, in the opinion of many enzymologists, the best available estimates of K_m and V_{max} are the medians of the values obtained from the intersections of these lines, provided there is no evidence from other plots of departures from linearity.

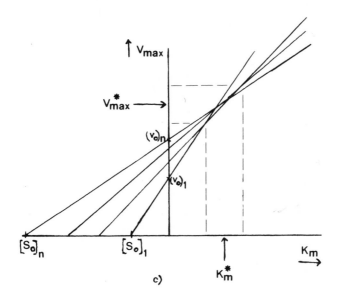

Fig. 7.3   (c) Typical Eisenthal–Cornish–Bowden plot for an enzyme-catalysed reaction at constant $[E_o]$. (V_{max}^* is the best estimate of V_{max}, K_m^* the best estimate of K_m.) See text for discussion.

7.1.7   The Haldane Relationship for Reversible Reactions

All reactions are to some degree reversible, and many enzyme-catalysed reactions can function in either direction within the cell. It is therefore of interest to compare the kinetics of the forward and back reactions.

Consider a single-substrate reaction, $S \rightleftharpoons P$, proceeding via the formation of a single intermediate complex: in the forward direction this would be regarded as an ES complex, but in the reverse direction the same complex would be called EP.

$$S + E \underset{k_{-1}}{\overset{k_1}{\rightleftharpoons}} ES/EP \underset{k_{-2}}{\overset{k_2}{\rightleftharpoons}} P + E.$$

The Michaelis-Menten equation in the forward direction, at fixed $[E_o]$, gives:

$$v_f = \frac{V_{max}^s[S_o]}{[S_o] + K_m^s}$$

where v_f is the initial velocity in the forward direction,
         V_{max}^s is the maximum initial velocity in the forward direction and
         $K_m^s = \dfrac{k_{-1} + k_2}{k_1}$.

The Michaelis-Menten equation for the reverse reaction (at constant $[E_o]$) gives:

$$v_b = \frac{V_{max}^P[P_o]}{[P_o] + K_m^P}$$

where v_b is the initial velocity in the back direction,
         V_{max}^P is the maximum initial velocity for the back reaction and
         $K_m^P = \dfrac{k_{-1} + k_2}{k_{-2}}$.

Haldane derived a useful relationship between the kinetic constants and the equilibrium constant of the reaction. At equilibrium the rate of the forward reaction equals the rate of the back reaction. Under these conditions:

$$k_{-1}[ES] = k_1[E][S].$$

$$\therefore \frac{[ES]}{[E]} = \frac{k_1}{k_{-1}}[S].$$

Also under these conditions:

$$k_2[ES] = k_{-2}[E][P].$$

$$\therefore \frac{[ES]}{[E]} = \frac{k_{-2}}{k_2}[P] = \frac{k_1}{k_{-1}}[S]$$

$$\therefore K_{eq} = \frac{[P]}{[S]} = \frac{k_1}{k_{-1}} \cdot \frac{k_2}{k_{-2}}$$

But $\quad V_{max}^s = k_2[E_o]$

and $\quad V_{max}^p = k_{-1}[E_o]$.

$$\therefore \frac{V_{max}^s}{V_{max}^p} = \frac{k_2}{k_{-1}}.$$

Also, $\quad \dfrac{K_m^s}{K_m^p} = \dfrac{(k_{-1} + k_2)}{k_1} \cdot \dfrac{k_{-2}}{(k_{-1} + k_2)} = \dfrac{k_{-2}}{k_1}$

$$\therefore K_{eq} = \frac{k_1 k_2}{k_{-1} k_{-2}} = \frac{V_{max}^s K_m^p}{V_{max}^p . K_m^s} \qquad \text{(the \textbf{Haldane relationship})}$$

If the equilibrium constant is known, this relationship can be used to check the validity of the kinetic constants which have been determined.

In general, it is likely that the K_m for the reaction in the metabolically-important direction will be less than that for the reaction in the opposite direction, since K_m often varies inversely with affinity (see Chapter 7.1.3). However, it should be realised that the direction of metabolic flow depends also on the concentrations of S and P present within the cell.

7.2 RAPID REACTION KINETICS

7.2.1 Pre-Steady-State Kinetics

Although investigations of steady-state kinetics are of great importance to the enzymologist, as discussed in Chapter 7.1, they have severe limitations. They allow the calculation of the kinetic constants K_m and k_{cat}, but the meaning of these constants depends on the assumptions made, such as the number of intermediates involved and the rates of their interconversion. These assumptions cannot be tested by ordinary "test-tube" methods. The turnover number, k_{cat}, of many enzymes is in the order of 100 s^{-1}, i.e. 100 molecules of product can be produced per second per molecule of enzyme, which means that the slowest step in the mechanism is likely to have a half-life of only a few milliseconds. However, continuous-flow and stopped-flow rapid-mixing methods (Chapter 6.5.2) can be designed to follow the build-up and decay of reaction intermediates over such a time-scale and they have been widely used for the study of enzyme-catalysed reactions. The experimental results can be compared with the theoretical results for a particular mechanism, which are often produced with the help of a computer. Gutfreund and others have obtained simulated results for a variety of reaction mechanisms.

The simplest possible reaction is a first-order reaction of the form:

$$A \xrightarrow{k} B.$$

The rate of reaction at time t is given by:

$$\frac{-d[A]}{dt} = \frac{+d[B]}{dt} = k[A].$$

Integration of this gives:

$$\log_e[A_0] - \log_e[A] = kt$$

(where $[A_0]$ is the initial concentration of A)

or $[A] = [A_0]e^{-kt}.$

The course of this reaction is depicted in Fig. 7.4.

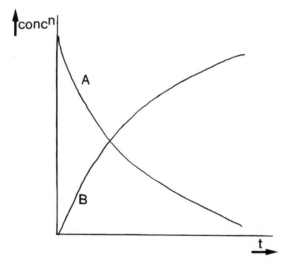

Fig. 7.4 Theoretical progress curves for a reaction of the form A → B.

The integrated equations for most reaction mechanisms are far more complicated than this. Let us examine the course of reaction for a single-substrate enzyme-catalysed reaction involving a single intermediate complex, where the initial substrate concentration is much greater than the initial enzyme concentration:

$$E + S \underset{k_{-1}}{\overset{k_1}{\rightleftharpoons}} ES \underset{k_{-2}}{\overset{k_2}{\rightleftharpoons}} E + P.$$

The rate of increase of [ES] at time t (within the initial period when [P] is negligible) is given by:

$$\frac{d[ES]}{dt} = k_1[E][S] - k_{-1}[ES] - k_2[ES]$$

$$= k_1([E_o] - [ES])([S_o] - [ES] - [P]) - k_{-1}[ES] - k_2[ES].$$

Since $[S_o] \gg [E_o]$,   then $([S_o] - [ES] - [P]) \simeq [S_o]$.

$$\therefore \frac{d[ES]}{dt} = k_1([E_o] - [ES])[S_o] - k_{-1}[ES] - k_2[ES].$$

This may be integrated to give an expression showing the change in [ES] with time. The integration is beyond the scope of this book, but the results are illustrated in Fig. 7.5. The changes in concentration with time for all the other participants in the reaction can be calculated in similar fashion and these are also shown in Fig. 7.5.

The linear part of the [P] against t graph represents the steady-state phase of the reaction and has a slope $= \dfrac{k_2[E_o][S_o]}{[S_o] + K_m}$ (obtained by substituting K_m for

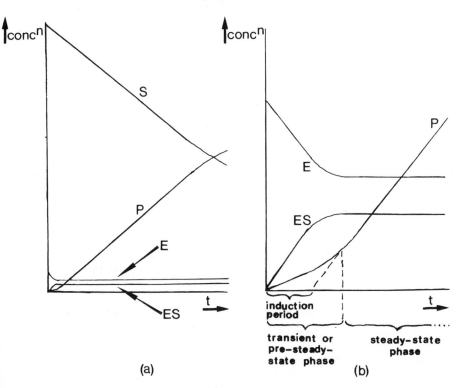

Fig. 7.5 (a) Theoretical progress curve for a reaction of the form $E + S \rightleftharpoons ES \rightarrow E + P$, where $[S_o] \gg [E_o]$; (b) the same, showing the initial period in greater magnification.

$\dfrac{k_{-1} + k_2}{k_1}$ in the integrated equation). If the linear, steady-state portion of this graph is extrapolated downwards it intercepts the t-axis where $t = \dfrac{1}{k_1[S_o] + K_m}$. This is called the **induction period**. If the experimental results are consistent with such a mechanism (e.g. no other reaction intermediate is implicated) and if K_m is known from steady-state investigations, then k_1, k_2 and hence k_{-1} can be calculated. Because of the finite time taken for mixing, the induction period must be greater than about 5 ms for meaningful results to be obtained. Enzymes with a turnover number greater than a few hundred per second catalyse reactions proceeding too quickly to be analysed by such methods.

The calculated progress curves for a slightly more complicated reaction:

$$E + S \underset{k_{-1}}{\overset{k_1}{\rightleftharpoons}} ES \underset{k_{-2}}{\overset{k_2}{\rightleftharpoons}} EP \underset{k_{-3}}{\overset{k_3}{\rightleftharpoons}} E + P$$

where the rate-limiting-step is $EP \to E + P$, are shown in Fig. 7.6.

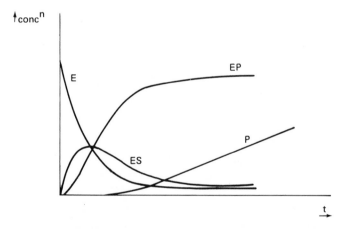

Fig. 7.6 Theoretical progress curves for a reaction of the form
$E + S \rightleftharpoons ES \rightleftharpoons EP \to E + P$, $EP \to E + P$ being the rate-limiting step.

As an example of the use of rapid-reaction methods in the investigation of reaction mechanisms, let us consider some findings for the enzyme **chymotrypsin**. This will catalyse the hydrolysis of p–nitrophenylacetate, giving the coloured p–nitrophenol as one of the products. Hartley and Kilby (1952) found that immediately after mixing the reagents there was an **initial burst** when one mole of p-nitrophenol per mole of enzyme was liberated much more rapidly than the subsequent steady-state liberation of p-nitrophenol. This suggested rapid acylation of the enzyme, the rate-limiting step of the overall reaction being the hydrolysis of this acyl–enzyme:

rate-limiting step

E + p-nitrophenylacetate→E-p-nitrophenylacetate→E-acetate ——→E + acetate.

p-nitrophenol

Gutfreund (1955) followed the liberation of p–nitrophenol using stopped-flow methods and confirmed these conclusions, as well as determining some rate constants.

From a study of many enzymes, Gutfreund concluded that the rate-limiting-step was usually one of the following: a change in conformation producing a reactive complex from the initial ES complex, the chemical interconversion of substrate to product, or the release of product from enzyme.

7.2.2 Relaxation Kinetics

Relaxation methods can also be used to investigate rapid reaction kinetics and since they start with the solutions already mixed and at equilibrium, they can be applied to processes taking place too quickly for investigation by pre-steady-state methods.

It can be shown that if a chemical reaction at equilibrium is subjected to any extremely rapid pressure or temperature change (e.g. $5°C$ in $5\ \mu s$) and the position of equilibrium changes slightly in consequence, the system will **relax** towards the new equilibrium position according to the relationship:

$$[\Delta A] = [\Delta A]_o e^{-\frac{t}{\tau}}$$

where $[\Delta A]$ is the difference between the concentration of A at time t and at the new equilibrium position, i.e. $[\Delta A] = [A] - [A_{eq}]$; and $[\Delta A]_o$ is the difference between the concentration of A immediately after the temperature or pressure change and at the new equilibrium position. τ is the **relaxation time**.

From the above equation:

$$\log_e[\Delta A] = \log_e[\Delta A]_o - \frac{t}{\tau}.$$

∴ the slope of a graph of $\log_e[\Delta A]$ against t gives $-\frac{1}{\tau}$.

This relationship holds for all chemical reactions where the perturbation in equilibrium is small, but the meaning of τ in terms of reaction constants depends on the actual reaction mechanism.

For the simplest reversible reaction $A \underset{k_{-1}}{\overset{k_1}{\rightleftharpoons}} B$, where the total concentration of A and B present is $[B_o]$, i.e. $[A] + [B] = [B_o]$, the rate of the formation reaction at time t is given by:

$$\frac{-d[A]}{dt} = \frac{+d[B]}{dt} = k_1[A] - k_{-1}[B]$$
$$= k_1[A] - k_{-1}([B_o] - [A])$$
$$= k_1[A] - k_{-1}[B_o] + k_{-1}[A]$$
$$= [A](k_1 + k_{-1}) - k_{-1}[B_o].$$

At equilibrium,

$$k_1[A_{eq}] = k_{-1}[B_{eq}].$$
$$\therefore k_1[A_{eq}] = k_{-1}([B_o] - [A_{eq}])$$
$$= k_{-1}[B_o] - k_{-1}[A_{eq}].$$
$$\therefore [A_{eq}](k_1 + k_{-1}) = k_{-1}[B_o].$$

Substituting for $k_{-1}[B_o]$ in the rate equation derived above:

$$\frac{-d[A]}{dt} = \frac{+d[B]}{dt} = [A](k_1 + k_{-1}) - [A_{eq}](k_1 + k_{-1})$$
$$= (k_1 + k_{-1})([A] - [A_{eq}])$$
$$= (k_1 + k_{-1})[\Delta A].$$

However, the rate of change of $[\Delta A]$ must be the same as the rate of change of $[A]$.

$$\therefore \frac{-d[A]}{dt} = \frac{-d[\Delta A]}{dt} = (k_1 + k_{-1})[\Delta A].$$
$$\therefore \frac{d[\Delta A]}{dt} = -(k_1 + k_{-1})[\Delta A].$$

Integrating,

$$\log_e[\Delta A] = \text{constant} - (k_1 + k_{-1})t.$$

At $t = 0$, $[\Delta A] = [\Delta A]_o$, so constant $= \log_e[\Delta A]_o$.

$$\therefore \log_e[\Delta A] = \log_e[\Delta A]_o - (k_1 + k_{-1})t.$$

This is of the same form as the general relaxation equation, so by comparison with this, $\tau = \dfrac{1}{k_1 + k_{-1}}$ for the reaction $A \rightleftharpoons B$.

Similarly, for a reaction $E + S \underset{k_{-1}}{\overset{k_1}{\rightleftharpoons}} ES$, it can be deduced that

$\dfrac{1}{\tau} = k_1([E_{eq}] + [S_{eq}]) + k_{-1}$. Reactions of this form include the binding of one substrate to an enzyme in the absence of other substrates, preventing the reaction from proceeding further; this has been used to investigate the binding of NADH to lactate dehydrogenase. A plot of $([E_{eq}] + [S_{eq}])$ against $1/\tau$ results in a straight line with the intercept equal to k_{-1} and the slope $= k_1$, so these constants can be calculated.

The meaning of τ for more complex reactions is, needless to say, very much more complicated and only in special cases can it be identified with individual steps and particular rate constants.

7.3 THE KING AND ALTMAN PROCEDURE

When more than one or two intermediates are present in a reaction sequence, the derivation of an overall rate equation becomes extremely complex, involving the solving of several simultaneous equations. However, King and Altman (1956) devised empirical rules which allow the rate equation for a particular mechanism to be written down by inspection as a function of individual rate constants, and this procedure was developed by Wong and Hanes (1962). The theoretical basis of these rules involves matrix algebra and is beyond the scope of this book. We will merely indicate the approach, using some simple examples.

First let us consider again the simplest possible enzyme-catalysed reaction under steady-state conditions:

$$E + S \underset{k_{-1}}{\overset{k_1}{\rightleftharpoons}} ES \xrightarrow{k_2} E + P.$$

The King and Altman procedure requires the reaction to be written as a cyclic process, showing the interconversions of the enzyme-forms involved (in this case E and ES). Each step must be described by a κ (kappa), which is the product of the rate constant and the concentration of free substrate involved in that step, or just the rate constant if no substrate is involved. Hence for the above sequence we have:

For each enzyme species we must then work out all the pathways by which that species may be synthesised. Each pathway must contain $n-1$ steps, where n is the number of enzyme species present (in this case 2) and the product of the kappas for the steps in each pathway is determined; each kappa product will contain $n-1$ terms. Pathways where two arrows arrive at a single enzyme species are permitted, but pathways where two arrows leave a single enzyme species are forbidden, so there are no closed loops. For the simple system being considered here (where $n = 2$) we have:

Enzyme species	Pathways forming enzyme species	Sum of kappa products
E	$E \xleftarrow{k_{-1}}$ and $E \, {k_2}$	$k_{-1} + k_2$
ES	$\xrightarrow{k_1[S_o]}$ ES	$k_1[S_o]$

For each enzyme species the following relationship then holds:

$$\frac{[\text{enzyme species}]}{[E_o]} = \frac{\text{sum of kappa products of that species}}{\text{sum of all kappa products}}.$$

Hence, for the mechanism under question,

$$\frac{[E]}{[E_o]} = \frac{k_{-1} + k_2}{k_{-1} + k_2 + k_1[S_o]} \quad \text{and} \quad \frac{[ES]}{[E_o]} = \frac{k_1[S_o]}{k_{-1} + k_2 + k_1[S_o]}.$$

The overall rate equation is given, as in Chapter 7.1.2, by

$$v_o = k_2[ES].$$

Substituting the expression for [ES] determined by the King and Altman procedure:

$$v_o = \frac{k_2.k_1[S_o][E_o]}{k_{-1} + k_2 + k_1[S_o]} = \frac{k_2[S_o][E_o]}{\left(\dfrac{k_{-1} + k_2}{k_1}\right) + [S_o]}$$

which is identical with the equation derived in Chapter 7.1.2.

Let us now use the King and Altman procedure to determine the rate equation for a slightly more complicated reaction sequence, that involving three enzyme species suggested for reactions catalysed by chymotrypsin (Chapter 7.2.1). Under steady-state conditions this may be written:

$$AX + E \underset{k_{-1}}{\overset{k_1}{\rightleftharpoons}} EAX \overset{k_2}{\underset{A}{\searrow}} EX \overset{k_3}{\longrightarrow} E + X$$

In cyclical form this is:

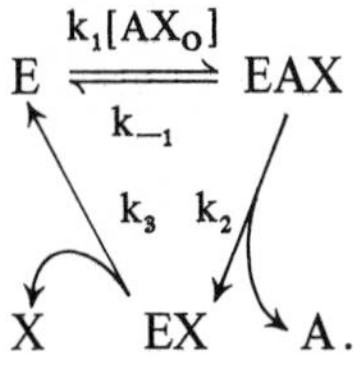

$$
\begin{array}{c}
E \underset{k_{-1}}{\overset{k_1[AX_0]}{\rightleftharpoons}} EAX \\
\text{(cyclic diagram with } k_3, k_2 \text{)} \\
X \quad EX \quad A.
\end{array}
$$

Enzyme species	*Pathways forming enzyme species*	*Sum of kappa products*
E	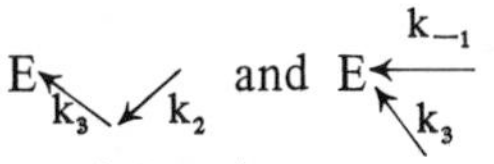 $E \overset{k_3}{\searrow} \overset{k_2}{\swarrow}$ and $E \overset{k_{-1}}{\underset{k_3}{\leftarrow}}$	$k_2 k_3 + k_{-1} k_3$
EAX	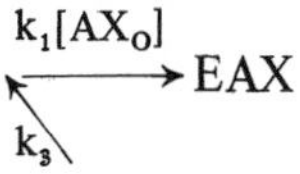 $\overset{k_1[AX_0]}{\underset{k_3}{\longrightarrow}} EAX$	$k_1 k_3 [AX_0]$
EX	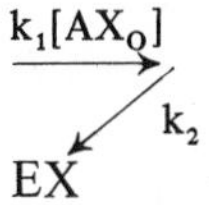 $\overset{k_1[AX_0]}{\underset{k_2}{\longrightarrow}} EX$	$k_1 k_2 [AX_0]$.

From this, for example,

$$\frac{[EX]}{[E_0]} = \frac{k_1 k_2 [AX_0]}{k_2 k_3 + k_{-1} k_3 + k_1 k_3 [AX_0] + k_1 k_2 [AX_0]}.$$

Under steady-state conditions the overall rate of reaction is given by:

$$v_0 = \frac{d[A]}{dt} = k_2 [EAX] = \frac{d[X]}{dt} = k_3 [EX].$$

If we substitute for either [EAX] or [EX] in the expression obtained by the King and Altman procedure we obtain the equation:

$$v_0 = \frac{k_1 k_2 k_3 [AX_0][E_0]}{k_2 k_3 + k_{-1} k_3 + k_1 k_3 [AX_0] + k_1 k_2 [AX_0]}$$

$$= \frac{k_1 k_2 k_3 [AX_0][E_0]}{k_3(k_2 + k_{-1}) + k_1[AX_0](k_2 + k_3)}.$$

SUMMARY OF CHAPTER 7

The hyperbolic graphs of v_0 against $[S_0]$ obtained experimentally for many enzyme-catalysed reactions can be obtained by the Michaelis–Menten equation. The original **equilibrium-assumption** of Michaelis and Menten is now seen as a special case of the **steady-state-assumption** of Briggs and Haldane. Steady-state kinetics can be used to calculate K_m and k_{cat}, which are characteristic of a particular enzyme. These calculations are facilitated by the use of linear plots derived from the Michaelis–Menten equation.

The kinetic mechanism of a reaction can be investigated in more detail by the use of rapid-reaction techniques. Continuous-flow, stopped-flow and relaxation methods give information as to the intermediates involved in the reaction and enable individual rate constants to be calculated.

A procedure which enables the rate equation for a complex reaction sequence to be written down by inspection has been described by King and Altman.

FURTHER READING

Colowick, S. P. and Kaplan, N. O. (eds.) (1979), *Methods in Enzymology,* **63**: Enzyme Kinetics and Mechanism, Academic Press.

Cornish-Bowden, A. (1979), *Fundamentals of Enzyme Kinetics* (Chapters 1-4, 9-10), Butterworth.

Dawes, E. A. (1980), *Quantitative Problems in Biochemistry,* 6th edition (Chapter 6 and Problems), Longman.

Dixon, M., Webb, E. C., Thorne, C. J. R. and Tipton, K. F. (1979), *Enzymes,* 3rd edition, (Chapter 4), Longman.

Gutfreund, H. (1972), *Enzymes: Physical Principles* (Chapters 6 and 8), Wiley.

Halford, S. E., Rapid Reaction Techniques, in Bull, A. T., Lagnado, J. R., Thomas, J. D. and Tipton, K. F. (eds.) (1974), *Companion to Biochemistry,* Longman.

Hammes, G. G. (1982), *Enzyme Catalysis and Regulation* (Chapters 3 and 4), Academic Press.

Morris, J. G. (1974), *A Biologist's Physical Chemistry,* 2nd edition (Chapter 11 and Problems), Arnold.

Price, N. C. and Stevens, L. (1982), *Fundamentals of Enzymology* (Chapter 4), Oxford University Press.

Roberts, D. V. (1977), *Enzyme Kinetics,* Cambridge University Press.

Wharton, C. W. and Eisenthal, R. (1981), *Molecular Enzymology* (Chapters 4, 6, 8 and 10), Blackie.

Williams, P. A. (1983), Enzpack: a microcomputer program to aid in the teaching of enzyme kinetics, *Biochemical Education,* **11** (pages 141-143).

PROBLEMS

7.1 The following results were obtained for an enzyme-catalysed reaction:

Substrate conc^n (mmol.l^{-1})	5.0	6.67	10.0	20.0	40.0
Initial velocity (μmol.l^{-1}.minute^{-1})	147	182	233	323	400.

Calculate K_m and V_{max}.

7.2 Malate dehydrogenase catalyses the reaction:

$$\text{L-malate} + NAD^+ \rightleftharpoons \text{oxaloacetate} + NADH + H^+.$$

The rate of the forward reaction was investigated in the presence of saturations of malate and a fixed concentration of enzyme. The following results were obtained:

NAD^+ conc^n	Absorbance (at 340 nm) at time t (minutes)					
(mmol.l^{-1})	$t = 0.5$	1.0	1.5	2.0	2.5	3.0
1.5	0.033	0.056	0.079	0.102	0.122	0.138
2.0	0.036	0.063	0.089	0.116	0.138	0.154
2.5	0.040	0.069	0.099	0.128	0.150	0.168
3.33	0.043	0.075	0.108	0.140	0.163	0.175
5.0	0.047	0.084	0.121	0.158	0.177	0.184
10.0	0.052	0.095	0.137	0.180	0.192	0.200.

Calculate K_m and V_{max}.

7.3 Ficin is a proteolytic enzyme which catalyses the hydrolysis of a variety of substrates. One such reaction was followed by rapid reaction techniques, and the following results were obtained:

Time (s)	0	0.1	0.2	0.3	0.4	0.5
Product conc^n (μmol.l^{-1})	0	4	8	17	27	37 .

Assuming that the reaction proceeds by the simplest possible mechanism, calculate the rate constants involved.

7.4 Use the King and Altman procedure to obtain a rate equation for the reaction:

$$E + S \underset{k_{-1}}{\overset{k_1}{\rightleftharpoons}} ES \underset{k_{-2}}{\overset{k_2}{\rightleftharpoons}} EP \overset{k_3}{\rightarrow} E + P.$$

CHAPTER 8

Enzyme Inhibition

8.1 INTRODUCTION

Inhibitors are substances which tend to decrease the rate of an enzyme-catalysed reaction. Although some act on a substrate or cofactor, we will restrict our discussion here to those which combine directly with an enzyme. **Reversible inhibitors** bind to an enzyme in a reversible fashion and can be removed by dialysis (or simply dilution) to restore full enzymic activity whereas **irreversible inhibitors** cannot be removed from an enzyme by dialysis. Sometimes it may be possible to remove an irreversible inhibitor from an enzyme by introducing another component to the reaction mixture, but this would not affect the classification of the original interaction.

Reversible inhibitors usually rapidly form an equilibrium system with an enzyme to show a definite degree of inhibition (depending on the concentration of enzyme, inhibitor and substrate) which remains constant over the period when initial velocity studies are normally carried out. In contrast, the degree of inhibition by irreversible inhibitors may increase over this period of time.

In this chapter we shall be concerned mainly with the inhibition of simple single-substrate enzyme-catalysed reactions. This group includes most single-substrate reactions obeying Michaelis–Menten kinetics (Chapter 7.1.3). Some aspects of the inhibition of two-substrate enzyme-catalysed reactions will be discussed in Chapter 9.3.2.

8.2 REVERSIBLE INHIBITION

8.2.1 Competitive Inhibition

Competitive inhibitors often closely resemble in some respects the substrates whose reactions they inhibit, and because of this structural similarity they may compete for the same binding-site on the enzyme. The enzyme-bound inhibitor then either lacks the appropriate reactive group or it is held in an unsuitable position with respect to the catalytic-site of the enzyme or to other potential substrates for a reaction to take place. In either case a **dead-end complex** is formed, and the inhibitor must dissociate from the enzyme and be replaced by a molecule of substrate before a reaction can take place at that particular enzyme molecule.

For example, **malonate** ($CO_2^-.CH_2.CO_2^-$) is a competitive inhibitor of the reaction catalysed by **succinate dehydrogenase**:

$$CO_2^-.CH_2.CH_2.CO_2^- \rightleftharpoons CO_2^-.CH = CH.CO_2^-.$$

    succinate   fumarate

Malonate has two carboxyl groups, like the substrate, succinate, and can fill the succinate-binding site on the enzyme. However the subsequent reaction involves the formation of a double bond, and since malonate, unlike succinate, has only one carbon atom between the carboxyl groups, it cannot react.

The effect of a competitive inhibitor depends on the inhibitor concentration, the substrate concentration and the relative affinities of the substrate and the inhibitor for the enzyme. In general, at a particular inhibitor and enzyme concentration, if the substrate concentration is low, the inhibitor will compete favourably with the substrate for the binding sites on the enzyme and the degree of inhibition will be great. However if, at this same inhibitor and enzyme concentration, the substrate concentration is high, then the inhibitor will be much less successful in competing with the substrate for the available binding sites and the degree of inhibition will be less marked. At very high substrate concentrations, molecules of substrate will greatly outnumber molecules of inhibitor and the effect of the inhibitor will be negligible. Hence V_{max} for the reaction is unchanged (Fig. 8.1(a)). However the apparent K_m, the substrate concentration when $v_o = \frac{1}{2}V_{max}$, is clearly increased as a result of the inhibition, and is given the symbol K_m'.

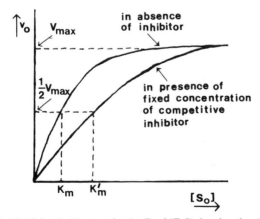

Fig. 8.1 (a) Michaelis–Menten plot (at fixed $[E_o]$) showing the effect of a competitive inhibitor.

Let us investigate the steady-state kinetics of a simple single-substrate single-binding-site single-intermediate enzyme-catalysed reaction in the presence of a competitive inhibitor, I.

$$E + S \underset{k_{-1}}{\overset{k_1}{\rightleftharpoons}} ES \xrightarrow{k_2} E + P$$

$$- I \, \big\Updownarrow + I$$

$$EI$$

The dissociation constant for the reaction between E and I is K_i, where $K_i = \dfrac{[E][I]}{[EI]}$.

In this context, K_i is called the **inhibitor constant**. Equilibrium between enzyme and inhibitor will normally be established almost instantaneously on mixing.

If we begin to derive the initial velocity equation using the steady-state assumption exactly as in Chapter 7.1.2 we reach the expression

$$\frac{[E][S]}{[ES]} = \frac{k_{-1} + k_2}{k_1} = K_m.$$

As before, we wish to make a substitution for $[E]$ in this equation, and here we have to take the inhibitor into account, since:

$$[E_o] = [E] + [ES] + [EI]$$

$$= [E] + [ES] + \frac{[E][I]}{K_i}$$

$$= [E] \left(1 + \frac{[I]}{K_i}\right) + [ES].$$

$$\therefore \ [E] = \frac{[E_o] - [ES]}{\left(1 + \dfrac{[I]}{K_i}\right)}.$$

If we now make this substitution for $[E]$:

$$\frac{([E_o] - [ES])[S]}{\left(1 + \dfrac{[I]}{K_i}\right)[ES]} = K_m.$$

$$\therefore \ \frac{([E_o] - [ES])[S]}{[ES]} = K_m \left(1 + \frac{[I]}{K_i}\right).$$

If we continue to develop the argument exactly as in Chapters 7.1.1 and 7.1.2 we reach the expression:

$$v_0 = \frac{V_{max}[S_0]}{[S_0] + K_m \left(1 + \frac{[I_0]}{K_i}\right)} .$$

This is an equation of the same form as the Michaelis-Menten equation, the only difference being that K_m has been increase by a factor $(1 + \frac{[I_0]}{K_i})$. (Note that the inhibitor concentration will usually be of the same order of magnitude as the substrate concentration and thus much greater than the enzyme concentration, so $[I] \doteq [I_0]$ just as $[S] \doteq [S_0]$). Therefore, for simple competitive inhibition V_{max} is unchanged but K_m is altered so that $K'_m = K_m(1 + \frac{[I_0]}{K_i})$, where K'_m is the apparent K_m in the presence of an initial concentration $[I_0]$ of competitive inhibitor. It can be seen that K_i is equal to the concentration of competitive inhibitor which apparently doubles the value of K_m.

The Lineweaver-Burk equation in the presence of a competitive inhibitor will be:

$$\frac{1}{v_0} = \frac{K'_m}{V_{max}} \cdot \frac{1}{[S_0]} + \frac{1}{V_{max}}$$

and Lineweaver-Burk plots showing the effect of competitive inhibition are shown in Figs. 8.1(b) and 8.1(c).

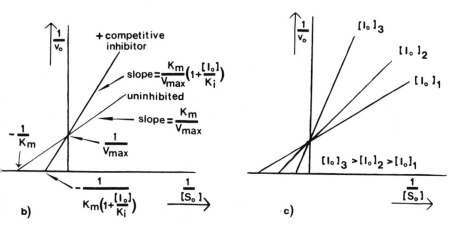

Fig. 8.1 (b) Lineweaver-Burk plot showing the effect of competitive inhibition; (c) the same, showing plots for several inhibitor concentrations at fixed enzyme concentrations.

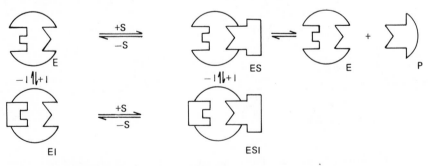

(a)   competitive inhibition, I binding to same site as S

(b)   competitive inhibition, I and S binding to different sites

(c)   uncompetitive inhibition

(d)   simple linear non-competitive inhibition

Fig. 8.2 Diagrammatic representation of some possible examples of reversible inhibition.

It must be pointed out that this identical expression would be obtained if the inhibitor-binding site was separate from the substrate-binding site, provided the binding of the substrate to the enzyme resulted in the blockage of the inhibitor-binding site by a conformational change or other mechanism (Fig. 8.2). In this situation the inhibitor could bind to E but not to ES, exactly as discussed above. For this reason it has become common to classify an inhibitor as competitive if, in its presence, a Lineweaver-Burk plot is obtained with changed K_m but unchanged V_{max}, *irrespective of the actual mechanism involved.*

Once the type of inhibitor has been established, it is desirable to determine the inhibitor constant K_i. It is obtained from the expression $K'_m = K_m (1 + \dfrac{[I_0]}{K_i})$, but a graphical method is prefered to a direct substitution of numbers to allow errors in individual determinations to be averaged out. From the above expression $K'_m = \dfrac{K_m}{K_i} [I_0] + K_m$, so a plot of K'_m (determined from the intercept on the $\dfrac{1}{[S_0]}$ axis of the **primary** Lineweaver-Burk plot) against $[I_0]$ will be linear, with the intercept on the $[I_0]$ axis giving $-K_i$ (Fig. 8.3(a)). Similarly, since the slope of the Lineweaver-Burk plot in the presence of a competitive inhibitor is $\dfrac{K_m}{V_{max}} (1 + \dfrac{[I_0]}{K_i})$, a graph of slope of the **primary** (Lineweaver-Burk) plot against $[I_0]$ will also be linear, the intercept on the $[I_0]$ axis giving $-K_i$ (Fig. 8.3(b)). These **secondary** plots of K'_m against $[I_0]$ and slope against $[I_0]$ must, of course, be constructed from data obtained at fixed $[E_0]$.

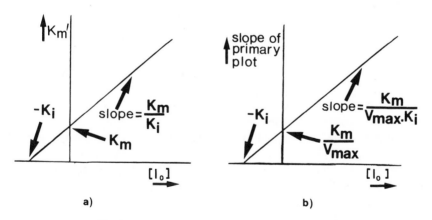

Fig. 8.3 Secondary plots for competitive inhibition.

An alternative graphical means of calculating K_i was suggested by Dixon (1953). The Lineweaver-Burk equation in the presence of a competitive inhibitor is:

$$\frac{1}{v_o} = \frac{K_m\left(1 + \dfrac{[I_o]}{K_i}\right)}{V_{max}} \cdot \frac{1}{[S_o]} + \frac{1}{V_{max}}$$

$$= \frac{K_m}{V_{max} \cdot [S_o]} \cdot \frac{[I_o]}{K_i} + \frac{K_m}{V_{max}[S_o]} + \frac{1}{V_{max}}.$$

Therefore, at fixed $[S_o]$, a plot of $\dfrac{1}{v_o}$ against $[I_o]$ (the **Dixon plot**) is linear. When $[I_o] = -K_i$ it can be seen that $\dfrac{1}{v_o} = \dfrac{1}{V_{max}}$ and is therefore independent of $[S_o]$ and so **Dixon plots** for different values of $[S_o]$ (but fixed $[E_o]$) intersect where $[I_o] = -K_i$ (Fig. 8.4).

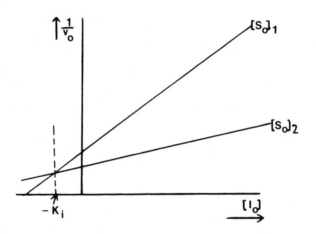

Fig. 8.4 Dixon plot for competitive inhibition.

The simplest forms of competitive inhibition considered above are sometimes termed **linear competitive inhibition** because both primary and secondary plots are linear. In more complicated systems the primary plots may be linear but the secondary plots non-linear. For example, if not one but two molecules of inhibitor can bind to the substrate binding site, then **parabolic competitive inhibition** is said to occur because of the shape of the secondary and Dixon plots. Similarly, if the inhibitor binds to a different site from the substrate and reduces the affinity of the enzyme for the substrate without altering the reaction characteristics of that substrate which does bind, then **hyperbolic competitive inhibition** results. In each case the primary plots are linear and the inhibition patterns are indistinguishable from those for **linear competitive inhibition**.

Competitive inhibitors, like other types of inhibitors, may be used to help elucidate metabolic pathways by causing accumulation of intermediates. In this way Krebs and colleagues used inhibition by malonate to investigate the tricarboxylic acid cycle, of which succinate dehydrogenase is a component.

Provided they are not dangerous to man, they may be used in medicine or agriculture as chemotherapeutic drugs, insecticides or herbicides to destroy or prevent growth of unwanted organisms. For example, sulphonamides such as sulphanilamide (H_2N-◯-SO_2NH_2), once widely used in medicine, are competitive inhibitors for bacterial enzymes involved in the biosynthesis of the coenzyme tetrahydrofolate (see Chapter 11.5.9) from p-aminobenzoic acid (H_2N-◯-CO_2H). This metabolic pathway is not found in man, so sulphonamides can be used to limit the growth of bacteria with relatively little risk to the patient.

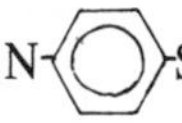

A detailed investigation of the binding characteristics of various competitive inhibitors which bind at the same site as a natural substrate can give useful information about factors governing the binding of the substrate. In two-substrate enzyme-catalysed reactions, competitive inhibition studies can help elucidate the reaction mechanism (Chapter 9.3.2). In this context and in general it should be noted that the product of a reaction often resembles the substrate and therefore may act as a competitive inhibitor. There are a few instances where competitive inhibition by a product may play an important role in metabolic regulation within the living cell. For example, **2,3-bisphosphoglycerate** inhibits its own formation from **3-phosphoglyceroyl phosphate**, a reaction catalysed by **bisphosphoglycerate mutase**.

8.2.2 Uncompetitive Inhibition

Uncompetitive inhibitors bind only to the enzyme-substrate complex and not to the free enzyme. Substrate-binding could cause a conformational change to take place in the enzyme and reveal an inhibitor binding-site (Fig. 8.2(c)), or the inhibitor could bind directly to the enzyme-bound substrate. In neither case does the inhibitor compete with the substrate for the same binding site, so the inhibition cannot be overcome by increasing the substrate concentration. Both K_m and V_{max} are altered, but a distinctive kinetic pattern emerges under steady-state conditions.

Once again let us consider the simplest situation:

$$E + S \rightleftharpoons ES \rightarrow E + P.$$
$$-I \updownarrow +I$$
$$ESI$$

ESI is a **dead-end complex**; the inhibitor constant $K_i = \dfrac{[ES][I]}{[ESI]}$.

Under steady-state conditions,

$$\frac{[E][S]}{[ES]} = K_m.$$

For this system,

$$[E_o] = [E] + [ES] + [ESI]$$

$$= [E] + [ES] + \frac{[ES][I]}{K_i}$$

$$= [E] + [ES]\left(1 + \frac{[I]}{K_i}\right).$$

$$\therefore [E] = [E_o] - [ES]\left(1 + \frac{[I]}{K_i}\right).$$

Substituting for [E] and continuing as in Chapter 7.1.1, the outcome is:

$$v_o = \frac{V_{max}[S_o]}{[S_o]\left(1 + \frac{[I_o]}{K_i}\right) + K_m}.$$

Dividing throughout by $(1 + \frac{[I_o]}{K_i})$ gives:

$$v_o = \frac{\dfrac{V_{max}}{\left(1 + \dfrac{[I_o]}{K_i}\right)} \cdot [S_o]}{[S_o] + \dfrac{K_m}{\left(1 + \dfrac{[I_o]}{K_i}\right)}}.$$

This is an equation of the same form as the Michaelis–Menten equation, the constants K_m and V_{max} both being divided by a factor $(1 + \frac{[I_o]}{K_i})$. Thus for uncompetitive inhibition,

$$V_{max}' = \frac{V_{max}}{\left(1 + \dfrac{[I_o]}{K_i}\right)} \quad \text{and} \quad K_m' = \frac{K_m}{\left(1 + \dfrac{[I_o]}{K_i}\right)}$$

where V_{max}' is the value of V_{max} in the presence of an initial concentration $[I_o]$ of uncompetitive inhibitor and K_m' is the apparent value of K_m under the same conditions. An inhibitor concentration equal to K_i will halve the values of both V_{max} and K_m.

The Lineweaver-Burk equation in the presence of an uncompetitive inhibitor is:

$$\frac{1}{v_o} = \frac{K_m'}{V_{max}'} \cdot \frac{1}{[S_o]} + \frac{1}{V_{max}'}$$

and the slope of a Lineweaver-Burk plot is equal to

$$\frac{K_m'}{V_{max}'} = \frac{K_m}{V_{max}} \frac{\left(1 + \dfrac{[I_o]}{K_i}\right)}{\left(1 + \dfrac{[I_o]}{K_i}\right)} = \frac{K_m}{V_{max}}.$$

In other words, the slope of a Lineweaver-Burk plot is not altered by the presence of an uncompetitive inhibitor, but both intercepts change (Fig. 8.5).

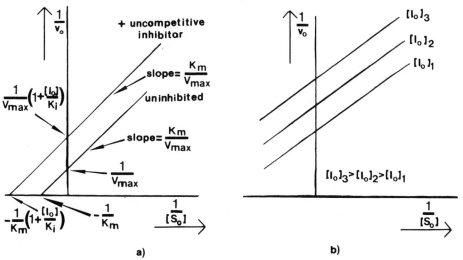

Fig. 8.5 (a) Lineweaver–Burk plots showing the effect of uncompetitive inhibition; (b) the same, showing plots for several inhibitor concentrations at fixed enzyme concentration.

As before, the inhibitor constant K_i can be determined using **secondary plots**. For uncompetitive inhibition, $\dfrac{1}{V_{max}'} = \dfrac{1}{V_{max}}(1 + \dfrac{[I_o]}{K_i})$ and $\dfrac{1}{K_m'} = \dfrac{1}{K_m}(1 + \dfrac{[I_o]}{K_i})$. Hence plots of $\dfrac{1}{V_{max}'}$ or $\dfrac{1}{K_m'}$ (obtained from intercepts on $\dfrac{1}{v_o}$ and $\dfrac{1}{[S_o]}$ axes respectively of the **primary plot**) against $[I_o]$ are linear, the intercept on the $[I_o]$ axes giving $-K_i$ (Fig. 8.6).

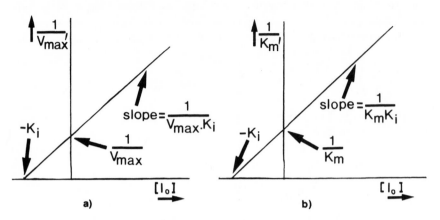

Fig. 8.6 Secondary plots for uncompetitive inhibition.

Uncompetitive inhibition of single-substrate enzyme catalysed reactions is a rare phenomenon, one of the few possible examples known being the inhibition of arylsulphatase by hydrazine. However uncompetitive inhibition patterns are seen with two-substrate reactions and this may help in the elucidation of the reaction mechanism (see Chapter 9.3.2).

8.2.3 Non-Competitive Inhibition

A non-competitive inhibitor can combine with an enzyme molecule to produce a **dead-end complex**, regardless of whether a substrate molecule is bound or not. Hence the inhibitor must bind at a different site from the substrate. We shall consider only the case where the inhibitor destroys the catalytic activity of the enzyme, either by binding to the catalytic site or as a result of a conformational change affecting the catalytic site, but does not affect substrate-binding. The situation for a simple single-substrate reaction will be as follows:

$$
\begin{array}{ccc}
\text{E} & \underset{-S}{\overset{+S}{\rightleftharpoons}} & \text{ES} \rightarrow \text{P}. \\[2mm]
-I \updownarrow +I & & -I \updownarrow +I \\[2mm]
\text{EI} & \underset{-S}{\overset{+S}{\rightleftharpoons}} & \text{ESI}
\end{array}
$$

Even this is a complex situation, for ES can be arrived at by alternative routes, making it impossible for an expression of the same form as the Michaelis–Menten equation to be derived using the general **steady-state-assumption**. However types of non-competitive inhibition consistent with a Michaelis–Menten-type equation and a linear Lineweaver-Burk plot can occur if the **equilibrium-assumption** (Chapter 7.1.1) is valid. In the simplest possible model, **simple linear non-competitive inhibition** (Fig. 8.2(d)), the substrate does not affect inhibitor-binding. Under these conditions the reactions $E + I \rightleftharpoons EI$ and $ES + I \rightleftharpoons ESI$ have an identical dissociation constant K_i, again called the **inhibitor constant**. The total enzyme concentration is effectively reduced by the inhibitor, decreasing the value of V_{max} but not altering K_m, since neither inhibitor nor substrate affect the binding of the other.

Let us once again derive an initial velocity equation, this time remembering we are only considering the special case where $K_m \backsimeq K_s$.

As before,

$$\frac{[E][S]}{[ES]} = K_m.$$

In the presence of a non-competitive inhibitor which will bind equally well to E or to ES, i.e. where $K_i = \dfrac{[E][I]}{[EI]} = \dfrac{[ES][I]}{[ESI]}$:

$$[E_o] = [E] + [ES] + [EI] + [ESI]$$

$$= [E] + [ES] + \frac{[E][I]}{K_i} + \frac{[ES][I]}{K_i}$$

$$= ([E] + [ES])\left(1 + \frac{[I]}{K_i}\right).$$

$$\therefore \ [E] + [ES] = \frac{[E_o]}{\left(1 + \dfrac{[I]}{K_i}\right)}$$

$$\therefore \ [E] = \frac{[E_o]}{\left(1 + \dfrac{[I]}{K_i}\right)} - [ES].$$

If we continue exactly as in Chapter 7.1.1 we conclude that, under these conditions:

$$v_o = \frac{V_{max}}{\left(1 + \dfrac{[I_o]}{K_i}\right)} \cdot \frac{[S_o]}{([S_o] + K_m)}.$$

This is of the form of the Michaelis–Menten equation, with V_{max} being divided by a factor $(1 + \frac{[I_o]}{K_i})$. Thus, for simple linear non-competitive inhibition, K_m is unchanged and V_{max} is altered so that

$$V_{max}{}' = \frac{V_{max}}{\left(1 + \frac{[I_o]}{K_i}\right)}, \quad \text{or} \quad \frac{1}{V_{max}{}'} = \frac{1}{V_{max}}\left(1 + \frac{[I_o]}{K_i}\right)$$

where $V_{max}{}'$ is the value of V_{max} in the presence of a concentration $[I_o]$ of non-competitive inhibitor. It follows that K_i for such a system is the inhibitor concentration which halves the value of V_{max}. The Lineweaver–Burk equation for simple linear non-competitive inhibition is:

$$\frac{1}{v_o} = \frac{K_m}{V_{max}{}'} \cdot \frac{1}{[S_o]} + \frac{1}{V_{max}{}'},$$

and Lineweaver–Burk plots showing the effect of such inhibition are shown in Fig. 8.7.

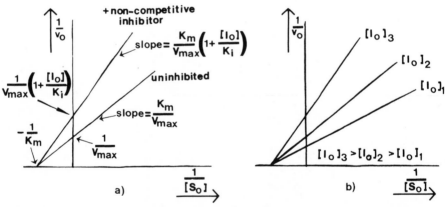

Fig. 8.7   (a) Lineweaver–Burk plot showing the effect of simple linear non-competitive inhibition; (b) the same, showing several inhibitor concentrations at fixed enzyme concentration.

Once the type of inhibition has been established, the inhibitor constant K_i may be determined using **secondary plots**. Since $\frac{1}{V_{max}{}'} = \frac{1}{V_{max}}(1 + \frac{[I_o]}{K_i})$ and slope of the primary Lineweaver-Burk plot $= \frac{K_m}{V_{max}}(1 + \frac{[I_o]}{K_i})$, it follows that plots of $\frac{1}{V_{max}{}'}$ against $[I_o]$ and of slope of the primary plot against $[I_o]$ are linear, the intercept on the $[I_o]$ axis giving $-K_i$ (Fig. 8.8).

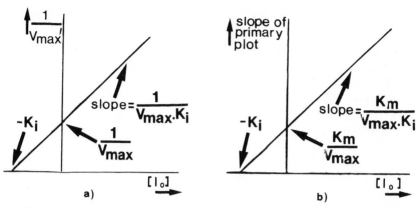

Fig. 8.8   Secondary plots (at constant $[E_o]$) for simple linear non-competitive inhibition.

A **Dixon plot** may also be used to determine K_i. From the Lineweaver–Burk equation:

$$\frac{1}{v_o} = \frac{K_m}{V_{max}}\left(1 + \frac{[I_o]}{K_i}\right) \cdot \frac{1}{[S_o]} + \frac{1}{V_{max}}\left(1 + \frac{[I_o]}{K_i}\right)$$

$$= \frac{K_m}{V_{max}\cdot[S_o]} + \frac{K_m[I_o]}{V_{max}\cdot K_i[S_o]} + \frac{1}{V_{max}} + \frac{[I_o]}{V_{max}\cdot K_i}$$

$$= \left(\frac{K_m}{V_{max}[S_o]} + \frac{1}{V_{max}}\right)\frac{[I_o]}{K_i} + \frac{K_m}{V_{max}[S_o]} + \frac{1}{V_{max}}.$$

Thus a Dixon plot of $\dfrac{1}{v_o}$ against $[I_o]$ will be linear at fixed $[E_o]$ and $[S_o]$ for simple linear non-competitive inhibition.

When $[I_o] = -K_i$, it can be seen that $\dfrac{1}{v_o} = 0$. Hence the intercept on the $[I_o]$ axis will give $-K_i$ (Fig. 8.9).

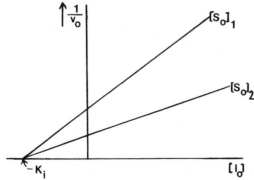

Fig. 8.9 Dixon plot for simple linear non-competitive inhibition.

Some enzymologists point out, with reason, that simple linear non-competitive inhibition is a special case of general non-competitive inhibition, and other forms of non-competitive inhibition may show different characteristics. However, since it is far easier to determine the inhibitor pattern than the actual mechanism of inhibition, and since the same pattern may be seen for different mechanisms (see Chapter 8.2.1), it has become established that inhibitors are classified according to the overall pattern observed. On this basis, non-competitive inhibition is only said to be present when the characteristics of simple linear non-competitive inhibition (i.e. linear Lineweaver-Burk plot, altered V_{max} but unchanged K_m) are demonstrated. If a linear Lineweaver-Burk plot is obtained but inhibition patterns characteristic of simple competitive inhibition (Fig. 8.1), uncompetitive inhibition (Fig. 8.5) or simple linear non-competitive inhibition (Fig. 8.7) are not observed, then **mixed inhibition** (Chapter 8.2.4) is said to occur, irrespective of the mechanism involved. In any discussion of enzyme inhibition, the basis for classification used should be made clear.

Few clear-cut instances of non-competitive inhibition of single-substrate enzyme-catalysed reactions are known. Hydrogen ions may be regarded as providing one of the simplest examples: some enzymes, e.g. chymotrypsin, where the catalytic site includes a proton acceptor may be inhibited by increasing hydrogen ion concentration (i.e. decreased pH), Lineweaver–Burk plots at different pH values over a relatively narrow range showing the characteristics of non-competitive inhibition. However it should be borne in mind that the effects of changing pH on enzyme activity are complex (Chapter 3.2.2 and 10.1.5).

Heavy-metal ions and organic molecules which bind to −SH groups of cysteine residues in the enzyme are sometimes quoted as being examples of non-competitive inhibitors, as are groups such as cyanide which bind to the metal ions of metalloenzymes and destroy enzyme activity. However in many cases such effects are irreversible, ruling out non-competitive inhibition, which must be reversible. The confusion may arise because irreversible inhibition can give kinetic patterns apparently characteristic of non-competitive inhibition, even though the two types of inhibition are otherwise quite distinct (Chapter 8.3). Nevertheless, it should be understood that the toxicity of substances such as cyanide, carbon monoxide, hydrogen sulphide and heavy metals is due to their action as enzyme inhibitors, whatever the precise mechanism in each case.

8.2.4 Mixed Inhibition

In Chapter 8.2.3 we obtained an expression for simple linear non-competitive inhibition which depended on the equilibrium-assumption (Chapter 7.1.1) being valid and further assumed that substrate-binding and inhibitor-binding were completely independent. Let us now consider the situation where the second assumption is *not* made.

There are two processes by which inhibitor may bind to the enzyme:

$$E + I \rightleftharpoons EI \quad \text{(inhibitor constant } K_i)$$

and $\quad ES + I \rightleftharpoons ESI$ (inhibitor constant K_I).

Hence $\quad K_i = \dfrac{[E][I]}{[EI]} \quad$ and $\quad K_I = \dfrac{[ES][I]}{[ESI]}$.

As in Chapter 8.2.2, for a single-substrate reaction:

$$\frac{[E][S]}{[ES]} = K_m$$

and $\quad [E_o] = [E] + [ES] + [EI] + [ESI]$.

If we develop the argument as before, but this time without assuming that K_i and K_I are identical:

$$[E_o] = [E] + [ES] + \frac{[E][I]}{K_i} + \frac{[ES][I]}{K_I}$$

$$= [E] \left(1 + \frac{[I]}{K_i}\right) + [ES] \left(1 + \frac{[I]}{K_I}\right).$$

$$\therefore [E] = \frac{[E_o] - [ES] \left(1 + \dfrac{[I]}{K_I}\right)}{\left(1 + \dfrac{[I]}{K_i}\right)}.$$

Substituting for [E] in the expression for K_m:

$$\frac{\left([E_o] - [ES] \left(1 + \dfrac{[I]}{K_I}\right)\right)[S]}{\left(1 + \dfrac{[I]}{K_i}\right)[ES]} = K_m.$$

$$\therefore [E_o][S] - [S][ES] \left(1 + \frac{[I]}{K_I}\right) = K_m[ES] \left(1 + \frac{[I]}{K_i}\right).$$

$$\therefore [ES] \left([S] \left(1 + \frac{[I]}{K_I}\right) + K_m \left(1 + \frac{[I]}{K_i}\right)\right) = [E_o][S].$$

$$\therefore [ES] = \frac{[E_o][S]}{[S] \left(1 + \dfrac{[I]}{K_I}\right) + K_m \left(1 + \dfrac{[I]}{K_i}\right)}.$$

Continuation as before gives:

$$v_0 = \frac{V_{max}[S_0]}{[S_0]\left(1 + \frac{[I_0]}{K_I}\right) + K_m\left(1 + \frac{[I_0]}{K_i}\right)}.$$

If numerator and denominator are both divided by $(1 + \frac{[I_0]}{K_i})$,

$$v_0 = \frac{\dfrac{V_{max}}{\left(1 + \dfrac{[I_0]}{K_I}\right)} \cdot [S_0]}{[S_0] + \dfrac{K_m\left(1 + \dfrac{[I_0]}{K_i}\right)}{\left(1 + \dfrac{[I_0]}{K_I}\right)}}.$$

This is of the same form as the Michaelis-Menten equation and can be written:

$$v_0 = \frac{V_{max}'[S_0]}{[S_0] + K_m'}$$

where $\qquad V_{max}' = \dfrac{V_{max}}{\left(1 + \dfrac{[I_0]}{K_I}\right)} \qquad$ and $\qquad K_m' = K_m \dfrac{\left(1 + \dfrac{[I_0]}{K_i}\right)}{\left(1 + \dfrac{[I_0]}{K_I}\right)}.$

Similarly the Lineweaver-Burk equation is:

$$\frac{1}{v_0} = \frac{K_m'}{V_{max}'} \cdot \frac{1}{[S_0]} + \frac{1}{V_{max}'}$$

and a Lineweaver-Burk plot will be linear. However, in general, K_m, V_{max} and slope, which equals $\frac{K_m'}{V_{max}'} = \frac{K_m}{V_{max}}(1 + \frac{[I_0]}{K_i})$ are all affected by the inhibitor. Thus plots at different inhibitor concentrations (at fixed $[E_0]$) will not intersect on either axis, nor will the slope be the same, so the pattern will be different from those characteristic of competitive, non-competitive and uncompetitive inhibition and is given the name mixed inhibition. It must be realised that this describes the overall pattern observed and does not imply that more than one type of inhibitor is present.

In the situation where $K_I > K_i$, the plots cross to the left of the $\frac{1}{v_o}$ axis but above the $\frac{1}{[S_o]}$ axis (Fig. 8.10(a)). This situation has been termed **competitive-non-competitive inhibition**, because the pattern observed lies between those for competitive (Fig. 8.1) and non-competitive (Fig. 8.7) inhibition.

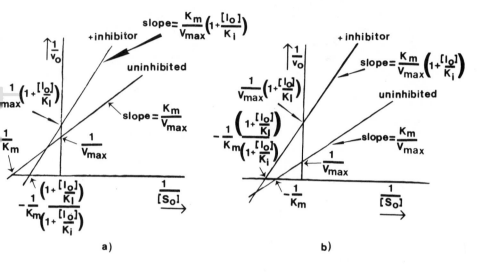

Fig. 8.10 Lineweaver–Burk plots showing the effect of mixed inhibition:
(a) $K_I > K_i$; (b) $K_I < K_i$.

In the situation where $K_I < K_i$, the plots cross to the left of the $\frac{1}{v_o}$ axis and below the $\frac{1}{[S_o]}$ axis (Fig. 8.10(b)). This form of mixed inhibition has been termed **non-competitive-uncompetitive inhibition** because the pattern is intermediate between those for non-competitive (Fig. 8.7) and uncompetitive (Fig. 8.5) inhibition.

In either case, K_i and K_I can be determined using **secondary plots**. For mixed inhibition, $\frac{1}{V_{max}'} = \frac{1}{V_{max}}(1 + \frac{[I_o]}{K_I})$ and slope for inhibited reaction = slope for uninhibited reaction $\times (1 + \frac{[I_o]}{K_i})$. Hence a secondary plot of $\frac{1}{V_{max}'}$ against $[I_o]$ will be linear, the intercept on the $[I_o]$ axis giving $-K_I$ (Fig. 8.11(a)); a graph of slope of primary plot against $[I_o]$ will also be linear, the intercept on the $[I_o]$ axis giving $-K_i$ (Fig. 8.11(b)).

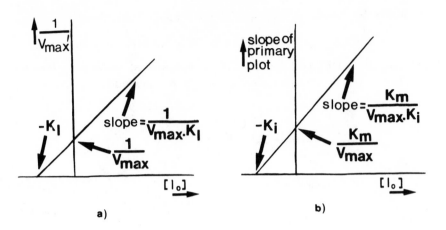

Fig. 8.11 Secondary plots for mixed inhibition.

However
$$K_m' = K_m \frac{\left(1 + \dfrac{[I_o]}{K_i}\right)}{\left(1 + \dfrac{[I_o]}{K_I}\right)}$$

which means that a graph of K_m' against $[I_o]$ will not be linear.

The equation for v_o derived above is a relatively general one, since no assumptions were made about the values of K_i and K_I, and it can be simplified for special cases. If *ESI* cannot be formed, then $K_I = \infty$ and the equation becomes that for competitive inhibition (Chapter 8.2.1), regardless of whether the substrate and inhibitor bind to the same or different sites. If the complex *ESI* can occur but not *EI*, then $K_i = \infty$ and the equation simplifies to that for uncompetitive inhibition (Chapter 8.2.2). When $K_i = K_I$, the equation reduces to that for simple linear non-competitive inhibition (Chapter 8.2.3).

8.2.5 Partial Inhibition

Hitherto we have considered only situations where enzyme–inhibitor complexes are dead-end ones, i.e. where no product can be formed from them. Let us now return to the general system described in Chapter 8.2.4 and consider what would happen if the inhibition was only partial and the *ESI* complex could break down to yield product according to the equation:

$$ESI \xrightarrow{k_2'} E + P + I.$$

Under these conditions, the overall initial velocity is given by:

$$v_o = k_2[ES] + k_2'[ESI]$$

$$= k_2[ES] + k_2' \frac{[ES][I]}{K_I}$$

$$= k_2[ES] \left(1 + \frac{k_2'[I]}{k_2 K_I}\right).$$

Using the same procedure as before, an expression of the same form as the Michaelis-Menten equation can be obtained:

$$v_o = \frac{V_{max}[S_o] \dfrac{\left(1 + \dfrac{k_2'[I_o]}{k_2 K_I}\right)}{\left(1 + \dfrac{[I_o]}{K_I}\right)}}{[S_o] + \dfrac{K_m\left(1 + \dfrac{[I_o]}{K_i}\right)}{\left(1 + \dfrac{[I_o]}{K_I}\right)}}.$$

Hence a Lineweaver-Burk plot would be linear. However **secondary plots** of intercept or slope against $[I_o]$ will not be linear, enabling **partial inhibition** to be distinguished from inhibition involving **dead-end complexes**.

8.2.6 Substrate Inhibition

A characteristic of enzyme-catalysed reactions, as we have seen, is that for a given enzyme concentration, the initial reaction velocity increases with increasing initial substrate concentration to a limiting value, V_{max} (Chapter 6.4). At still higher substrate concentrations, the initial velocity is sometimes found to be less than the maximum value. In some instances the observations may be explained away on the basis of interaction between the detecting system and excess substrate, but in other cases it appears that the substrate, in very high concentrations, really can inhibit its own conversion to product.

Let us consider one possible mechanism for substrate inhibition at high substrate concentrations, in relation to the reaction catalysed by succinate dehydrogenase. For a reaction to take place, both carboxyl groups of the substrate have to bind to the enzyme:

$$
\begin{array}{c}
\text{\textbar} \cdots \cdots ^-O_2C \\
\qquad | \\
\qquad CH_2 \\
\qquad | \\
\qquad CH_2 \\
\text{\textbar} \cdots \cdots ^-O_2C
\end{array}
\longrightarrow
\quad
\begin{array}{c}
\text{\textbar} \\
\text{\textbar} \\
\text{\textbar}
\end{array}
+
\begin{array}{c}
^-O_2C \\
| \\
CH \\
\| \\
CH \\
| \\
^-O_2C
\end{array}
+ \; 2H.
$$

E   succinate   E   fumarate

At very high substrate concentrations there is an increased possibility of carboxyl groups from two separate substrate molecules binding to the same enzyme:

$$
\begin{array}{l}
\text{\textbar} \cdots \cdots ^-O_2C.CH_2.CH_2.CO_2^- \\
\\
\text{\textbar} \cdots \cdots ^-O_2C.CH_2.CH_2.CO_2^-. \\
E
\end{array}
$$

If this happens, a reaction cannot take place until one of them has dissociated away again.

The characteristic features of substrate inhibition are shown in Fig. 8.12.

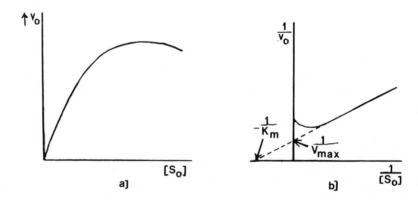

Fig. 8.12 (a) Michaelis–Menten and (b) Lineweaver–Burk plots, showing the effects of substrate inhibition.

In general it can be seen that substrate inhibition occurs when a molecule of substrate binds to one site on the enzyme and then another molecule of substrate binds to a separate site on the enzyme to form a dead-end complex. This can be regarded as a form of uncompetitive inhibition, the extra substrate molecule being the inhibitor. As shown in Chapter 8.2.2, the initial velocity equation for uncompetitive inhibition is:

$$v_o = \frac{\dfrac{V_{max}}{\left(1 + \dfrac{[I_o]}{K_i}\right)} \cdot [S_o]}{[S_o] + \dfrac{K_m}{\left(1 + \dfrac{[I_o]}{K_i}\right)}}$$

If the inhibitor is identical to the substrate, this becomes:

$$v_o = \frac{\dfrac{V_{max}}{\left(1 + \dfrac{[S_o]}{K_i}\right)} \cdot [S_o]}{[S_o] + \dfrac{K_m}{\left(1 + \dfrac{[S_o]}{K_i}\right)}} = \frac{V_{max}[S_o]}{[S_o]\left(1 + \dfrac{[S_o]}{K_i}\right) + K_m}$$

This is consistent with the plots in Fig. 8.12. At low $[S_o]$ the term $\dfrac{[S_o]}{K_i}$ is negligible and the expression reduces to the normal Michaelis-Menten equation. When $[S_o]$ is very high, then $[S_o](1 + \dfrac{[S_o]}{K_i}) + K_m \simeq [S_o](1 + \dfrac{[S_o]}{K_i})$ and the equation simplifies to:

$$v_o = \frac{V_{max}}{1 + \dfrac{[S_o]}{K_i}}$$

Under these circumstances, v_o decreases as $[S_o]$ increases, as observed for substrate inhibition at high substrate concentrations.

8.2.7 Allosteric Inhibition

The forms of inhibition considered previously in this chapter have many uses (Chapter 8.2.1) but in general they do not play a major role in the normal functioning of the living cell. In contrast, allosteric inhibition plays a vital role in metabolic regulation. Consider a biosynthetic pathway:

$$A \rightarrow B \rightarrow C \rightarrow D \rightarrow E \rightarrow F.$$

Unnecessary production of excess F may be prevented, and supplies of A conserved, by feedback inhibition, where the end-product F acts as an allosteric inhibitor of an early enzyme in the pathway, e.g. that catalysing the reaction $A \rightarrow B$.

An allosteric inhibitor, by definition, binds to the enzyme at a site distinct from the substrate-binding site. Therefore, some of the types of inhibition we have considered previously may be regarded as forms of allosteric inhibition. However the term allosteric inhibition is usually reserved for the situation where the inhibitor, rather than forming a **dead-end complex** with the enzyme, influences **conformational changes** which may alter the **binding characteristics** of the enzyme for the substrate or the subsequent **reaction characteristics** (or both) (Chapter 13.2.3). The Michaelis-Menten plot becomes less hyperbolic and more sigmoidal (S-shaped) (Fig. 13.1), which means that the rate of reaction is reduced at low substrate concentrations but not necessarily at others.

If the binding characteristics alone are affected, V_{max} will usually remain unchanged, so the inhibition pattern could be regarded as competitive. Similarly, other forms of allosteric inhibition, where V_{max} is altered, could be regarded as giving non-competitive or mixed inhibition, depending on whether K_m (the substrate concentration where $v_o = \frac{1}{2} V_{max}$) is changed or not. However, in most cases the Michaelis–Menten equation is not obeyed in the presence of allosteric inhibitors, nor are linear Lineweaver–Burk plots obtained, so the terms competitive, non-competitive and mixed inhibition are not strictly applicable.

Almost all enzymes known to be subject to end-product (feedback) inhibition are oligomeric proteins, so the mechanism of allosteric inhibition may involve interactions between the enzyme sub-units. In support of this, allosteric control, but not catalytic activity, of some enzymes has been shown to be lost when the oligomer is separated into its monomeric units.

The subject of allosteric inhibition is discussed in more detail in Chapters 13 and 14.

8.3 IRREVERSIBLE INHIBITION

An irreversible inhibitor binds to the active site of the enzyme by an irreversible reaction:

$$E + I \rightarrow EI,$$

and hence cannot subsequently dissociate from it. A covalent bond is usually formed between inhibitor and enzyme. The inhibitor may act by preventing substrate-binding or it may destroy some component of the catalytic site. Compounds which irreversibly denature the enzyme protein or cause non-specific inactivation of the active site are not usually regarded as irreversible inhibitors.

Of course in practice no process is totally irreversible, but an inhibitor which shows great affinity for the enzyme (dissociation constant in the order of 10^{-9} mol.1^{-1}) is regarded as irreversible. Unlike reversible inhibition, where an equilibrium is quickly set up between inhibitor and enzyme, making the system suitable for investigation by initial velocity studies, irreversible inhibition is progressive and will increase with time until either all the inhibitor or all the enzyme present has been used up in forming enzyme-inhibitor complex.

Regardless of whether the reaction between enzyme and irreversible inhibitor has gone to completion before initial velocity studies are commenced, these will give little or no information about the characteristics of the inhibitor (see below). It is much more useful to describe the system in terms of the rate constant for the binding of inhibitor to substrate. If these react on a 1:1 basis then the time taken to reduce the enzyme activity to 50% of its original value will be inversely proportional to the initial inhibitor concentration (at fixed total enzyme concentration).

If a molar excess of the inhibitor is present, all molecules of enzyme present will eventually become bound to inhibitor and catalytic activity will be reduced to a residual level or completely lost. (In general, the binding of such an inhibitor to a catalytic site will totally destroy catalytic activity, but binding to a substrate-binding site may not prevent a small amount of catalysis taking place, provided the catalytic site remains active and accessible to substrate molecules approaching it by random movement in free solution). Hence irreversible inhibitors may be titrated against an enzyme, particularly where the reaction between enzyme and inhibitor proceeds rapidly. In contrast, reversible inhibitors, like substrates (Chapter 7.1.1), can be present in great excess without saturating the available enzyme.

Irreversible inhibitors effectively reduce the concentration of enzyme present. An inhibitor of initial concentration $[I_o]$ will reduce the concentration of active enzyme from an initial value of $[E_o]$ to $[E_o] - [I_o]$, assuming the inhibitor is not in excess. If a substrate is introduced after the reaction between inhibitor and enzyme has gone to completion, a system which obeys the Michaelis-Menten equation in the absence of inhibitor will still do so. The value of K_m will be the same as for the uninhibited reaction, but V_{max} will be reduced (to V_{max}').

In the absence of inhibitor, $\quad V_{max} = k_{cat}[E_o]$.

In the presence of inhibitor, $\quad V_{max}' = k_{cat}([E_o] - [I_o])$.

$$\therefore \frac{V_{max}'}{V_{max}} = \frac{[E_o] - [I_o]}{[E_o]}.$$

$$\therefore V_{max}' = V_{max} \cdot [E_o] \left(1 - \frac{[I_o]}{[E_o]}\right).$$

Similar results would be obtained even if the reaction between enzyme and inhibitor had not gone to completion, provided the degree of inhibition was relatively constant over the period when initial velocity studies were being carried out.

Therefore, patterns resembling those for reversible non-competitive inhibition, with unchanged K_m and reduced V_{max}, may be obtained with irreversible inhibitors, even when the inhibitor binds to the same site as the substrate. However the very real differences between the two forms of inhibition should be clear from the above discussion, and any attempt to calculate K_i from initial velocity measurements would be a totally meaningless exercise, since the relationship between V_{max}' and V_{max} does not involve K_i in this case. Hence, if a pattern of non-competitive inhibition is obtained in the investigation of a system, it is important to establish whether the inhibition is reversible or irreversible before the results can be interpreted.

Many irreversible inhibitors attack $-SH$ groups (in cysteine side chains) which are often found at the active sites of enzymes. Important examples are alkylating agents, such as iodoacetate and iodoacetamide, which form covalent linkages with essential $-SH$ groups:

$$E - SH + ICH_2.CO_2^- \rightarrow E - S - CH_2.CO_2^- + HI.$$
$$\text{enzyme} \qquad \text{iodoacetate}$$

Another well known group are the organophosphorus compounds which react with essential $-OH$ groups (in serine side chains) of some enzymes. An example is **diisopropylphosphofluoridate (DFP)**, which is a nerve poison since one of the enzymes it inactivates is acetylcholinesterase, important in nerve function:

$$E - OH + F-\overset{\displaystyle OCH(CH_3)_2}{\underset{\displaystyle OCH(CH_3)_2}{P}}=O \quad \rightarrow \quad E-O-\overset{\displaystyle OCH(CH_3)_2}{\underset{\displaystyle OCH(CH_3)_2}{P}}=O \quad + HF.$$
$$\text{enzyme} \qquad\qquad \text{DFP}$$

Irreversible inhibitors are useful in the investigation of the active site of an enzyme, since the inhibitor, unlike the substrate, will remain firmly bound to one of the amino-acids of the enzyme and thus act as a marker to enable it to be identified (Chapter 10). Some organophosphorus compounds are also used as insecticides.

SUMMARY OF CHAPTER 8

Competitive inhibitors usually compete with the substrate for the same binding site on the enzyme. In the characteristic form, Michaelis–Menten kinetics are obeyed, K_m is increased and V_{max} unchanged.

Uncompetitive inhibitors bind to a site other than the substrate-binding site on the enzyme-substrate complex, altering the K_m and V_{max} but not the slope of the Lineweaver-Burk plot.

Non-competitive inhibitors bind to a site other than the substrate-binding site on the enzyme and enzyme-substrate complex. In the characteristic form, Michaelis-Menten kinetics are obeyed, K_m is unchanged and V_{max} decreased.

Forms of inhibition obeying Michaelis-Menten kinetics but not giving patterns characteristic of competitive, uncompetitive or non-competitive inhibition are usually termed **mixed inhibition**, irrespective of the actual mechanism.

Secondary plots and **Dixon plots** enable the inhibitor constant, K_i, to be calculated and help distinguish between mechanisms which give identical **primary Lineweaver-Burk plots**.

These types of inhibition are used in biochemical research and have applications in medicine and agriculture. **Allosteric inhibition**, which results in more sigmoidal reaction characteristics, plays an important role in metabolic regulation in the living cell. All these forms of inhibition are **reversible**, but **irreversible inhibition** is also known. Irreversible inhibitors have been used to identify amino-acids in the active centres of enzymes.

FURTHER READING

As for Chapter 7, plus:

Brodbeck, U. (ed.) (1980), *Enzyme Inhibitors*, Verlag-Chemie.

Dixon, M., Webb, E. C., Thorne, C. J. R. and Tipton, K. F. (1979), *Enzymes*, Third Edition (Chapter 8), Longman.

Jain, M. K. (1982), *Handbook of Enzyme Inhibitors*, Wiley.

Webb, J. L. (1963), *Enzyme and Metabolic Inhibitors*, 1, Academic Press.

PROBLEMS

8.1 An enzyme catalysed reaction was found to be affected by two inhibitors A and B. The following results were obtained at fixed total enzyme concentration:

Substrate concn (mmol. 1^{-1})	Initial velocity (absorbance units per minute)		
	Uninhibited	With 1 mmol. 1^{-1} A	With 1 mmol. 1^{-1} B
50	0.684	–	–
20	1.08	–	–
10	1.43	1.01	0.653
5	1.02	0.649	0.468
3.3	0.798	0.476	0.363
2.5	0.657	0.374	0.296
2.0	0.549	0.311	0.250

Comment on these results.

8.2 The system investigated in problem 7.1 was investigated again under identical conditions but in the presence of an inhibitor, giving the following data:

Substrate conc^n (mmol.l^{-1})   5.0   6.67   10.0   20.0   40.0

Initial velocity (μmol.l^{-1}.minute^{-1})   100   122   156   222   278 .

Determine the type of inhibition. If K_i for this system is 2.9 mmol.l^{-1}, calculate the inhibitor concentration present.

8.3 The system investigated in problem 7.2 was investigated again under identical conditions but in the presence of oxaloacetate (initial concentration 2.0 mmol.l^{-1} in each case). These results were obtained:

NAD$^+$ conc^n	Absorbance (at 340 nm) at time t (minutes)					
(mmol.l^{-1})	t = 0.5	1.0	1.5	2.0	2.5	3.0
1.5	0.026	0.042	0.058	0.074	0.088	0.102
2.0	0.028	0.045	0.062	0.080	0.096	0.112
2.5	0.029	0.047	0.066	0.084	0.102	0.117
3.33	0.030	0.050	0.070	0.090	0.108	0.123
5.0	0.032	0.053	0.074	0.097	0.115	0.130
10.0	0.033	0.057	0.080	0.103	0.124	0.140

What type of inhibition is exhibited? What would be the value of V_{max}' in presence of 3.0 mmol.l^{-1} oxaloacetate?

8.4 A single-substrate enzyme-catalysed reaction was investigated in the presence of 1.0 mmol.l^{-1} inhibitor and in the absence of inhibitor, the initial enzyme concentration being constant throughout. The following results were obtained:

Substrate conc^n	Product conc^n (μmol.l^{-1}) at time t					
(mmol.l^{-1})	t = 0	60 s	120 s	180 s	240 s	300 s
5.0 inhibited	0	110	221	333	430	480
5.0 uninhibited	0	161	320	482	598	662
6.67 inhibited	0	142	281	420	531	598
6.67 uninhibited	0	194	388	581	745	796
10.0 inhibited	0	183	367	549	705	752
10.0 uninhibited	0	263	525	789	998	1120
20.0 inhibited	0	279	558	837	1050	1170
20.0 uninhibited	0	400	798	1200	1520	1760
50.0 inhibited	0	398	798	1190	1520	1760
50.0 uninhibited	0	576	1150	1730	2170	2460

Determine the type of inhibition and calculate the values of K_m' and V_{max}' in the presence of 3.0 mmol.l^{-1} inhibitor.

8.5 The following results were obtained for a single-substrate enzyme-catalysed reaction (at fixed initial enzyme concentration):

Substrate conc^n (mmol.l^{-1})	Initial velocity (absorbance units. minute^{-1}) at initial inhibitor concentration $[I_o]$ (mmol.l^{-1})				
	$[I_o] = 1.0$	2.0	3.0	4.0	5.0
2.0	0.432	0.396	0.365	0.339	0.317
2.5	0.485	0.448	0.417	0.389	0.365
3.33	0.552	0.516	0.485	0.457	0.432
5.0	0.642	0.609	0.579	0.552	0.528.

What can you deduce about the nature of the inhibition?

8.6 A single-substrate enzyme-catalysed reaction was investigated at fixed initial enzyme concentration and the following data obtained:

Substrate conc^n (mmol.l^{-1})	Initial velocity (absorbance units. minute^{-1}) at initial inhibitor concentration $[I_o]$ (mmol.l^{-1})				
	$[I_o] = 1.0$	2.0	3.0	4.0	5.0
2.0	0.400	0.351	0.317	0.291	0.271
2.5	0.447	0.394	0.356	0.328	0.306
3.33	0.506	0.447	0.405	0.374	0.350
5.0	0.584	0.521	0.475	0.440	0.413.

What can be deduced from these results about the nature of the inhibition?

CHAPTER 9

Kinetics of Multi-Substrate Enzyme-Catalysed Reactions

9.1 EXAMPLES OF POSSIBLE MECHANISMS

9.1.1 Introduction

Most biochemical reactions involve at least two substrates, so it is necessary to consider the kinetics of such reactions. This is a vast and extremely complex topic and, for reasons of simplicity, we will restrict our discussion to some specific examples of **two-substrate two-product (bi-bi)** reactions. These are often transfer reactions of one type or another (including oxidation/reduction reactions) and can best be represented as:

$$AX + B \rightleftharpoons BX + A.$$

The reaction mechanism may be a **sequential** one, where both substrates bind to the enzyme to form a **ternary complex** before the first product is formed, or it may be **non-sequential**.

9.1.2 Ping-Pong Bi-Bi Mechanism

An example of non-sequential mechanism is the ping-pong bi-bi or double-displacement mechanism:

$$AX + E \rightleftharpoons E.AX \rightleftharpoons EX.A \rightleftharpoons EX + A$$
$$EX + B \rightleftharpoons EX.B \rightleftharpoons E.BX \rightleftharpoons E + BX.$$

AX first binds to the enzyme E, forming a binary complex $E.AX$ (X is usually a small group and does not participate in the reaction as a free molecule, so it is not regarded as a separate reactant). An intramolecular reorganisation takes place, the bond $E-X$ being formed and the $X-A$ bond being broken. The first product, A, then leaves before the second substrate arrives. B cannot bind to the enzyme E but can bind to the modified enzyme EX. Since only one substrate is present on the enzyme at any one time there may only be a single binding site.

Another intramolecular rearrangement takes place, the bond B—X being formed and the bond E—X being broken. The second product, B, is then liberated, leaving the enzyme in its original form.

Cleland has devised a diagrammatic representation which shows this sequence of events as follows:

$$
\begin{array}{ccccccc}
\text{AX} & & \text{A} & \text{B} & & \text{BX} & \\
\downarrow & & \uparrow & \downarrow & & \uparrow & \\
\text{E} \overline{\hspace{4em}} & & & & & & \text{E.} \\
\text{E.AX} & \text{EX.A} & \text{EX} & & \text{EX.B} & \text{E.BX} &
\end{array}
$$

9.1.3 Random-Order Mechanism

A random-order mechanism is one in which any substrate can bind first to the enzyme and any product can leave first. It is a sequential mechanism and for a two-substrate reaction involves the formation of a **ternary complex** (one involving enzyme and both substrates):

$$
\begin{array}{lcl}
\text{E} + \text{AX} \rightleftharpoons \text{E.AX} & & \text{EA} \rightleftharpoons \text{E} + \text{A} \\
\quad\quad\quad\quad\searrow\!{\scriptstyle+B} & \quad {\scriptstyle-BX}\!\nearrow & \\
\quad\quad\quad \text{E.AX.B} \rightleftharpoons \text{E.A.BX} & & \\
\quad {\scriptstyle+AX}\!\nearrow & \quad {\scriptstyle-A}\!\searrow & \\
\text{E} + \text{B} \rightleftharpoons \text{EB} & & \text{E.BX} \rightleftharpoons \text{E} + \text{BX.}
\end{array}
$$

There will be two separate binding sites on the enzyme, one for A/AX and one for B/BX.

9.1.4 Compulsory-Order Mechanism

A compulsory-order (or simply ordered) mechanism is a sequential mechanism where the order of binding to and leaving the enzyme is compulsory. For a two-substrate reaction, a **ternary complex** will be involved. The precise order must be specified,

e.g. $\text{E} + \text{AX} \rightleftharpoons \text{E.AX} \overset{+B}{\rightleftharpoons} \text{E.AX.B} \rightleftharpoons \text{E.A.BX} \overset{-BX}{\rightleftharpoons} \text{EA} \rightleftharpoons \text{E} + \text{A}$

or $\text{E} + \text{B} \rightleftharpoons \text{EB} \overset{+AX}{\rightleftharpoons} \text{E.B.AX} \rightleftharpoons \text{E.BX.A} \overset{-A}{\rightleftharpoons} \text{E.BX} \rightleftharpoons \text{E} + \text{BX.}$

As before, the enzyme will have a binding site for A/AX and a separate one for B/BX.

9.2 STEADY-STATE KINETICS

9.2.1 The General Rate Equation of Alberty

Many two-substrate enzyme-catalysed reactions obey the Michaelis–Menten equation with respect to one substrate at constant concentrations of the other substrate. This applies both to reactions catalysed by enzymes with just one binding site per substrate and by those with several binding sites per substrate, provided there is no interaction between the binding sites. For such reactions, Alberty (1953) derived the general equation:

$$v_0 = \frac{V_{max}[AX_o][B_o]}{K_m^B[AX_o] + K_m^{AX}[B_o] + [AX_o][B_o] + K_s^{AX}.K_m^B}$$

where V_{max} is the maximum possible v_o when AX and B are both saturating,
K_m^{AX} is the concentration of AX which gives $\frac{1}{2}V_{max}$ when B is saturating,
K_m^B is the concentration of B which gives $\frac{1}{2}V_{max}$ when AX is saturating,
and K_s^{AX} is the dissociation constant for $E + AX \rightleftharpoons EAX$.

The total enzyme concentration is constant and much smaller than the concentrations of the two substrates.

At very large $[B_o]$ the general equation simplifies to:

$$v_0 = \frac{V_{max}}{1 + \frac{K_m^{AX}}{[AX_o]}} = \frac{V_{max}[AX_o]}{[AX_o] + K_m^{AX}}$$

(which is the Michaelis–Menten equation).

Similarly at very large $[AX_o]$,

$$v_0 = \frac{V_{max}}{1 + \frac{K_m^B}{[B_o]}} = \frac{V_{max}[B_o]}{[B_o] + K_m^B}.$$

At constant but not saturating $[B_o]$, the general equation can be rearranged to give:

$$v_0 = \frac{V_{max}.K_1[AX_o]}{[AX_o] + K_2}$$

(which is of the form of the Michaelis–Menten equation)

where $K_1 = \dfrac{[B_o]}{K_m^B + [B_o]}$ and $K_2 = \dfrac{K_s^{AX}.K_m^B + K_m^{AX}[B_o]}{K_m^B + [B_o]}$.

At constant but not saturating $[AX_o]$, a similar expression can be obtained, also of the form of the Michaelis–Menten equation.

The reason for the mixed $K_s^{AX} \cdot K_m^B$ term (rather than $K_m^{AX} \cdot K_m^B$) can be seen if we consider a compulsory-order mechanism of the form:

$$E + AX \rightleftharpoons EAX \rightleftharpoons EAXB \rightarrow \rightarrow products.$$

As $[B_o]$ tends to zero, there will be very little formation of EAXB from EAX, so $E + AX \rightleftharpoons EAX$ will be very close to equilibrium and K_m^{AX} tends to K_s^{AX}. This is consistent with the Alberty equation, for at a constant but very low concentration of B,

$$K_1 \simeq \frac{[B_o]}{K_m^B} \quad \text{and} \quad K_2 \simeq \frac{K_s^{AX} \cdot (K_m^B)}{(K_m^B)}$$

giving $v_o = \dfrac{V_{max}[B_o][AX_o]}{K_m^B([AX_o] + K_s^{AX})}$, an expression involving K_s^{AX} but not K_m^{AX}.

For a compulsory-order mechanism where B binds first to the enzyme, the $K_s^{AX} \cdot K_m^B$ term would be replaced in the general equation by $K_s^B \cdot K_m^{AX}$.

For a random-order mechanism either term could be used, and would be justified on grounds similar to the above.

Many random- and compulsory-order reactions involving ternary complexes obey the general rate equation of Alberty, particularly where the rate-limiting step is the interconversion of the ternary complexes ($E.AX.B \rightleftharpoons E.A.BX$, the main chemical reaction taking place, involving the breakage of the $A-X$ bond and the formation of the $B-X$ bond in the forward direction). A similar observation was made for single-substrate reactions in Chapter 7.1.3, for these obeyed the Michaelis–Menten equation if $ES \rightarrow EP$ was the rate-limiting step. A random-order mechanism where all the steps except the interconversion of the ternary complexes are rapid is sometimes said to have a **random-order rapid-equilibrium bi-bi mechanism**.

For a ping-pong bi-bi mechanism, the liberation of A from the enzyme in the initial period of the reaction will be irreversible because the concentration of product A present will be negligible. Hence $K_s^{AX} = 0$ and so $K_s^{AX} \cdot K_m^B = 0$, giving this mechanism a simpler rate equation:

$$v_o = \frac{V_{max}[AX_o][B_o]}{K_m^B[AX_o] + K_m^{AX}[B_o] + [AX_o][B_o]}.$$

9.2.2 Primary Plots for Mechanisms Following the General Rate Equation

Two-substrate reactions obeying the general rate equation of Alberty also obey the Michaelis–Menten equation with respect to one substrate, provided the concentration of the other substrate is maintained fixed. It follows, therefore, that the corresponding double-reciprocal (i.e. Lineweaver–Burk) plot should be linear. Such **primary plots** are drawn for reactions investigated with the fixed substrate concentration being in one of two categories: either saturating the enzyme, or approximately half-saturating it. The theoretical results are shown in Fig. 9.1, together with some of the characteristics of the graphs.

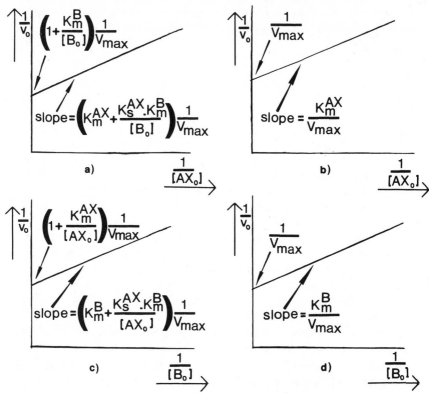

Fig. 9.1 Primary plots for reactions with mechanisms following the general rate equation: (a) $\frac{1}{v_0}$ against $\frac{1}{[AX_0]}$, where $[B_0]$ is constant and $\simeq K_m^B$, (b) $\frac{1}{v_0}$ against $\frac{1}{[AX_0]}$, where $[B_0]$ is saturating (i.e. pseudo single-substrate with regard to AX, (c) $\frac{1}{v_0}$ against $\frac{1}{[B_0]}$, where $[AX_0]$ is constant and $\simeq K_m^{AX}$, (d) $\frac{1}{v_0}$ against $\frac{1}{[B_0]}$, where $[AX_0]$ is saturating (i.e. pseudo single-substrate with respect to B).

9.2.3 The General Rate Equation of Dalziel

Dalziel (1957) gave the general rate equation in the form:

$$\frac{[E_0]}{v_0} = \phi_0 + \frac{\phi_{AX}}{[AX_0]} + \frac{\phi_B}{[B_0]} + \frac{\phi_{AXB}}{[AX_0][B_0]}.$$

The ϕ terms, called **kinetic coefficients**, can be obtained by drawing primary and secondary plots as follows. Primary plots of $\dfrac{[E_o]}{v_o}$ against $\dfrac{1}{[AX_o]}$ at constant $[B_o]$ are drawn for a series of different values of $[B_o]$. The slopes and the intercepts on the $\dfrac{[E_o]}{v_o}$ axis for each graph are determined, and then **secondary replots** are drawn as in Fig. 9.2. From the secondary replots, the values of ϕ_o, ϕ_{AX}, ϕ_B and ϕ_{AXB} can be determined. The whole procedure could be repeated from primary plots of $\dfrac{[E_o]}{v_o}$ against $\dfrac{1}{[B_o]}$ at constant $[AX_o]$, but the same results should be obtained.

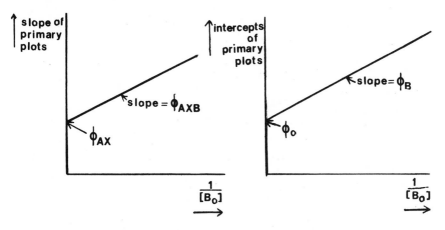

Fig. 9.2 Secondary replots to enable calculation of the kinetic coefficients of Dalziel.

The Dalziel equation can be arranged in the same form as the Alberty equation and the relationship between the respective constants determined. Thus V_{max} in the Alberty equation is $\dfrac{[E_o]}{\phi_o}$ in the Dalziel equation, whereas

$$K_m^{AX} = \frac{\phi_{AX}}{\phi_o}, \quad K_m^B = \frac{\phi_B}{\phi_o} \quad \text{and} \quad K_s^{AX} = \frac{\phi_{AXB}}{\phi_B}.$$

9.2.4 Rate Constants and the Constants of Alberty and Dalziel

The general rate equation of Alberty bears an obvious resemblance to the Michaelis–Menten equation and so serves as a useful introduction to the subject of two-substrate kinetics. However the Dalziel equation may be of more value to an enzyme kineticist.

The kinetic constants in the general rate equations are, of course, functions involving the individual rate constants of the steps involved in the reaction.

Relationships can be derived for particular mechanisms and it is found that the Dalziel kinetic coefficients give more straightforward relationships between rate constants than do the Alberty constants. In either case the mechanisms involved in calculating these relationships can be extremely laborious, so the **King and Altman procedures** (Chapter 7.3) prove extremely useful.

Let us consider the ping-pong bi-bi mechanism (Chapter 9.1.2) under steady-state conditions. Since the initial concentrations of both products are zero, the release of these products from the enzyme will be effectively irreversible. The reaction sequence, in cyclic form, is:

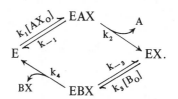

Enzyme
Species   *Pathways forming enzyme species*   *Sum of kappa products*

E

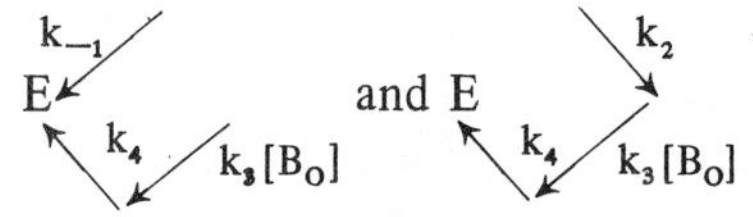

$$k_{-1}k_3k_4[B_o] + k_2k_3k_4[B_o]$$
$$= k_3k_4[B_o](k_{-1} + k_2)$$

EAX

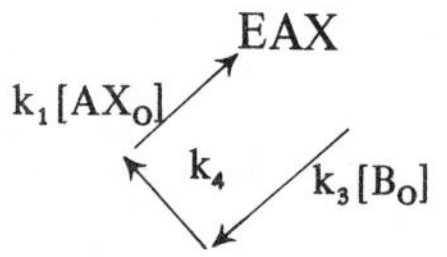

$$k_1k_3k_4[AX_o][B_o]$$

EX

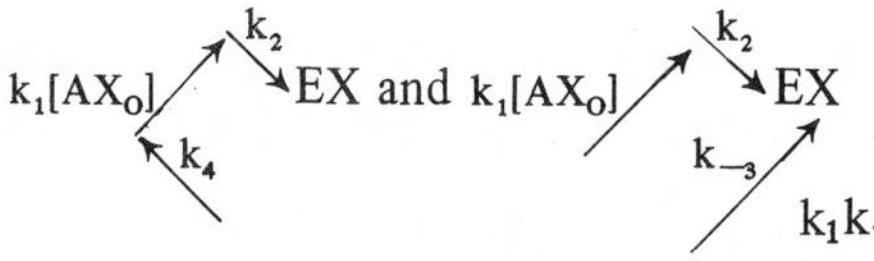

$$k_1k_2k_4[AX_o] + k_1k_2k_{-3}[AX_o]$$
$$= k_1k_2[AX_o](k_{-3} + k_4)$$

EBX

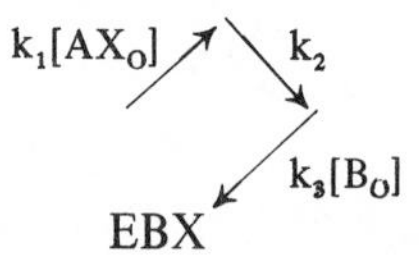

$$k_1k_2k_3[AX_o][B_o]$$

Hence, for example:

$$\frac{[EAX]}{[E_o]} = \frac{k_1k_3k_4[AX_o][B_o]}{k_1k_3[AX_o][B_o](k_2 + k_4) + k_3k_4[B_o](k_{-1} + k_2) + k_1k_2[AX_o](k_{-3} + k_4)}.$$

The overall rate of reaction, $v_o = k_2[EAX] = k_4[EBX]$.

Substituting for [EAX] and rearranging,

$$\frac{[E_o]}{v_o} = \frac{k_1k_3[AX_o][B_o](k_2+k_4)+k_3k_4[B_o](k_{-1}+k_2)+k_1k_2[AX_o](k_{-3}+k_4)}{k_1k_2k_3k_4[AX_o][B_o]}$$

$$= \frac{k_2+k_4}{k_2k_4} + \frac{k_{-1}+k_2}{k_1k_2}\cdot\frac{1}{[AX_o]} + \frac{k_{-3}+k_4}{k_3k_4}\cdot\frac{1}{[B_o]}.$$

This is in the form of the Dalziel equation. If the expression is rearranged into the form of the Alberty equation we obtain:

$$v_o = \frac{\left(\dfrac{k_2k_4}{k_2+k_4}\right)[E_o][AX_o][B_o]}{\dfrac{k_2}{k_3}\left(\dfrac{k_{-3}+k_4}{k_2+k_4}\right)[AX_o]+\dfrac{k_4}{k_1}\left(\dfrac{k_{-1}+k_2}{k_2+k_4}\right)[B_o]+[AX_o][B_o]}.$$

Steady-state rate equations for other mechanisms can be obtained similarly.

9.3 INVESTIGATION OF REACTION MECHANISMS USING STEADY-STATE METHODS

9.3.1 The Use of Primary Plots

Reaction mechanisms which obey the general rate equation of Alberty (and that of Dalziel) give linear primary plots of $\frac{1}{v_o}$ against $\frac{1}{[AX_o]}$ at constant $[B_o]$, and of $\frac{1}{v_o}$ against $\frac{1}{[B_o]}$ at constant $[AX_o]$. As shown in Fig. 9.1, the expressions for the intercepts on the $\frac{1}{v_o}$ axis and the slopes of these graphs both include the concentration of the fixed substrate at non-saturating concentrations, so both intercept and slope change if the experiments are repeated with the fixed substrate concentration set at a different value. The primary plots for such reactions, e.g. those proceeding by compulsory-order ternary-complex mechanisms or by random-order rapid-equilibrium bi-bi mechanisms, will be of the form shown in Fig. 9.3(a).

In contrast, the equation for reaction proceeding by a ping-pong bi-bi mechanism is simpler in that $K_s^{AX}.K_m^B = 0$. From Fig. 9.1 it can be seen that this results in the slope of the primary plot being independent of the concentration of the fixed substrate. A series of primary plots obtained at different concentrations of the fixed substrate would thus be parallel if a ping-pong bi-bi mechanism operates (Fig. 9.3(b)).

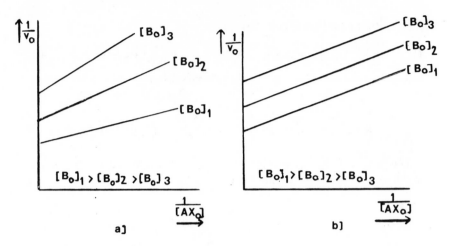

Fig. 9.3   Primary plots obtained at non-saturating concentrations of both substrates: (a) for compulsory-order and random-order ternary-complex mechanisms; (b) for ping-pong bi-bi mechanisms. (In each case similar results would be obtained for plots of $\frac{1}{v_0}$ against $\frac{1}{[B_0]}$ at fixed $[AX_0]$.)

Compulsory-order and random-order ternary-complex mechanisms can be distinguished from ping-pong bi-bi mechanisms, but not from each other, by the use of primary plots.

For example, Hersh and Jencks (1967) obtained a graph of the form of Fig. 9.3(b) for **3-ketoacid CoA-transferase** and concluded that the mechanism was ping-pong bi-bi, probably proceeding as follows:

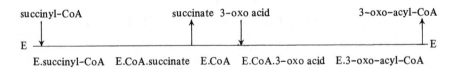

They confirmed this mechanism by isolating the E−CoA intermediate.

The drawing of primary plots is not in itself sufficient basis for concluding that a particular mechanism is ping-pong bi-bi, because it is difficult to be certain whether lines drawn from experimental results really are parallel. Fromm and co-workers (1962) concluded from primary plots that yeast **hexokinase** had a ternary-complex mechanism but brain hexokinase probably proceeded by a ping-pong bi-bi mechanism involving an E−P intermediate. Although isoenzymes do not necessarily act by identical mechanisms, further studies (1969) showed that the plots for brain hexokinase met far below the horizontal axis, so this enzyme too forms a ternary-complex.

9.3.2   The Use of Inhibitors which Compete with Substrates for Binding Sites

The use of inhibitors which compete with one of the substrates for a site on the enzymes can give useful information as to the mechanism of the reaction. A product of the reaction, for example, if present at the start may compete with a substrate for a binding site on the enzyme and thus slow down the rate of the forward reaction.

Because of the complex nature of two-substrate reactions, the fact that a particular inhibitor acts in a competitive way will not necessarily result in a characteristic competitive inhibition pattern (Fig. 8.1) being seen for the overall reaction. For double-reciprocal (Lineweaver–Burk) primary plots (Fig. 9.1) drawn for experiments performed at a single concentration of the fixed substrate but at a series of different inhibitor concentrations, the overall pattern will only be competitive if the intercept on the $\frac{1}{v_o}$ axis is unaffected by the inhibitor. If, on the other hand, the intercept does depend on the inhibitor concentration but the slope does not, then a series of parallel lines will be obtained at different inhibitor concentrations and the overall inhibition pattern will be uncompetitive (Fig. 8.5). If both intercept and slope are influenced by inhibitor concentration and the plots at different inhibitor concentrations converge to the left of the $\frac{1}{v_o}$ axis, then the overall inhibition pattern is mixed (Fig. 8.10), except in the special case of non-competitive inhibition (Fig. 8.7) where the plots converge at the horizontal axis. All of these inhibition patterns may be seen with two-substrate reactions where the inhibition is essentially competitive in nature.

Cleland has formulated a **series of rules** which enable the inhibition patterns for a particular mechanism to be predicted.

One rule states that **the intercept on the $\frac{1}{v_o}$ axis of a double-reciprocal plot is affected only by an inhibitor which binds reversibly to an enzyme-form other than that to which the variable substrate combines.** The explanation for this is straightforward. For there to be no change in intercept as a result of the presence of an inhibitor, the variable substrate in saturating concentrations must be able to prevent the inhibitor binding. *Therefore, for a characteristic pattern of competitive inhibition to occur, the inhibitor must compete with the substrate whose concentration is being varied for a site on the same enzyme-form.*

Another rule is that **the slope of a double-reciprocal plot is affected by an inhibitor which binds to the same enzyme-form as the variable substrate, or to an enzyme-form which is connected by a series of reversible steps to that with which the variable substrate combines.** The explanation for this rule is as follows. Most enzyme-catalysed reaction obey the equation:

$$v_o = \frac{V_{max}[S_o]}{K_m^S + [S_o]}$$

where S is the only substrate, or the only substrate whose concentration is being varied. When $[S_o]$ is low, $K_m^S + [S_o] \simeq K_m^S$ and therefore $v_o = \dfrac{V_{max}}{K_m^S}.[S_o]$. Hence the apparent first-order constant when $[S_o]$ is low is $\dfrac{V_{max}}{K_m^S}$, and this is the reciprocal of the slope of the Lineweaver–Burk plot (Fig. 7.2). The substrate S binds to the enzyme E according to the equation $E + S \rightleftharpoons ES$, so any factor which increases the concentration of ES relative to E will reduce $\left(\dfrac{V_{max}}{K_m^S}\right)$ and hence increase the slope of the double-reciprocal (Lineweaver–Burk) plot. Similarly anything which decreases the concentration of ES relative to E will increase $\left(\dfrac{V_{max}}{K_m^S}\right)$ and decrease the slope of the double-reciprocal plot. Hence, for there to be no change in slope in the presence of an inhibitor, i.e. *for a pattern characteristic of uncompetitive inhibition to be seen, there must be no reversible link between inhibitor and variable substrate.* Most of the steps involved in enzyme-catalysed reactions are reversible, but irreversible steps which in certain circumstances may give rise to uncompetitive inhibition occur where a substrate is present in saturating concentrations or where an inhibitor forms a dead-complex with the enzyme. It should be realised that in the case of uncompetitive inhibition of a single-substrate reaction (Chapter 8.2.2) there is no reversible link between inhibitor and substrate because the inhibitor binds to ES to form a dead-end complex.

We will now consider product inhibition of the ping-pong bi-bi mechanism and ternary-complex mechanisms obeying the general rate equation (Chapter 9.2.1). Non-saturating concentrations of the fixed-substrate will be assumed unless otherwise stated.

Let us look first at a **random-order ternary-complex mechanism**:

$$E + AX \rightleftharpoons E.AX \qquad\qquad\qquad\qquad EA \rightleftharpoons E + A$$

$$\begin{array}{ccc} & +B & \qquad\qquad -BX & \\ & E.AX.B \rightleftharpoons E.A.BX & \end{array}$$

$$+AX \qquad\qquad\qquad\qquad\qquad -A$$

$$E + B \rightleftharpoons EB \qquad\qquad\qquad\qquad E.BX \rightleftharpoons E + BX.$$

If the forward reaction is investigated under steady-state conditions at fixed $[AX_o]$ and variable $[B_o]$ in the presence of the product BX, then a competitive inhibition pattern will result: the variable substrate B will compete with BX for the same site on the same enzyme-form (E); a graph of the form of Fig. 9.4(a) will result if the experiment is performed at a series of concentrations of the product BX. Similar results will be obtained for inhibition by the product A at fixed $[B_o]$ and variable $[AX_o]$. However the other two combinations give less clear-cut predictions. Particularly in the case of a rapid-equilibrium mechanism, there may be competition between substrate and inhibitor for the same enzyme-form, even where the competition is not for the same *site*. Consider, for example,

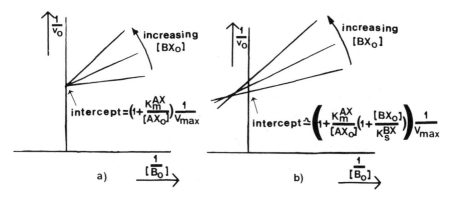

Fig. 9.4 Plots of $\dfrac{1}{v_o}$ against $\dfrac{1}{[B_o]}$ at constant and non-saturating $[AX_o]$ and constant $[E_o]$ showing; (a) a pattern characteristic of competitive inhibition and (b) a pattern characteristic of mixed inhibition. The inhibitor is the product, BX.

inhibition by the product A where the variable substrate is B. Saturating concentrations of B will remove all free E from the system, thus preventing A binding to it, a characteristic of **competitive** inhibition. However excess B cannot prevent the possibility of A binding to EB to form a dead-end complex EBA; if such a complex can be formed, a **mixed** inhibition pattern would result (Fig. 9.4(b)).

Similarly, if BX is the inhibitor and AX the varying substrate, the inhibitor pattern will be characteristic of **competitive** inhibition unless the dead-end complex $E.AX.BX$ can be formed (a less likely possibility than EBA because of congestion at the binding sites).

· Product inhibition patterns of the type discussed above have been found for reactions with rapid-equilibrium random-order mechanisms (e.g. that catalysed by **creatine kinase**) and with many other reactions with random-order mechanisms where the rate-limiting step is not solely the interconversion of the ternary complexes (e.g. reactions catalysed by **hexokinase** enzymes).

For a **compulsory-order mechanism** where AX binds first:

$$AX + E \rightleftharpoons E.AX \rightleftharpoons E.AX.B \rightleftharpoons E.A.BX \rightleftharpoons EA \rightleftharpoons E + A,$$

A and AX will compete for a site on the enzyme-form E; therefore, overall **competitive** inhibition (as in Fig. 9.4(a)) will result from inhibition by the product A where AX is the variable substrate. No other combination of inhibitor and variable substrate will give a pattern characteristic of **competitive** inhibition for this mechanism: inhibition by the product BX where B is the variable substrate will give **mixed** inhibition (as in Fig. 9.4(b)) since BX and B do not bind to the same enzyme-form, BX binding to EA and B to $E.AX$.

Similarly for a **compulsory-order mechanism** of the form:

$$B + E \rightleftharpoons EB \rightleftharpoons E.B.AX \rightleftharpoons E.BX.A \rightleftharpoons E.BX \rightleftharpoons E + BX,$$

only inhibition by the product BX where B is the variable substrate will give a pattern characteristic of **competitive** inhibition.

For a **ping-pong bi-bi mechanism**:

$$AX + E \rightleftharpoons E.AX \rightleftharpoons EX.A \rightleftharpoons EX + A$$

$$EX + B \rightleftharpoons EX.B \rightleftharpoons E.XB \rightleftharpoons E + BX,$$

AX and BX compete for a site on the enzyme-form E, whereas A and B compete for a site on the enzyme-form EX. Therefore inhibition by BX when AX is the variable substrate will give overall **competitive** inhibition, as will inhibition by A when B is the variable substrate. However this only applies to ping-pong bi-bi mechanisms where there is a single binding site, e.g. many reactions catalysed by **transaminase (aminotransferase)** enzymes.

In all of the situations discussed above, there is a reversible link between inhibitor and variable substrate at non-saturating concentrations of the fixed substrates, so the possibility of uncompetitive inhibition does not arise. The overall inhibition patterns under these conditions may be summarised as follows:

Mechanism	Product used as inhibitor	Inhibition pattern observed with respect to varying $[Ax_0]$	with respect to varying $[B_0]$
compulsory-order ternary-complex (AX binding first)	A	competitive	mixed
	BX	mixed	mixed
compulsory-order ternary-complex (B binding first)	A	mixed	mixed
	BX	mixed	competitive
random-order ternary-complex	A	competitive	competitive or mixed
	BX	competitive or mixed	competitive
ping-pong bi-bi	A	mixed	competitive
	BX	competitive	mixed

Although these general conclusions may not apply in certain instances because of the effect of individual rate constants, a systematic study will in most cases give good evidence as to the mechanism of the reaction.

Further information may be obtained by performing inhibition studies in the presence of saturating concentration of the fixed substrate. Of particular interest is the situation where the saturating fixed substrate is the second substrate to be bound in a compulsory-order mechanism. For example, if $[B_o]$ is saturating for a mechanism where AX must bind first to the enzyme, the following occurs:

$$E \overset{+AX}{\rightleftharpoons} E.AX \overset{+B}{\rightleftharpoons} E.AX.B \rightleftharpoons E.A.BX \underset{+BX}{\rightleftharpoons} EA \rightleftharpoons E + A;$$

inhibition by BX will give an **uncompetitive** pattern with respect to varying AX because there is no reversible link between inhibitor and variable substrate.

Dead-end inhibitors which are analogues of the products will give generally similar inhibitor patterns. However there is a fundamental difference between the two types of inhibition. In the case of product inhibition the enzyme–inhibitor complex is an intermediate normally present but which in the circumstances is non-productive for the reaction in the forward direction. In the case of dead-end inhibitors, the enzyme–inhibitor complex is not present in the uninhibited reaction and so effectively removes some of the enzyme from the reacting system. Therefore, there may be slight differences between the inhibition patterns in the two cases. For example, for a compulsory-order mechanism where AX binds first, inhibition by the product BX gives a **mixed** inhibition pattern with respect to varying AX at fixed but non-saturating concentrations of B; however an analogue of BX which binds in the same way but forms a dead-end complex will give overall **uncompetitive** inhibition because there is no reversible link between inhibitor and variable substrate.

9.4 INVESTIGATION OF REACTION MECHANISMS USING NON-STEADY-STATE METHODS

9.4.1 Isotope Exchange at Equilibrium

Boyer, in 1959, first suggested the investigation of isotope exchange at chemical equilibrium as a means of investigating reaction mechanisms, and he and his co-workers have done much to develop the technique.

If isotope exchange can be demonstrated between a reactant and product in the absence of other reactants and products, then a **ping-pong mechanism** must be indicated. For example, isotope exchange takes place between the substrate orthophosphate and the product glucose 1–phosphate in the presence of the enzyme **sucrose phosphorylase** but in the absence of the other substrate, sucrose, and the other product, fructose. Also, isotope can be exchanged between sucrose and fructose in the absence of orthophosphate and glucose 1–phosphate. This and other evidence suggests the following mechanism for the forward reaction:

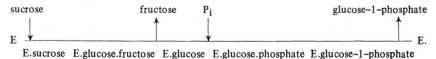

sucrose   fructose P_i glucose–1–phosphate

E   E.

E.sucrose E.glucose.fructose E.glucose E.glucose.phosphate E.glucose–1–phosphate

Of more general application is the investigation of the change in the rate of equilibrium isotope exchange when the concentrations of reactants and products are altered without changing the position of equilibrium. Consider a reaction of the general form:

$$AX + B \rightleftharpoons BX + A.$$

At equilibrium, the rate of forward reaction equals the rate of back reaction and $\dfrac{[BX][A]}{[AX][B]} = K_{eq}$.

A small amount of radioactively labelled B is introduced (an amount too small to significantly affect the equilibrium) and the rate of formation of labelled BX is measured. Then the concentrations of A and AX are increased, keeping the ratio $[A]:[AX]$ constant so as not to alter the position of equilibrium. The equilibrium concentrations of B and BX will remain unchanged, but the rate of isotope exchange between B and BX will be affected.

If the reaction has a compulsory-order mechanism of the form:

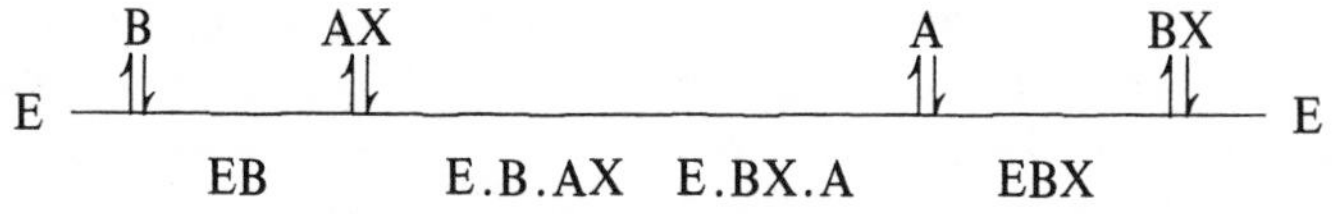

slight increases in the concentrations of A and AX may increase the rate of isotope exchange between B and BX. However substantial increases in the concentrations of A and AX will force the enzyme towards formation of the ternary-complexes $E.B.AX$ and $E.BX.A$, making it more difficult for B to dissociate from EB and BX to dissociate from EBX, and thus reducing the $B \rightarrow BX$ exchange (Fig. 9.5(a)).

Different results are found for other mechanisms. If the reaction has a compulsory-order mechanism with AX binding first:

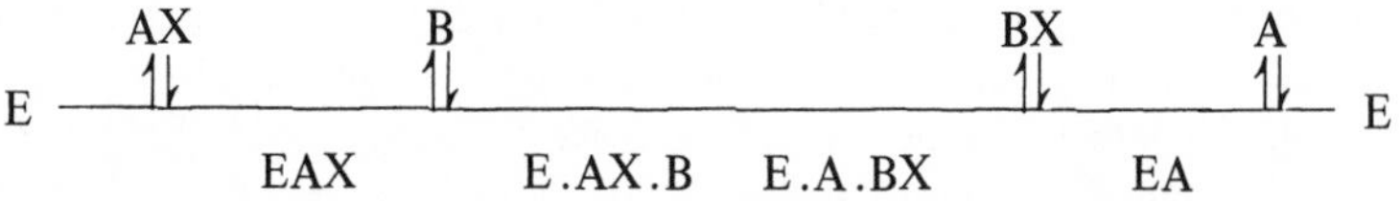

as the concentrations of AX and A increase, the free enzyme is forced into the EAX and EA forms. Since EAX is the form which reacts with B while EA does not affect the initial velocity of liberation of BX from $E.A.BX$, the rate of isotope exchange between B and BX will increase in a hyperbolic manner with increasing $[AX]$ (Fig. 9.5(b)). Similar results will be obtained if the reaction has a random-order ternary-complex mechanism.

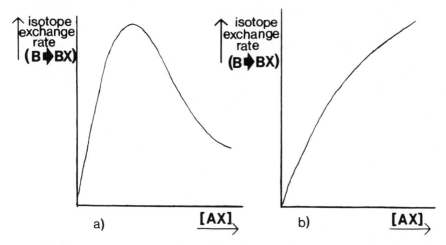

Fig. 9.5 Plots of equilibrium isotope exchange from B to BX against [AX] at constant [AX]/[A]: (a) for a compulsory-order mechanism, B binding first and (b) for a compulsory-order mechanism, AX binding first, or for a random-order mechanism.

A decreased rate of equilibrium isotope exchange with increased reactant and product concentration in an experiment of this type is diagnostic of a compulsory-order mechanism where the isotope exchange rate is being measured between the first substrate to be bound and the last product to be released.

An interesting example is the reaction catalysed by horse liver **alcohol dehydrogenase**:

$$CH_3CH_2OH + NAD^+ \rightleftharpoons CH_3CHO + NADH (+ H^+).$$

Increased concentrations of CH_3CH_2OH and CH_3CHO, keeping the ratio of one to the other constant, slows down the rate of isotope exchange between NAD^+ and NADH. It can therefore be concluded that this reaction has a compulsory-order ternary-complex mechanism, with NAD^+ binding first and NADH being the last product released. However, this is not the whole story, as will be seen in Chapter 9.4.2.

9.4.2 Rapid-Reaction Studies

Rapid-reaction techniques may also be used to investigate the mechanism of two-substrate reactions. Theorell and Chance (1951) investigated the reaction catalysed by horse liver **alcohol dehydrogenase** (Chapter 9.4.1) using stopped-flow techniques with a double-beam detector. There is a difference between the absorbance of NADH and E.NADH at 350 nm but not at 328 nm. Therefore, monitoring the differences between the absorbances at 350 nm and 328 nm gives the rate of E.NADH → Ė + NADH conversion. Also, the rate of change of absorbance at 328 nm gives the rate of reduction of NAD^+ (i.e. $E.NAD^+ \to E.NADH$ conversion). Theorell and Chance found that the interconversion of the ternary-complexes took place extremely rapidly and that the rate-limiting step of the overall reaction was the dissociation of NADH from the E.NADH complex. This type of compulsory-order mechanism, characterised by very rapid interconversion of the ternary complexes, has been termed the **Theorell–Chance mechanism** and is represented as follows:

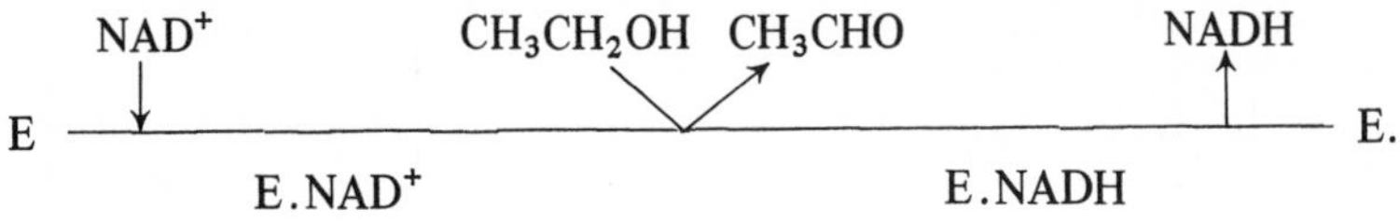

SUMMARY OF CHAPTER 9

Two-substrate enzyme-catalysed reactions may proceed by a variety of mechanisms, including the ping-pong bi-bi, compulsory-order ternary-complex and random-order ternary-complex mechanisms. General rate equations have been derived by Alberty and Dalziel for two-substrate reactions which obey the Michaelis-Menten equation with respect to one substrate at fixed concentrations of the other. The Alberty equation has the more obvious resemblance to the Michaelis-Menten equation but the constants of the Dalziel equation give a simpler relationship between the rate constants for the individual steps involved in the reaction. These relationships may be determined for a particular mechanism by the use of King and Altman procedures.

The mechanism of a two-substrate reaction may be investigated by steady-state methods, including product inhibition studies, and by non-steady-state methods, such as equilibrium isotope exchange and rapid-reaction techniques.

FURTHER READING

Cleland, W. W. (1970), Steady state kinetics, in Boyer, P. D. (ed.), *The Enzymes*, 3rd edition, 2 (pages 1–65), Academic Press.

Dalziel, K. (1975), Kinetics and mechanism of nicotinamide-dinucleotide-linked dehydrogenases, in Boyer, P. D. (ed.), *The Enzymes*, 3rd edition, 11 (pages 1-60), Academic Press.

Engel, P. C. (1977), *Enzyme Kinetics* (Chapters 5 and 6), Chapman and Hall.

Montgomery, R. and Swenson, C. A. (1976), *Quantitative Problems in the Biochemical Sciences*, 2nd edition (Chapter 11 and problems), Freeman.

Tipton, K. F. (1974), Enzyme Kinetics, in Bull, A. T., Lagnado, J. R., Thomas, J. O. and Tipton, K. F. (eds.), *Companion to Biochemistry* (pages 227-252), Longman.

Wharton, C. W. and Eisenthal, R. (1981), *Molecular Enzymology* (Chapter 7), Blackie.

PROBLEMS

9.1 For the enzyme-catalysed transfer reaction:

$$AX + B \rightleftharpoons BX + A$$

a series of experiments were performed at fixed total enzyme concentration. The results are summarised in the following table, which shows the initial velocity of the reaction at different initial concentrations of AX, B and I (I being an inhibitor resembling BX in structure).

$[AX_0]$ (mmol.l^{-1})	$[I_0]$ (mmol.l^{-1})	Initial rate of reaction (μmol.minute^{-1}) where:					
		$[B_0]$ (mmol.l^{-1}) = 20.0	10.0	5.0	3.33	2.50	2.0
3.0	0	1460	1190	881	694	575	488
4.0	0	1720	1410	1030	806	667	564
6.0	0	2080	1690	1230	961	794	668
8.0	0	2380	1920	1360	1060	870	735
8.0	1.0	2250	1740	1200	913	738	617
8.0	2.0	2080	1590	1060	800	641	532.

Further, product inhibition studies showed that the presence of A leads to competitive inhibition patterns with varying $[AX_0]$ at fixed $[B_0]$. What can you conclude about the mechanism of the reaction?

9.2 Malate dehydrogenase catalyses the reaction:

$$malate + NAD^+ \rightleftharpoons oxaloacetate + NADH + H^+.$$

A malate dehydrogenase enzyme was investigated at pH 8 and 298°K as follows. The initial velocity of the reaction, in the direction of oxaloacetate formation, was determined spectrophotometrically by the appearance of NADH at 340 nm at different initial concentrations of malate and NAD^+. The effect of inhibition by the product oxaloacetate was also investigated. No NADH was present initially in any experiment, and the total enzyme concentration was the same in each case. The following results were obtained.

NAD^+ conc^n (mmol.l^{-1})	oxaloacetate conc^n (mmol.l^{-1})	Initial rate of reaction (absorbance change per minute at 340 nm) at malate conc^n (mmol.l^{-1}):						
		1.25	1.5	2.0	2.5	3.33	5.0	10.0
2.0	0	0.019	0.021	0.025	0.028	0.032	0.038	0.045
2.0	1.0	0.016	0.018	0.021	0.024	0.027	0.032	0.038
2.0	2.0	0.014	0.015	0.019	0.021	0.024	0.028	0.033
3.0	0	0.024	0.026	0.032	0.036	0.041	0.049	0.059
4.0	0	0.027	0.031	0.037	0.042	0.048	0.057	0.069
6.0	0	0.031	0.036	0.043	0.049	0.057	0.067	0.083.

The equilibrium rate of conversion of NAD^+ to NADH was investigated by isotope exchange studies in the presence of different concentrations of malate, but keeping the ratio of [malate] to [oxaloacetate] constant at 100:1 and the enzyme concentrations the same in each case. The following results were obtained:

Malate conc^n (mmol.1^{-1})	50	100	150	200	250	300
equilibrium reaction rate, $NAD^+ \rightarrow NADH$ (μmol.1^{-1}.minute^{-1})	41	58	21	15	12	10.

What can you conclude about the reaction mechanism from the above data?

9.3   An enzyme-catalysed transfer reaction of the form:

$$AX + B \rightleftharpoons BX + A$$

was investigated as follows.

(a) The effect of a dead-end inhibitor similar in structure to BX was investigated at varying initial concentrations of AX but fixed initial concentrations of B and enzyme. Temperature and pH were maintained constant, and no BX or A were present at zero time. The following results were obtained:

Initial concentration of AX (mmol.1^{-1})	Initial velocity of the forward reaction (μmol.1^{-1}.minute^{-1})		
	In absence of inhibitor	In presence of 1.0 mmol.1^{-1} inhibitor	In presence of 2.0 mmol.1^{-1} inhibitor
2.0	14.3	12.5	11.1
2.5	16.7	14.3	12.5
3.3	19.9	16.6	14.2
5.0	24.9	20.1	16.7
10.0	33.4	25.1	20.0.

(b) The effect of inhibition of the forward reaction by the product A at a concentration of 1.0 mmol.1^{-1} was investigated at varying initial concentrations of AX under identical conditions to those used in part (a). These results were obtained:

Initial concentration of AX (mmol.1^{-1})	2.0	2.5	3.3	5.0	10.0
Initial velocity of the forward reaction (μmol.1^{-1}.minute^{-1})	10.5	12.5	15.3	20.1	28.6.

(c) The forward reaction was investigated at varying initial concentrations of B and at a series of fixed (but non-saturating) initial concentrations of AX. Initial enzyme concentration, temperature and pH were the same for each experiment. No product A was present at zero time in any experiment, and no dead-end inhibitor was present at any time. In some experiments the product BX was also initially absent, but in others the effect of product inhibition by BX was investigated. The following results were obtained:

Initial concentration of:		Initial velocity of the forward reaction $(\mu mol.l^{-1}.minute^{-1})$ at initial concentration of B $(mmol.l^{-1}) = [B_o]$				
AX $(mmol.l^{-1})$	BX $(mmol.l^{-1})$	$[B_o] = 2.0$	2.5	3.3	5.0	10.0
2.0	0	9.5	11.0	13.0	15.9	20.4
2.5	0	10.8	12.4	14.7	18.1	23.4
3.3	0	12.3	14.3	17.0	21.1	27.6
2.0	1.0	8.2	9.5	11.2	13.8	17.9
2.0	2.0	7.3	8.4	9.9	12.3	16.0 .

(d) Equilibrium isotope exchange studies, performed at fixed enzyme concentration and constant $[A]:[AX]$ ratio gave these results:

Concentration of AX $(mmol.l^{-1})$	50	100	150	200	250	300
Equilibrium reaction rate $(B \rightarrow BX)$ $(\mu mol.l^{-1}.minute^{-1})$	25	54	60	64	68	70 .

What can be concluded about the reaction mechanism from the above data?

9.4 Deduce the constants of the Dalziel equation in terms of rate constants for the reaction mechanism:

$$E \underset{k_{-1}}{\overset{k_1}{\rightleftharpoons}} E.AX \underset{k_{-2}}{\overset{k_2}{\rightleftharpoons}} E.AX.B \underset{k_{-3}}{\overset{k_3}{\rightleftharpoons}} E.A.BX \overset{k_4}{\rightarrow} EA \overset{k_5}{\rightarrow} E .$$

9.5 Lysine-oxoglutarate reductase (E.C. 1.5.1.8) catalyses the first step in lysine catabolism in mammalian liver:

$$L\text{-lysine} + 2\text{-oxoglutarate} + NADPH \rightleftharpoons L\text{-saccharopine} + NADP^+ .$$

Purified bovine enzyme was subjected to product inhibition studies with saccharopine. In each case, the rate of reaction was determined by monitoring the conversion of NADPH to $NADP^+$ at 340 nm at pH 7.4 and 35°C. In each case, no $NADP^+$ was present at zero time, and the concentrations of two of the three substrates were fixed. When the concentration of L-lysine was fixed, its value was 10 mmol.1^{-1}; when the concentration of 2-oxoglutarate was fixed, its value was 4 mmol.1^{-1}; and when the concentration of NADPH was fixed its value was 0.4 mmol.1^{-1}. The following data (adapted from M. Ameen, PhD Thesis, Trent Polytechnic, 1982) were obtained.

Saccharopine conc^n (mmol.1^{-1})	Initial velocity (absorbance units per minute) at L-lysine concentration (mmol.1^{-1}) =				
	1.0	1.25	1.67	2.5	5.0
1.1	0.143	0.167	0.200	0.250	0.333
0.80	0.167	0.193	0.232	0.288	0.385
0.28	0.246	0.286	0.338	0.417	0.541
0	0.278	0.323	0.382	0.463	0.592

Saccharopine conc^n (mmol.1^{-1})	Initial velocity (absorbance units per minute) at 2-oxoglutarate concentration (mmol.1^{-1}) =				
	0.25	0.30	0.40	0.50	1.0
1.1	0.124	0.143	0.178	0.207	0.313
0.80	0.156	0.178	0.217	0.250	0.357
0.28	0.223	0.250	0.294	0.330	0.435
0	0.278	0.308	0.353	0.385	0.476

Saccharopine conc^n (mmol.1^{-1})	Initial velocity (absorbance units per minute) at NADPH concentration (mmol.1^{-1}) =				
	0.10	0.125	0.167	0.25	0.50
1.1	0.177	0.195	0.217	0.246	0.284
0.80	0.189	0.208	0.235	0.270	0.313
0.28	0.227	0.257	0.299	0.357	0.439
0	0.242	0.278	0.327	0.398	0.500

What can be deduced from thes results about the mechanism of the reaction catalysed by lysine-oxoglutarate reductase?

CHAPTER 10

The Investigation of Active Site Structure

10.1 THE IDENTIFICATION OF BINDING SITES AND CATALYTIC SITES

10.1.1 Trapping the Enzyme–Substrate Complex

The reversible character of the steps involved in enzyme-catalysed reactions makes the determination of each substrate-binding site less than straightforward. At steady-state, a constant amount of enzyme–substrate complex is known to be present, but if an attempt is made to isolate this and hydrolyse it so as to identify the amino-acid to which the substrate is attached, the effort will not usually be rewarded with success: this is because the substrate will dissociate from the enzyme during the procedures involved. However, if the enzyme–substrate complex can be trapped in a modified form by some chemical process so that the substrate is no longer able to dissociate from the enzyme, then it may be possible to identify the substrate-binding site.

Consider, for example, the reversible reaction catalysed by **fructose-bisphosphate aldolase**:

dihydroxyacetone phosphate + glyceraldehyde 3-phosphate $\rightleftharpoons$ fructose 1,6-bisphosphate.

Horecker and colleagues showed in 1962 that if the reaction mixture was treated with sodium borohydride, a strong reducing agent, then an inactive complex was produced. On hydrolysis of this complex, ϵ-N-glyceryl lysine was found among the products. From this it was concluded that dihydroxyacetone-phosphate normally binds to the ϵ (i.e. side chain) amino group of a lysine residue in the enzyme by a **Schiff's base** ($-N=CH-$) linkage. The presumed sequences for the normal reaction and for the procedures used to identify the substrate-binding site are as follows:

$$E-NH_2 \;+\; O{=}\underset{\underset{CH_2OPO_3^{2-}}{|}}{\overset{\overset{CH_2OH}{|}}{C}} \;\rightleftharpoons\; E-N{=}\underset{\underset{CH_2OPO_3^{2-}}{|}}{\overset{\overset{CH_2OH}{|}}{C}} \;\overset{+\text{ glyceraldehyde}\atop -3\text{-P}}{\rightleftharpoons}\; \text{fructose-1,6-bisP}$$

enzyme   dihydroxyacetone-   enzyme-substrate
            phosphate   complex

$$\Big\downarrow \begin{array}{l} 2H \\ \text{(borohydride)} \end{array}$$

$$E-NH.\underset{\underset{CH_2OPO_3^{2-}}{|}}{\overset{\overset{CH_2OH}{|}}{CH}} \;\xrightarrow{\text{hydrolysis}}\; \text{lysine}-NH.\underset{\underset{CH_2OPO_3^{2-}}{|}}{\overset{\overset{CH_2OH}{|}}{CH}}$$

inactive complex   ε-N-glyceryl lysine

In general, once an enzyme-substrate complex has been trapped as an inactive complex, it may be subjected to partial hydrolysis and the amino-acid sequence for a few residues in each side of the binding site determined. It may also be possible to determine the complete primary structure of the inactivated complex (as in Chapter 2.4) and hence the substrate-binding site may be precisely located.

10.1.2 The Use of Substrate Analogues

An alternative way of producing a complex more stable than the normal enzyme-substrate complex is to replace the natural substrate by an analogue which binds to the same site on the enzyme but is then less readily removed.

An example was discussed in Chapter 7.2.1: the first step in the hydrolysis of p-nitrophenyl acetate and other acyl esters by **chymotrypsin** is the rapid splitting of the ester to yield the first product (the alcohol) and form an **acyl-enzyme**; the subsequent liberation of the acyl group from the enzyme is extremely slow, allowing the structure of the complex to be investigated. It is found, as might be expected, that the acyl group binds by an ester linkage to the —OH group of a serine residue in the enzyme.

Enzymes form even stronger linkages with **irreversible inhibitors**, since in this case there is no subsequent reaction at all. Thus **DFP** (Chapter 8.3) binds to the serine residue at the active site of chymotrypsin and other serine proteases to form very stable complexes. Partial hydrolysis of each enzyme-inhibitor complex gives a series of peptide fragments, which can be separated from each other and analysed; the amino-acid sequence of any fragment containing DFP must be the primary structure of part of the active site. In this way it can be shown that the amino-acid sequence around the essential serine residue of chymotrypsin is -Gly-Asp-Ser-Gly-Gly-Pro-, and the serine residue in question identified as serine-195. An identical sequence is found for trypsin.

The assumption that an irreversible inhibitor binds at the active site of an enzyme is particularly valid when the inhibitor resembles a substrate. For example, tosyl (N-toluenesulphonyl)-L-phenylalanine chloromethyl ketone (TPCK) resembles esters which are hydrolysed by **chymotrypsin**, but TPCK itself acts as an irreversible inhibitor of this enzyme by alkylating the histidine-57 residue.

$$CH_3 - \bigcirc - SO_2 - NH . \underset{\underset{CH_2}{|}}{CH} - CO.CH_2Cl \qquad CH_3 - \bigcirc - SO_2 - NH . \underset{\underset{CH_2}{|}}{CH} - COOR$$

tosyl-L-phenylalanine chloromethyl
ketone (TPCK)
(an inhibitor of chymotrypsin)

ester of tosyl-L-phenylalanine

(a substrate for chymotrypsin)

Thus there is evidence that both serine-195 and histidine-57 are present at the active site of chymotrypsin. It would appear that the binding of TPCK to the enzyme brings the reactive $-Cl$ group into close proximity to the histidine-57 residue and facilitates the formation of a covalent $C-N$ bond between inhibitor and imidazole side chain.

The subject of enzyme inactivation by irreversible inhibitors and other agents is developed in the next section (Chapter 10.1.3).

10.1.3 Enzyme Modification by Chemical Procedures Affecting Amino-Acid Side Chains

If an enzyme is modified by the conversion of a particular amino-acid side chain to a different form (e.g. by the action of an irreversible inhibitor) and this modification results in a loss of catalytic activity, then it is possible that the amino-acid concerned is a component of the active site of the enzyme. However it is also possible that the loss of activity is due to a change in tertiary structure resulting from a modification to an amino-acid residue not present at the active site. The two situations may be distinguished by attempting to carry out the same modification in presence of excess amounts of substrate, which should protect the substrate-binding site and amino-acid residues in the neighbouring region from being modified. Similar protection should be provided by excess amounts of reversible inhibitors which bind to the substrate-binding site, e.g. most competitive inhibitors. Thus, if an enzyme loses activity when one of its amino-acid side chains (R) is modified by a particular treatment in the absence of substrate (or competitive inhibitor):

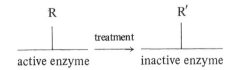

but retains full activity if subjected to the same treatment in the presence of saturating amounts of the substrate (or competitive inhibitor):

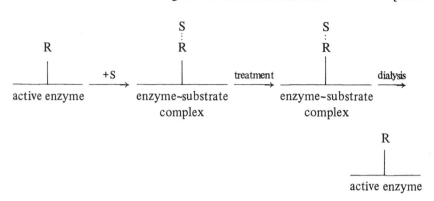

then the amino-acid residue must be present at the active site of the enzyme. It should be realised that the presence of the substrate would protect not only the substrate-binding site (as illustrated above) but also neighbouring residues, which are likely to include the catalytic sites.

As a rough generalisation, modification of a catalytic site would be expected to result in a complete loss of enzyme activity: the substrate would still be able to bind to the enzyme, but no subsequent catalysed reaction would occur. On the other hand, modification of a binding site might leave a residual catalytic activity: the substrate would no longer be able to bind to the enzyme, but might still be able to approach the catalytic site by random motion. This situation could possibly result in V_{max} remaining unchanged, although the substrate concentration required to achieve it would be very high indeed. Regardless of this, the absence of substrate-binding would result in a great increase in K_m.

Enzyme modification experiments have revealed **histidine, cysteine, serine, methionine, tyrosine, aspartate, glutamate, lysine** and **tryptophan** residues at the active sites of enzymes.

The imidazole side chain of **histidine** might be expected to be an important contributor towards catalytic activity since it is the only amino-acid side chain to have a pK_a in the pH range at which most enzymes function: it can thus act as both a proton donor and a proton acceptor. As mentioned in Chapter 10.1.2, tosyl-L-phenylalanine chloromethyl ketone (TPCK) inhibits **chymotrypsin** irreversibly by alkylating histidine-57; this inactivation is prevented by the presence of excess concentrations of the competitive inhibitor benzamide, which supports the view that TPCK acts at the active site of the enzyme. TPCK does not inhibit **trypsin**, which has a different specificity to chymotrypsin, hydrolysing bonds adjacent to lysine rather than aromatic amino-acid residues; this enzyme is inactivated by tosyl-L-lysine chloromethylketone (TLCK), again by alkylation of an essential histidine residue (Schoellman and Shaw, 1963).

Alkylation of histidine residues by iodoacetate was shown by Crestfield, Stein and Moore (1963) to cause inactivation of **ribonuclease**:

$$E-CH_2 \overset{N \diagup\diagdown NH}{\underset{\diagdown\diagup}{\rule{0pt}{0pt}}} + ICH_2CO_2^- \rightarrow E-CH_2 \overset{N \diagup\diagdown N-CH_2CO_2^-}{\underset{\diagdown\diagup}{\rule{0pt}{0pt}}} + HI.$$

enzyme   iodoacetate   carboxymethyl-enzyme

Histidine-119 or, to a lesser degree, histidine-12, may be alkylated, but never both in the same ribonuclease molecule, suggesting that these two histidine residues are found close together at the active site. Iodoacetamide, a similar alkylating agent, has no effect on the enzyme, so it appears that the negative charge of the iodoacetate may form an electrostatic link with the protonated form of one of the histidine residues prior to the carboxymethylation of the other residue. In fact iodoacetate itself does not usually attack histidine residues under the conditions used (pH 5.5), which suggests that histidine-12 and histidine-119 are in a special environment. The inactivation of ribonuclease by alkylation is prevented by the presence of excess phosphate, which binds to the active site.

Another technique which may be used to inactivate enzymes by modification of amino-acid side chains is **photo-oxidation**, i.e. oxidation by activated oxygen in the presence of a photosensitiser such as methylene blue or rose bengal. This method is non-specific and may oxidise histidine, tryptophan, methionine and cysteines residues. However, some degree of specificity may be obtained by careful choice of photosensitising dye and pH. With certain enzymes and under certain conditions, it may be found that only a single amino-acid residue is photo-oxidised. Thus Westhead (1965) found that photo-oxidation of **enolase** in the presence of rose bengal affected only a single histidine residue, but this was sufficient to inactivate the enzyme; magnesium ions, which are essential for the activation of enolase, could protect it from the effects of photo-oxidation. The oxidation product of a histidine residue depends on a variety of factors, including the environment of the residue in the protein: in some cases, aspartate may be the major product.

The thiol group of **cysteine** residues may be alkylated by halogeno-compounds of the type which also alkylate histidine residues. For example, cysteine-25 of the proteolytic enzyme **papain** is alkylated by iodoacetamide, with resulting inactivation of the enzyme. Enzymes containing essential cysteine residues may also be inhibited by unsaturated compounds such as N-ethylmaleimide:

$$\underset{\text{enzyme}}{\overset{SH}{\underset{E}{|}}} + \underset{\text{N-ethylmaleimide}}{\overset{CH.C \diagup\diagdown O}{\underset{CH.C \diagdown\diagup O}{\| \quad \diagdown N.C_2H_5}}} \rightarrow \overset{CH_2.C \diagup\diagdown O}{\underset{\underset{E}{\overset{|}{S}}}{\overset{CH.C}{\| \quad \diagdown N.C_2H_5}}} \diagdown O.$$

Heavy metal ions have a great affinity for thiol groups, and papain is inactivated by p-chloromercuribenzoate (PCMB) and similar mercurial compounds. Such reagents tend to attack −SH groups in a non-specific fashion, so it should not be concluded that the cysteine residues in question are at the active site unless there is further evidence for this (as there is in the case of papain − see above).

The essential **serine** residue of **chymotrypsin** (serine-195) was modified by Koshland and co-workers (1963). Both the tosyl-derivative and its elimination product were found to be inactive:

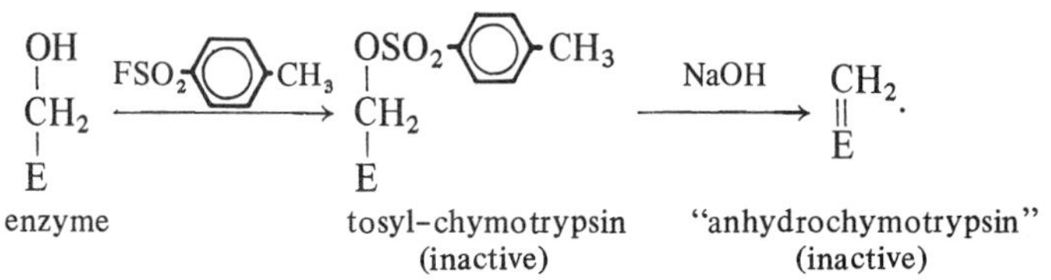

enzyme   tosyl-chymotrypsin   "anhydrochymotrypsin"
                    (inactive)   (inactive)

Methionine residues, in common with those of other amino-acids discussed in this chapter, may be modified by alkylation or photo-oxidation. Hence, it is important to determine precisely which amino-acids are affected in each modification experiment. Koshland and colleagues (1962) showed that photo-oxidation of **chymotrypsin** under certain conditions led to the oxidation of only a single methionine residue (methionine-192), which formed the corresponding sulphoxide; this modification led to only partial inactivation of the enzyme, as did alkylation of the same methionine residue. It appears that methionine-192 is present at the active site of chymotrypsin and may play a part in substrate-binding, but is not otherwise involved in catalytic activity.

A **tyrosine** residue (tyrosine-248) has been demonstrated at the active site of **carboxypeptidase A** by a variety of modification techniques: Vallee and colleagues (1963) showed that iodination or nitration of the benzene ring, or acylation of the phenolic −OH group, all inactivate the enzyme.

The presence of **aspartate** and **glutamate** at the active site of certain enzymes may be demonstrated by the production of esters of the side chain carboxyl groups. For example, Erlanger and others (1966) have shown that p-bromophenacyl bromide can inactivate **pepsin** by forming an ester linkage with an aspartate residue:

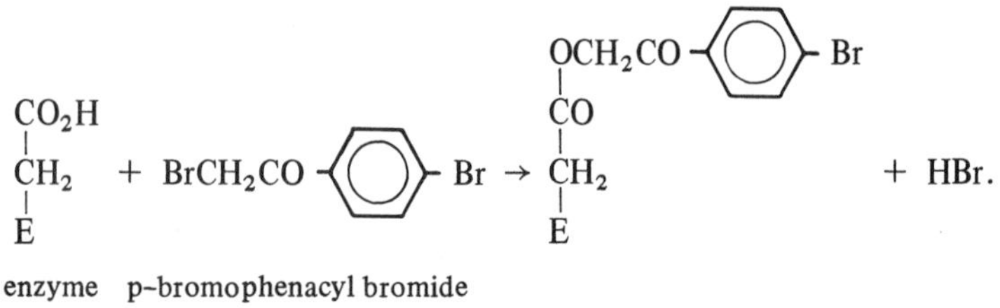

enzyme   p-bromophenacyl bromide

This modification can be prevented by the presence of excess substrate, showing that the aspartate residue is indeed at the active site.

Lin and Koshland (1969) treated **lysozyme** with aminomethanesulphonic acid and a carbodiimide, causing modification of all carboxyl groups present and a total loss of activity of the enzyme:

$$\underset{\text{enzyme}}{\overset{\displaystyle \text{C}\!\!\overset{\displaystyle O}{\diagup}}{\underset{\displaystyle E}{\underset{\displaystyle |}{}}}\text{OH}} \;+\; \text{H}_2\text{NCH}_2\text{SO}_3\text{H} \;\rightarrow\; \underset{E}{\overset{\displaystyle \text{C}\!\!\overset{\displaystyle O}{\diagup}}{\underset{\displaystyle |}{}}}\text{NHCH}_2\text{SO}_3\text{H} \;+\; \text{H}_2\text{O}.$$

enzyme   aminomethane-
sulphonic acid

Modification in the presence of the substrate protected aspartate-52, with the result that 50% of total enzyme activity was retained, suggesting that this residue is a component of the active site.

A **lysine** residue (lysine-41) has been demonstrated to be present at the active site of **ribonuclease** by modification with dinitrofluorobenzene:

$$\underset{E}{\overset{}{\underset{|}{}}}\text{NH}_2 \;+\; \text{F}\!-\!\underset{\text{NO}_2}{\overset{\text{NO}_2}{\bigcirc}}\!-\!\text{NO}_2 \;\rightarrow\; \underset{E}{\overset{}{\underset{|}{}}}\text{HN}\!-\!\underset{}{\overset{\text{NO}_2}{\bigcirc}}\!-\!\text{NO}_2 \;+\; \text{HF}.$$

enzyme   dinitrofluorobenzene

Photo-oxidation of **tryptophan** residues in **lysozyme** inactivates the enzyme, suggesting the presence of these residues at the active site.

10.1.4 Enzyme Modification by Treatment with Proteolytic Enzymes

Just as enzyme-proteins may be modified by chemical procedures which affect the structures of amino-acid side chains (Chapter 10.1.3), so their primary structure may be modified by the action of proteolytic enzymes. In certain circumstances, this may give information as to the structure of the active site of the enzyme. For example, the removal of C-terminal tyrosine residues results in a reduced activity of **fructose-bisphosphate aldolase** towards its fructose 1,6-bisphosphate substrate. This, and other evidence, suggests that the tyrosine residue binds the 6-phosphate group.

Another well-known example is the action of **subtilisin** on **ribonuclease**. This splits the molecule into two portions: one is called the **S-peptide** (comprising the first 20 residues in the primary structure of ribonuclease) and the other is the **S-protein** (the other 104 residues of the ribonuclease molecule). Separately, the S-peptide and S-protein have no catalytic activity, but Richards and colleagues (1959) showed that if they are mixed in equimolar proportions, full enzyme activity is restored. This suggests that both contribute to the active site of the enzyme, agreeing with evidence discussed in Chapter 10.1.3 that residues 12, 41 and 119 are all required for enzymic activity.

10.1.5 The Effect of Changing pH

Since the characteristics of ionisable side chains of amino-acids depend on pH, enzymic activity usually varies with pH changes (Chapter 3.2.2). At extremes of pH, the tertiary structure of the protein may be disrupted and the protein denatured; even at moderate pH values, where the tertiary structure is not disrupted, enzyme activity may depend on the degree of ionisation of certain amino-acid side chains, and the pH profile of an enzyme may suggest the identity of those residues (as investigated by Von Euler and colleagues, 1924; and Dixon, 1953).

Consider an enzyme whose catalytic activity depends on a certain amino-acid side chain (Y^-) being able to act as a proton acceptor, according to the general equation:

$$E.YH \rightleftharpoons H^+ + E.Y^-.$$

protonated   deprotonated
enzyme   enzyme

The titration graph of pH against added NaOH for such a system, as predicted by the **Henderson–Hasselbalch equation**, is shown in Fig. 3.8. Since the amount of $E.Y^-$ present is directly proportional to the amount of NaOH added, a graph of pH against the fraction of the enzyme in the form $E.Y^-$ will be exactly the same shape as Fig. 3.8. This graph, with the axes reversed, is shown in Fig. 10.1(a).

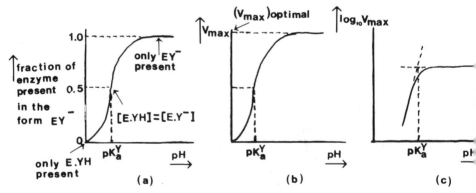

Fig. 10.1 Graphs showing the effect of dissociation of the enzyme form E.YH, according to the Henderson–Hasselbalch equation: (a) graph of $\dfrac{[EY^-]}{[E_0]}$ against pH, (b) graph of V_{max} against pH, at constant $[E_0]$, where catalytic activity depends on the presence of $E.Y^-$, and (c) the corresponding graph of $\log_{10} V_{max}$ against pH.

The initial reaction velocity is given by $v_o = k_{cat}[ES]$, where $k_{cat} = V_{max}/[E_o]$ (Chapter 7.1.3). At each given pH, k_{cat} (and hence $V_{max}/[E_o]$) will be proportional to the fraction of enzyme in the form $E.Y^-$, if this form is required for catalytic activity.

$$\therefore \frac{V_{max}}{[E_o]} = k\frac{[EY^-]}{[E_o]}, \quad \text{where k is a constant.}$$

$$\therefore V_{max} = k[EY^-].$$

At the optimal pH, where all of the enzyme present is in the required form and $V_{max} = (V_{max})$ optimal:

$$(V_{max}) \text{ optimal} = k[E_o].$$

So, in general, $V_{max} = (V_{max})$ optimal. $\frac{[E.Y^-]}{[E_o]}$.

Hence a graph of V_{max} against pH at constant $[E_o]$ (Fig. 10.1(b)) will be identical in shape to that shown in Fig. 10.1(a). The pK_a of the ionisable group (the amino-acid side chain) can be obtained from the point of inflection of the curve (where $V_{max} = \frac{1}{2}(V_{max})$ optimal), thus making it possible to identify the amino-acid residue in question. In plotting such graphs, it is important to ensure that the enzyme is fully saturated with substrate at each pH investigated, for the concentration required to achieve this may also vary with pH (see later).

The pK_a value may also be obtained by an alternative treatment of this data. The dissociation constant, K_a^Y, of E.YH is given by:

$$K_a^Y = \frac{[H^+][E.Y^-]}{[E.YH]}.$$

Also, $[E_o] = [E.Y^-] + [E.YH].$

Hence $[E_o] = [E.Y^-] + \dfrac{[H^+][E.Y^-]}{K_a^Y} = [E.Y^-]\left(1 + \dfrac{[H^+]}{K_a^Y}\right).$

$$\therefore \frac{[E.Y^-]}{[E_o]} = \frac{1}{\left(1 + \dfrac{[H^+]}{K_a^Y}\right)}.$$

Substituting for $\dfrac{[E.Y^-]}{[E_o]}$ in the expression for V_{max} obtained above:

$$V_{max} = \frac{(V_{max}) \text{ optimal}}{\left(1 + \dfrac{[H^+]}{K_a^Y}\right)}.$$

Taking logarithms,

$$\log_{10} V_{max} = \log_{10}(V_{max})\text{optimal} - \log_{10}\left(1 + \frac{[H^+]}{K_a^Y}\right).$$

When $[H^+] \gg K_a^Y$, $\left(1 + \frac{[H^+]}{K_a^Y}\right) \simeq \frac{[H^+]}{K_a^Y}$, and

the equation reduces to:

$$\log_{10} V_{max} = \log_{10}(V_{max})\text{optimal} - \log_{10}[H^+] + \log_{10}K_a^Y$$

$$= \log_{10}(V_{max})\text{optimal} + pH - pK_a^Y.$$

Hence a plot of $\log_{10} V_{max}$ against pH, where $[H^+] \gg K_a^Y$, will be linear and of slope $= +1$.

When $[H^+] \ll K_a^Y$, the expression for $\log_{10} V_{max}$ reduces to:

$$\log_{10} V_{max} = \log_{10}(V_{max})\text{optimal} - \log_{10} 1$$

$$= \log_{10}(V_{max})\text{optimal}.$$

Hence a plot of $\log_{10} V_{max}$ against pH, where $[H^+] \ll K_a^Y$, will be horizontal.

By combining the equations obtained under the two extreme conditions, it can be seen that the two linear graphs, when extrapolated, intersect where pH $= pK_a^Y$ (Fig. 10.1(c)).

Let us now consider the situation of an enzyme whose catalytic activity depends on a certain amino-acid side chain ($-ZH$) being able to act as a proton donor, according to the general equation:

$$E.ZH \rightleftharpoons H^+ + E.Z^-.$$

From the Henderson–Hasselbalch equation, the relationship between the proportion of enzyme in the form $E.Z^-$ and the pH will be exactly as shown for $E.Y^-$ in Fig. 10.1(a). However, in this instance, $E.ZH$ and not $E.Z^-$ is required for catalytic activity. Since $[E_o] = [E.ZH] + [E.Z^-]$, it readily follows that the relationship between the proportion of enzyme in the form $E.ZH$ and the pH will be as in Fig. 10.2(a), and a graph of V_{max} against pH, at constant $[E_o]$, for such an enzyme will be as in Fig. 10.2(b). The pK_a of the ionisable group is the pH where $[E.ZH] = [E.Z^-]$, so is the same regardless of whether $[E.ZH]$ or $[E.Z^-]$ is plotted against pH and is given by the point of inflection of either graph. Therefore it is again possible to identify the amino-acid residue in question from a graph of V_{max} against pH, or alternatively of $\log_{10} V_{max}$ against pH (Fig. 10.2(c)).

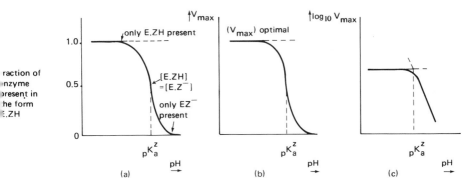

Fig. 10.2 Graphs showing the effects of dissociation of E.ZH according to the Henderson–Hasselbalch equation: (a) a graph of $\dfrac{[EZH]}{[E_0]}$ against pH, (b) a graph of V_{max} against pH, at constant $[E_0]$, where catalytic activity depends on the presence of E.ZH, and (c) the corresponding graph of $\log_{10} V_{max}$ against pH.

Finally, let us consider an enzyme whose catalytic activity depends on one amino-acid side chain $(-Y^-)$ being able to act as a proton acceptor and another $(-ZH)$ being able to act as a proton donor. In this situation, a graph of V_{max} against pH at constant $[E_0]$ will be a combination of Figs. 10.1(b) and 10.2(b); the two points of inflection give the pK_a values of the two amino-acid side chains involved. If these pK_a values are far apart (Fig. 10.3(a)), their determination from a plot of V_{max} against pH is relatively straightforward, for the points of inflection will occur where $V_{max} = \frac{1}{2}(V_{max})$ optimal at a particular enzyme concentration. If, however, the two pK_a values are less than 2 pH units apart (Fig. 10.3(b)), then the theoretical (V_{max}) optimal will never be achieved

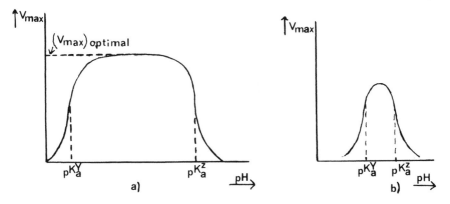

Fig. 10.3 Graphs of V_{max} against pH, at constant $[E_0]$, where catalytic activity depends on the simultaneous presence of E.Y and E.ZH: (a) where pK_a^Y and pK_a^Z are more than 2 units apart, and (b) where pK_a^Y and pK_a^Z are less than two units apart.

and the pK_a values cannot be accurately determined from such a graph or one of $\log_{10} V_{max}$ against pH. Nevertheless, reasonable estimates can be made, and more accurate values obtained from further mathematical treatment of the results, to give some indication as to the probable amino-acid residues involved. It will be noted that this bell-shaped curve of V_{max} against pH is obtained for many enzyme-catalysed reactions.

So, in general, a plot of V_{max} against pH at fixed $[E_0]$ for a particular enzyme-catalysed reaction may enable one or two pK_a values to be obtained for processes which are essential for catalytic activity; a comparison of these values with known pK_a values for amino-acid side chains (Fig. 3.9) may identify the amino-acid residues involved.

However, it is necessary at this point to register a few words of caution: a straightforward interpretation of the results is only valid if the assumptions discussed above are true of a particular enzyme-catalysed reaction and no other pH-dependent factors are involved. In practice, for example, the degree of ionisation of the substrate may vary with pH. Also, V_{max} (or, more accurately, k_{cat}) may be a composite function of several rate constants, not all of which may be affected by pH to the same degree. Thirdly, the microenvironment in which a particular amino-acid residue exists as part of an enzyme may affect the pK_a value of its ionisable side chain by as much as 3 pH units from that found in free solution.

For these, and other reasons, results of pH studies have to be treated with caution, unless confirmed by other methods. Nevertheless, with this proviso, such studies are useful since they are easy to perform and can give the first indication of which amino-acid residues should be considered as essential components of a particular active site.

So, for example, hydrolysis of chitin (poly N–acetylglucosamine) by **lysozyme** shows a bell-shaped curve of activity against pH, the pH optimum being about 5 and the two component pK_a values approximately 4 and 6. The lower of these suggests the presence of an essential carboxyl group, which had been confirmed by other studies and shown to belong to aspartate–52. Less predictably, the other essential ionisable amino-acid side chain has also been shown to be a carboxyl group, this time belonging to glutamate–35: in contrast to aspartate–52, glutamate–35 is located in a non-polar region and is thus protonated at pH 5.

In general, the variation of V_{max} with changing pH will indicate the essential ionising groups of the **enzyme–substrate complex**, i.e. those involved in the catalytic process. It is also possible that some ionising groups might be involved in substrate-binding: these can be identified from the pK_a values obtained by plotting $\dfrac{1}{K_m}$ against pH, provided the substrate is non-ionisable. However, only where the Michaelis–Menten equilibrium-assumption (Chapter 7.1.1) is valid will the results be independent of ionisations involved in the catalytic process.

Further information may be obtained by plotting $\dfrac{V_{max}}{K_m}$ against pH, at constant $[E_o]$: this will indicate the essential ionising groups of the **free enzyme**, i.e. those involved in both the binding and catalytic processes. Ionisable groups which are not involved in either of these processes will not usually contribute to changes in K_m or V_{max}.

Ribonuclease has been shown by Rabin and his colleagues (1961) to have two active site ionisable groups, their pK_a values being 5.2 and 6.8 in the free enzyme, but perturbed by the binding of substrate to 6.3 and 8.1 in the enzyme-substrate complex. From this and other data it has been concluded that both of the ionisable groups are histidine side chains (histidine-12 and histidine-119), one required to act as a proton donor and the other as a proton acceptor.

Two distinct ionisable groups have been demonstrated by Bender and others to influence the reaction catalysed by **chymotrypsin**. One of these causes an increase in K_m at high pH, without a corresponding change in V_{max}: this has been attributed to an N-terminal isoleucine residue (isoleucine-16, see Chapter 5.1) whose α-amino group must be in a protonated form to maintain the enzyme in a structure capable of binding substrate. The other ionisation, which affects V_{max}, shows a pK_a of about 6.8: this is apparently associated with aspartate-102, the anomolous pK_a value being due to its hydrophobic environment. The plot of $\dfrac{V_{max}}{K_m}$ against pH, at constant $[E_o]$, is consistent with these results, showing a bell-shaped curve with a maximum at about pH 8 and pK_a values of about 6.8 and 9.0.

10.2 THE INVESTIGATION OF THE THREE-DIMENSIONAL STRUCTURES OF ACTIVE SITES

From the above discussion, it must be clear that the active site of an enzyme is more than the sum of its parts: in order to understand how an enzyme works it is necessary to know not only the identities of the amino-acid residues which make up the substrate-binding sites and catalytic sites, but also the environment in which each site exists and the arrangement in space of the sites with respect to each other.

The technique which has given far more information than any other about the three-dimensional structure of enzymes is x-ray crystallography (see Chapter 2.5). Of course it is the structure in solution which determines the properties of enzymes, but all available evidence suggests that structures determined in crystals can exist in solution; some crystals of enzymes, e.g. ribonuclease, even retain catalytic activity.

If this technique is applied to investigate the binding of the substrate to the active site of an enzyme, a major difficulty is encountered: x-ray diffraction analysis requires the structure under investigation to be static, but enzyme-substrate complexes break down rapidly to give reaction products. Hence, the structure of an enzyme-substrate complex is usually inferred from that obtained when the substrate is replaced by a slow-reacting analogue or an inhibitor (see Chapter 10.1.2).

Enzyme crystals often consist of about 50% water, so it may be possible to diffuse these ligands (substrate analogue or inhibitor) into an existing crystal; alternatively the enzyme-ligand complex may be crystallised from solution. For example, Blow and his colleagues (1967) prepared **tosyl-chymotrypsin** (see Chapter 10.1.3) by diffusing tosyl-fluoride into a crystal of α-chymotrypsin; they then used x-ray crystallography to elucidate its structure, the findings supporting much of the data discussed in Chapter 10.1. The tosyl group was found to be bound to serine-195, which was in close proximity to histidine-57 and methionine-192 on the surface of the molecule (Fig. 10.4); the imidazole ring of histidine-57 was shown to be linked by a hydrogen bond to the side chain carboxyl group of aspartate-102, itself located in the hydrophobic interior of the molecule; also, the protonated α-amino group of the N-terminal isoleucine residue (isoleucine-16) was found to form an electrostatic linkage with the side chain of aspartate-194.

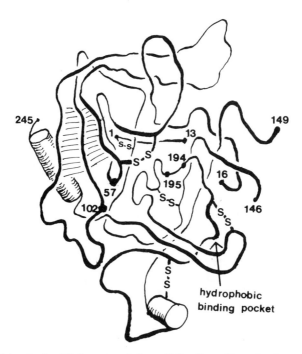

Fig. 10.4   A simplified representation of the three-dimensional structure of α-chymotrypsin, as revealed by the x-ray diffraction studies of Blow and colleagues (1967). (Conventions as for Fig. 2.8.)

Further examples of x-ray crystallography in the elucidation of enzyme structure and function are discussed in Chapter 11.

Spectroscopy in its various forms can give further information about enzyme–substrate interactions. The nmr technique, for example, detects atoms with an odd number of protons in their nuclei, so giving them a magnetic moment. If a kilogauss magnetic field is applied, such nuclei will precess around it with a frequency depending on the magnetic moment and the magnitude of the applied field; if a radiofrequency (around 100 MHz) field is then generated so that its magnetic vector rotates perpendicular to the kilogauss field and the conditions adjusted, resonance will occur when the frequency of oscillation of the field corresponds with the precession of a nuclear dipole. This resonance is detected as the absorption of energy by the nucleus from the radiofrequency field. The intensity of a resonance absorption line is directly proportional to the number of nuclei in an identical environment. However, chemical shifts in resonance frequency for identical nuclei in different electronic environments can be detected, as can the splitting of resonance peaks into multiplets of fine structure because of the interaction of neighbouring nuclear spins, while relaxation times can give a measurement of the exchange of energy with neighbouring nuclei of the same kind. Nuclei that can be investigated in this way include those of ^1H, ^{13}C and ^{15}N. Of course, before isotopes such as ^{13}C and ^{15}N, which do not occur naturally, can be used in the investigation of enzyme structure, they must first be incorporated into the enzyme, for example by utilising bacteria to synthetise it in a medium rich in the isotope. Studies by nmr can give much information about interactions between particular atoms in an enzyme–substrate complex, and if carried out at various pH values can indicate the pK_a values of individual groups (see Chapter 11.3.2).

Various techniques have indicated that segments of a protein may be flexible and free to move within a crystal: flexible, or disordered, segments are "invisible" to x-ray diffraction analysis, since only regular arrangements of atoms give rise to well-defined patterns. A disordered region often starts at a glycine residue, the absence of a side chain allowing maximum flexibility, and usually contains no aromatic residues.

Huber and his colleagues (1977) have shown that **trypsinogen** contains several disordered regions, but the active enzyme, **trypsin**, does not: the inactivity of trypsinogen may be due to the inability of the binding sites in the disordered regions to form bonds of sufficient strength with the substrate. The electrostatic linkage occurring in chymotrypsin between isoleucine-16 and aspartate-194 is also found in trypsin, but not in trypsinogen or chymotrypsinogen, and it appears to act as a clamp, holding the enzyme in the rigid structure required for acitivity.

In other enzyme systems, e.g. **acetyl–CoA carboxylase** (Chapter 11.5.8), flexibility is essential for activity, since it appears that an intermediate may be transferred from one site to another on a flexible protein arm. In some enzymes, e.g. carboxypeptidase A (Chapter 11.4.4), important conformational changes have been shown to take place during the binding of a substrate. In general, proteins are likely to be more flexible in solution than when crystalline.

SUMMARY OF CHAPTER 10

A major problem in the identification of the amino-acid residues involved in substrate-binding and catalysis is posed by the instability of the enzyme-substrate complex: substrate and product molecules will easily dissociate from the enzyme during investigation, making their positions of attachment difficult to determine. This problem may be overcome by the use of chemical procedures to prevent the breakdown of the enzyme-substrate complex, or by the use of substrate analogues, including irreversible inhibitors, which bind readily to the enzyme but do not easily dissociate from it.

If the chemical modification of a particular amino-acid side chain results in a loss of catalytic activity, then it may be assumed that the residue in question is located at the active site of the enzyme, provided the modification can be prevented by the presence of excess substrate or competitive inhibitor. The identity of amino-acid residues involved in binding or catalysis may also be indicated by the activity profile of an enzyme at different pH values.

However, the technique which has so far given most information about the structure of enzymes and the nature of enzyme-substrate interactions is x-ray crystallography.

FURTHER READING

As for Chapters 2, 4 and 5.

PROBLEMS

10.1 The hydrolysis of an ester substrate by papain was investigated at fixed temperature and enzyme concentration, but at a series of pH values. The following results were obtained:

Initial velocity (μmol.l^{-1}.min^{-1}) at [S_o] (mol.l^{-1}) =

pH	2.0	2.5	3.33	5.0	10.0
2.0	1	1	1	2	2
3.0	11	12	14	16	20
4.0	83	93	105	122	143
5.0	156	172	196	222	263

pH	2.0	2.5	3.33	5.0	10.0
6.0	157	173	197	223	264
7.0	156	173	196	223	263
8.0	139	156	179	208	250
9.0	43	52	66	89	141
10.0	5	7	9	13	25 .

What can you deduce about the possible involvement of ionisable amino-acid side chains?

CHAPTER 11

The Chemical Nature of Enzyme Catalysis

11.1 AN INTRODUCTION TO REACTION MECHANISMS IN ORGANIC CHEMISTRY

A covalent chemical bond involves the sharing of a pair of electrons between the atoms (X:Y). If the bond breaks by **homolytic fission**, each atom separates as a highly reactive **free radical**, possessing an unpaired electron $(X \cdot + \cdot Y)$. Such reactions are very uncommon in aqueous solutions, where most bonds are broken by **heterolytic fission**, leaving one of the atoms with both electrons. If both electrons are retained by a carbon atom, a **carbanion** is produced:

$$R_3C:X \rightleftharpoons R_3C:^- + X^+.$$
$$\text{carbanion}$$

If neither electron is retained by the carbon atom, a **carbonium ion** is produced:

$$R_3C:X \rightleftharpoons R_3C^+ + :X^-.$$
$$\text{carbonium ion}$$

Carbanions and carbonium ions may be stabilised by **delocalisation** of the charge over other atoms in the molecule. For example:

$$CH_2 = CH - \overset{+}{C}H_2 \longleftrightarrow \overset{+}{C}H_2 - CH = CH_2.$$

The curved arrow indicates the flow of a pair of electrons which would take place if one of these **canonical structures** of the carbonium ion was converted to the other. The actual structure is somewhere in between the two canonical structures written down: it should not be imagined that the two forms are being rapidly interconverted. **Conjugated** double bond systems (i.e. where double and single bonds occur alternately) are of particular importance for stabilisation, for a charge can be delocalised over the entire system.

The formation of carbanions and carbonium ions is likely to occur as part of a complete reaction sequence. Organic compounds can participate in four main types of reaction: displacement (or substitution) reactions, addition reactions, elimination reactions and rearrangements.

Substitution reactions may be **nucleophilic**, when the group attacking the carbon atom is an electron donor called a nucleophile since it is attracted to nuclei; alternatively they may be **electrophilic**, when the attacking group is an electron acceptor, or electrophile. Electrophilic substitution reactions often involve the displacement of hydrogen. For example:

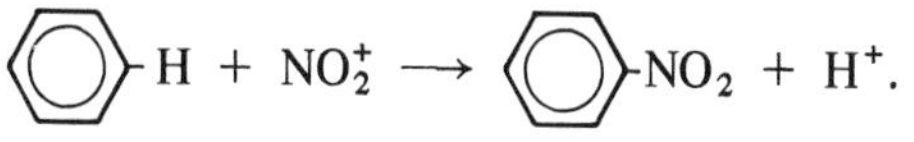

$$\langle\bigcirc\rangle\text{-H} + \text{NO}_2^+ \longrightarrow \langle\bigcirc\rangle\text{-NO}_2 + \text{H}^+.$$

benzene   nitrobenzene

In **nucleophilic substitution**, an atom other than hydrogen is usually displaced. Such reactions may have a unimolecular or a bimolecular mechanism.

In **unimolecular nucleophilic substitution (S_N1)** reactions, the rate-limiting step is the ionisation of a single molecule to form a carbonium ion which then reacts with a nucleophile, as in the following example:

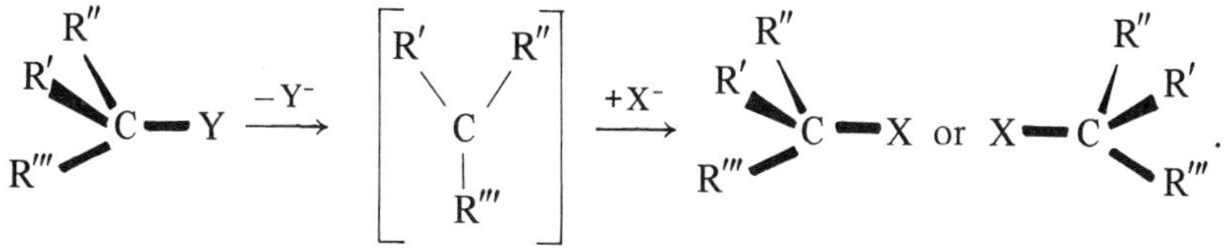

The carbonium ion is planar and , if the resulting product has four different groups attached to that particular carbon atom (as above), a racemic mixture of optical isomers will be obtained. The overall rate of the reaction is determined by the rate of ionisation, and thus on the basicity of the leaving group (Y^-): weak bases are good leaving groups and depart readily; strong bases are poor leaving groups. The strength of the attacking nucleophile does not usually affect the rate of the reaction.

In **bimolecular nucleophilic substitution (S_N2)** reactions, the attacking nucleophile adds to the carbon atom at a point diametrically opposite the leaving group, which it displaces in one rapid step:

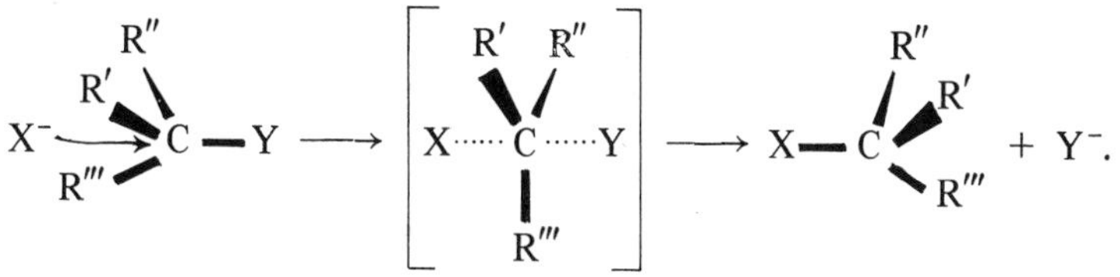

transition-state

It can be seen that the reaction involves an inversion of configuration at the carbon atom. The rate of the reaction depends, among other factors, on the relative strengths of X^- and Y^- as nucleophiles.

It is unnecessary here to go into details about the other types of reaction. **Addition reactions** involve the addition of an electrophile or a nucleophile to form a relatively stable compound without any group being displaced; **elimination reactions** involve the removal of a group, usually in a strongly basic solution, without it being replaced by another group; **rearrangements** are reactions where bonds are formed and broken within a single molecule.

Information about the mechanism of a particular reaction can be obtained from the **stereochemistry** of the products, as discussed above for the S_N1 and S_N2 reactions. Another useful technique is the investigation of the effect of **isotopic substitution** on the reaction rate. For example, it is usually found that a C—D bond is broken more slowly than a C—H bond, so the use of deuterium-labelled reactants can give information as to which C—H bonds are normally broken as part of the rate-limiting step.

11.2 MECHANISMS OF CATALYSIS

11.2.1 Acid-Base Catalysis
Acids can catalyse reactions by temporarily donating a proton; bases can do the same by temporarily accepting a proton. In Chapter 6.2.3 we saw how a base could catalyse the hydrolysis of an ester by stabilising the transition-state. Bases may also increase reaction rates by increasing the nucleophilic character of the attacking groups. Thus, hydroxide ions displace halides from alkyl halides more rapidly than do neutral water molecules. Similarly, acids facilitate the removal of leaving groups, where these are strong bases. For example, the following reaction would normally proceed extremely slowly:

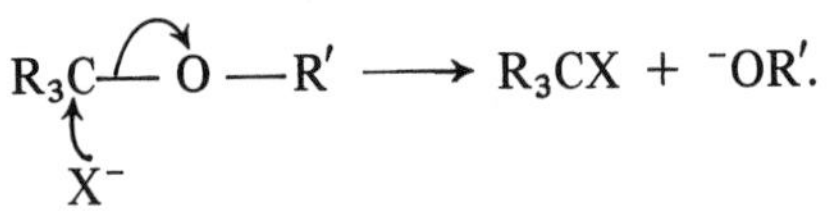

$$R_3C \overset{\frown}{—} O—R' \longrightarrow R_3CX + {}^-OR'.$$
$$\underset{X^-}{\big\uparrow}$$

In the presence of an acid, however, conditions are much more favourable:

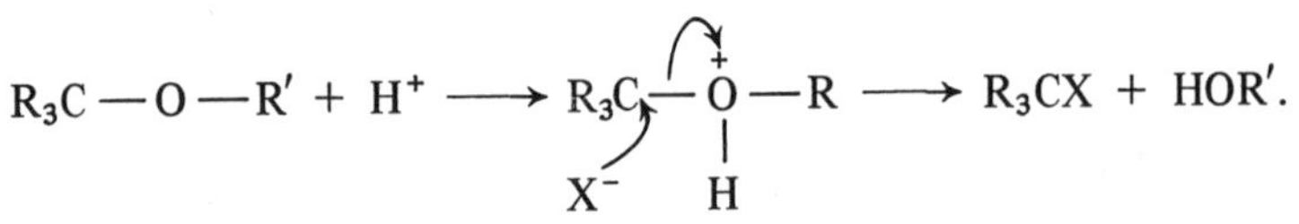

$$R_3C—O—R' + H^+ \longrightarrow R_3C \overset{\frown}{\underset{X^-}{—}} \overset{+}{\underset{H}{O}}—R \longrightarrow R_3CX + HOR'.$$

In general, acid or base catalysis may be shown to be operating by determining the rate constants of a reaction at different concentrations of acid or base.

11.2.2 Electrostatic Catalysis

A transition-state may be stabilised by electrostatic interaction between its charged groups and charged groups on a catalyst. Thus, the positive charge on a carbonium ion can be stabilised by interaction with a negatively charged carboxylate ion; similarly, the negative charge on an oxyanion can be stabilised by a positively charged metal ion. For example, the hydrolysis of glycine esters:

$$H_2N.CH_2\overset{\overset{\displaystyle O}{\|}}{C}.OCH_3 + H_2O \rightleftharpoons H_2N.CH_2\overset{\overset{\displaystyle O}{\|}}{C}.OH + CH_3OH$$

may be catalysed by cupric ions, the mechanism probably involving the following steps:

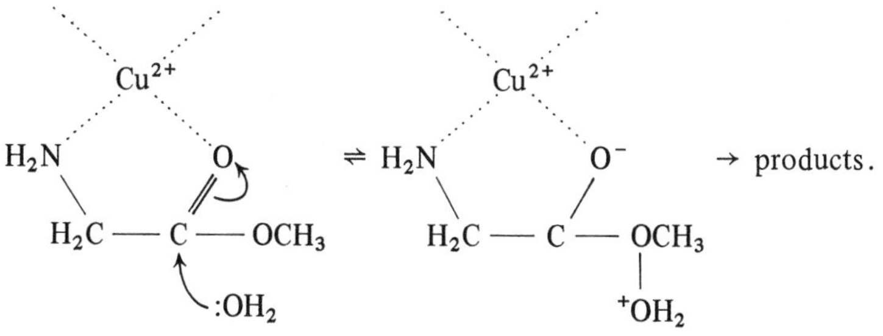

11.2.3 Covalent Catalysis

In contrast to acid-base and electrostatic catalysis, where the transition-state is merely modified, covalent catalysis introduces a different reaction mechanism and is sometimes termed **alternative pathway catalysis**. In the case of nucleophilic catalysis, the catalyst is more nucleophilic than the normal attacking groups and so rapidly forms an intermediate which itself rapidly breaks down to give the products. For example, Bender and co-workers (1957) have shown that a variety of tertiary amines catalyse the hydrolysis of esters, as follows:

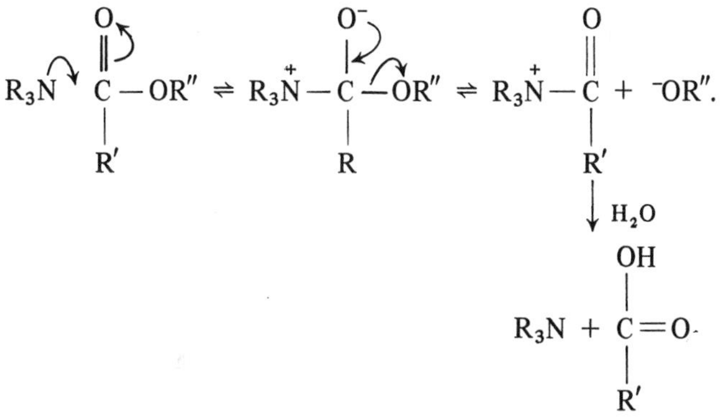

In contrast, electrophilic catalysts act by withdrawing electrons from the reaction centre of an intermediate and may be termed **electron sinks**. This type of catalysis is found in the reactions of the coenzymes **thiamine pyrophosphate** (Chapter 11.5.6) and **pyridoxal phosphate** (Chapter 11.5.7).

11.2.4 Enzyme Catalysis

All of the above mechanisms of catalysis are seen in enzyme-catalysed reactions. However, enzymes, because of their great size and range of properties are able to impose their presence on a reaction to a far greater extent than most catalysts.

All catalysts act by reducing the energy of activation, or more correctly, the free energy of activation ($\Delta G^{\ddagger}$), of the reaction being catalysed (see Chapters 6.2.3 and 6.6). As we saw in Chapter 6.1.2, free energy is made up of an entropic and an enthalpic component, since $\Delta G = \Delta H - T\Delta S$. Hence $\Delta G^{\ddagger} = \Delta H^{\ddagger} - T\Delta S^{\ddagger}$, where $\Delta H^{\ddagger}$ is the enthalpy of activation and $\Delta S^{\ddagger}$ is the entropy of activation, and so anything which decreases $\Delta H^{\ddagger}$ or makes ΔS more positive will lower the value of $\Delta G^{\ddagger}$ and help the reaction proceed.

There is a great loss of entropy when reactants leave their random existence in free solution to become bound in the transition-state, so $-T\Delta S^{\ddagger}$ will generally have a large positive value, possibly making up about half of the total free energy of activation where there is more than one reactant. In enzyme-catalysed reactions there is inevitably a similar loss of entropy at some stage, but it occurs largely in the binding steps when enzyme–substrate complexes are formed, and not to the same degree in the actual conversion of substrates to products: the binding of substrate molecules in close proximity to each other on the enzyme surface effectively increases their concentrations and reduces the entropy loss for the subsequent formation of a transition state; this has been called the **proximation effect**, the **approximation effect** and the **propinquity effect** by Bruice, Jencks and others. The enzyme may also ensure that the reacting groups of the bound substrates approach each other with their electronic orbitals correctly orientated, thus ensuring that the reaction takes place under optimal conditions; Koshland and others have termed this property of enzymes **orbital steering**.

The contribution of the catalytic sites on the enzyme to the reaction will generally be to the enthalpic factor by stabilising the transition-state; it appears that the stability of the enzyme-bound transition-state may be the most important single factor in determining whether the reaction proceeds (Chapter 4.6). It is possible that several different catalytic processes may be involved in the same enzyme-catalysed reaction, as may be seen in the examples discussed in Chapter 11.3–11.5. Another way in which the enthalpy factor may be reduced occurs if the positions of the binding sites on the enzyme do not correspond exactly to those on a substrate: as the substrate binds, some of its bonds may be distorted and weakened, resembling those found in the transition-state; this is the

Haldane and Pauling concept of **strain**, discussed in Chapter 4.5. Alternatively, the enzyme–substrate complex may be under stress from internal interactions which are absent from the enzyme-bound transition-state complex: again, this would facilitate the forward reaction (Chapter 4.5). A distortion of the protein to fit the substrate (the Koshland induced-fit hypothesis, Chapter 4.4) would not have any affect on enthalpy, but would not preclude stress factors from playing a part.

Some reactions proceed better in organic rather than aqueous solutions. Most enzymes contain regions of non-polar character (Chapter 2.5.2) and these may be of importance in the catalysis of such reactions.

So, enzymes create a suitable environment for the reaction to proceed; they reduce the total free energy of activation required and also spread out the free energy requirement over several stages. Nevertheless, some activation energy is still required for each stage of the reaction sequence. This cannot come from the substrate because the translational energy that these possess in free solution is lost the moment they become bound to the enzyme. It seems likely that the required energy is obtained from collisions between solute molecules and the enzyme–substrate complex, but the mechanism for this is far from clear.

11.3 MECHANISMS OF REACTIONS CATALYSED BY ENZYMES WITHOUT COFACTORS

11.3.1 Introduction

Enzymes which operate without cofactors tend to be relatively small and have relatively straightforward reaction mechanisms. For these reasons, such enzymes were amongst the first to be investigated in detail. Some examples are given below.

11.3.2 Chymotrypsin

Chymotrypsin is formed by the cleavage of several peptide bonds in the inactive monomeric protein chymotrypsinogen, which is synthesised and secreted by mammalian pancreas; the active enzyme thus produced consists of three non-identical polypeptide chains (Chapter 5.1.2).

Chymotrypsin catalyses the cleavage of peptide bonds at the carboxyl side of aromatic amino-acid (phenylalanine, tyrosine or tryptophan) residues; it will also hydrolyse a variety of amides and esters, and these artificial substrates have been used to investigate the enzyme in great detail. In Chapter 7.2.1 we discussed kinetic evidence showing that the chymotrypsin-catalysed hydrolysis of an ester proceeded via the formation of an acyl–enzyme. This is also true for the hydrolysis of amides, and we can write the reaction in general terms as follows:

$$E + R.CO.Y \longrightarrow E.R.CO.Y \xrightarrow[YH]{} E.CO.R \xrightarrow{H_2O} E + RCOOH.$$
$$\text{ester or} \qquad\qquad\qquad \text{acyl}$$
$$\text{amide} \qquad\qquad\qquad\quad \text{enzyme}$$

In Chapters 10.1.2, 10.1.3 and 10.1.5 we discussed evidence that the substrate can bind to serine-195; and histidine-57 was also implicated in the reaction mechanism. X-ray diffraction studies (see Chapter 10.2 and Fig. 10.4) revealed a hydrophobic binding pocket at the active site for aromatic side chains, and showed that aspartate-102, buried in the interior of the molecule, could be closely linked to the action of histidine-57 and serine-195, possibly setting up what has been termed a charge relay system, with aspartate-102 removing a proton from histidine-57, making it easier for the latter to remove a proton from serine-195 during the course of the reaction. However, detailed ^{1}H, ^{13}C and ^{15}N nmr studies have shown that histidine-57 has a relatively normal pK_a value near 7, at least in the free enzyme, casting doubt on the charge relay theory.

An outline reaction mechanism for the hydrolysis of an ester or amide by chymotrypsin (or other serine protease) is as follows. Histidine-57 acts as a base catalyst to enable the oxygen of serine-195 make a nucleophilic attack on the carboxyl group of the enzyme-bound substrate. An unstable tetrahedryl intermediate is formed whose negatively charged oxygen atom may be stabilised by hydrogen bonding with the backbone —NH of glycine-193. The imidazole group of histidine-57 then acts as an acid catalyst to facilitate the liberation of the first product (YH), leaving behind the acyl enzyme. (Note the covalent bond linking the acyl group to serine-195.)

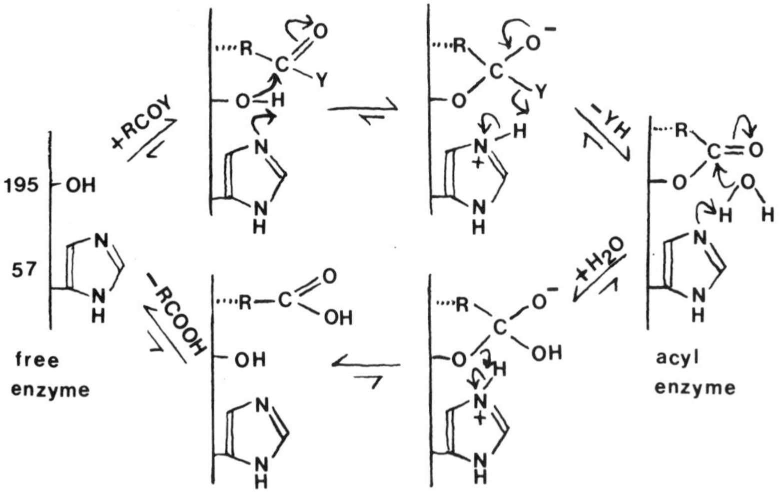

The second stage of the reaction, like the first, is initiated by a nucleophilic attack, this time by water; the liberation of the product (RCOOH) is again assisted by acid catalysis.

The same mechanism is thought to occur for peptide substrates R'.NH. C(R'').CO.NH.R''': if the substrate is again written in the form R.CO.Y, to be compatible with the illustration above, R— represents R'.NH.C(R'') − (where R is the side chain which fits into the binding pocket, the −NH forming a hydrogen bond with the backbone carboxyl group of serine-214) and −Y represents −NH.R'''.

11.3.3 Ribonuclease

Bovine pancreatic ribonuclease A has been studied in great detail by enzymologists and was the first enzyme to have its complete amino-acid sequence determined (Smyth, Stein and Moore, 1963). It catalyses the cleavage of the phosphodiester backbone of ribonucleic acids by a reaction involving transfer of a phosphate group from the 5'-position of one nucleotide to the 3''-position of the next nucleotide in the chain (see Chapter 3.1.1).

In Chapters 10.1.3 and 10.1.5 we discussed evidence that histidine-12 and histidine-119 both contribute to the catalytic process, one possibly acting as a proton donor and the other as a proton acceptor; similar investigations have suggested that lysine-41 is also involved. The three-dimensional structure of the enzyme, as revealed by the x-ray diffraction studies of Kartha and colleagues (1967), shows these three amino-acids are sited close together near a substrate-binding cleft in the molecule (Fig. 11.1).

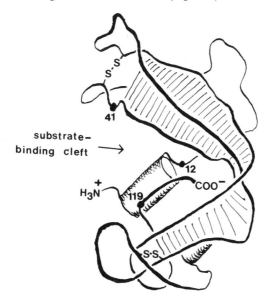

Fig. 11.1   A simplified representation of the three-dimensional structure of ribonuclease A, as revealed by the x-ray diffraction studies of Kartha and colleagues (1967). (Conventions as for Fig. 2.8.)

In 1962, Rabin and co-workers suggested the following mechanism for the reaction:

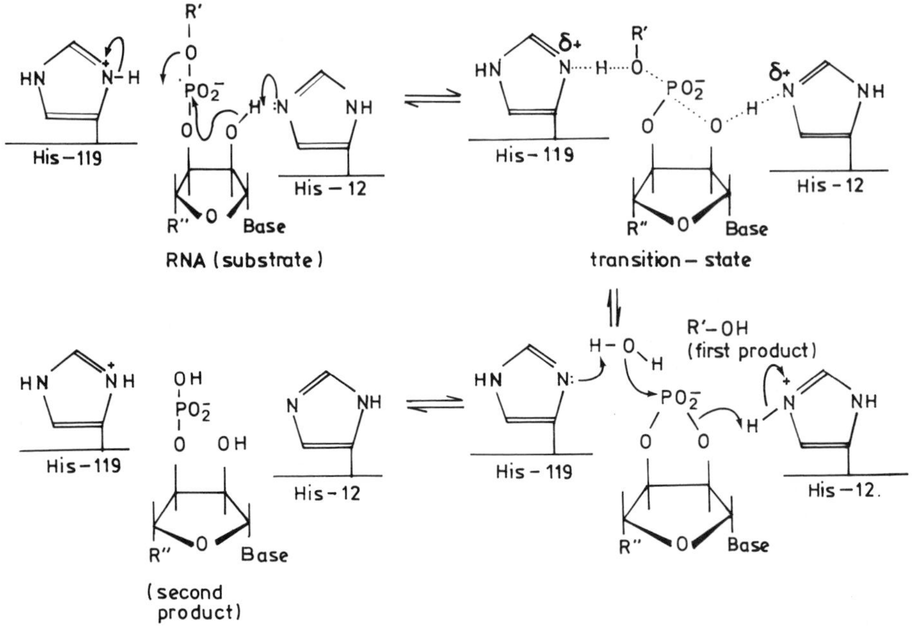

Histidine-12 removes a proton from the substrate, producing a nucleophilic oxygen atom which attacks the *P*. Histidine-119 donates a proton to the complex, and the first product, R'OH, is liberated. The rest of the reaction is the reverse of the first part, with water replacing the R'OH.

This mechanism is consistent with the evidence now available. It is likely that lysine-41 helps stabilise the complex by an electrostatic interaction with the negative charge of the phosphate.

11.3.4 Lysozyme

Lysozyme catalyses the hydrolysis of the polymer consisting of alternating N-acetylglucosamine (NAG) and N-acetylmuramic acid (NAM) units which gives shape and rigidity to bacterial cell walls.

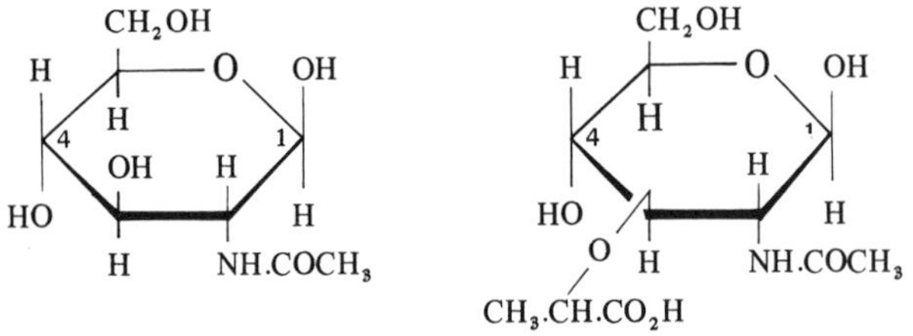

The lysozyme which has been most studied is that from hen egg-white; this was the first enzyme to have its complete three-dimensional structure determined by x-ray crystallography (Blake, Phillips and colleagues 1965). These and subsequent investigations revealed that the enzyme has a large cleft which can accommodate six of the substrate's amino-sugar units (see Fig. 2.8). This may be represented diagrammatically as follows:

position of cleavage

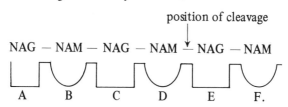

NAG — NAM — NAG — NAM — NAG — NAM

A   B   C   D   E   F.

Each of the six amino-sugars is bound to the enzyme by hydrogen bonds, e.g. those in sites B and C to aspartate-101, that in site B to tryptophan-62 and tryptophan-63, and that in site F to asparagine-37. NAM, because of its large lactyl side chain, cannot bind to sites A, C or E, where there is restricted space, and during or immediately after binding to site D its hexose ring is distorted from the chair to the the less-stable half-chair conformation.

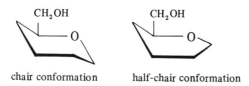

chair conformation   half-chair conformation

The hydrolysis takes place between the units in sites D and E, and therefore must always involve the β-1,4 glycosidic link between C1 (the reducing end) of NAM and C4 of NAG. The amino-acids near this bond are glutamate-35 and aspartate-52; evidence that these are involved in the reaction has been discussed in Chapters 10.1.3 and 10.1.5.

Vernon (1967) suggested the following mechanism for the reaction (many bonds and atoms not involved in the reaction being omitted from the diagram for the sake of clarity):

The substrate binds to the enzyme and ring D is distorted to the half-chair conformation, which facilitates the formation of a carbonium ion also having the half-chair conformation. Glutamate-35, which is in a non-polar environment and thus protonated at pH 5, can act as a general acid catalyst and donates its proton to the oxygen of the glycosidic bond, causing the bond to break. Aspartate-52 is in a more polar environment than glutamate-35 and is negatively charged at pH 5; therefore, it can stabilise the carbonium ion by electrostatic interaction. The first product leaves, and the reaction is completed by a nucleophilic attack on the carbonium ion by water.

11.3.5 Triose phosphate isomerase

Muscle triose phosphate isomerase is a dimeric enzyme which catalyses the interconversion of glyceraldehyde 3-P(G3P) and dihydroxyacetone-P (DHAP). X-ray crystallography has revealed that each sub-unit has an inner cylinder of eight strands of parallel β-pleated sheet linked by predominantly helical regions which form an outer cylinder. Each of the two identical binding sites contains lysine-13, histidine-95, glutamate-165 and glycine-232 from one sub-unit and phenylalanine-74 from the other (Fig. 11.2). The reaction mechanism apparently involves glutamate-165 acting as a base catalyst and histidine-95 as an acid one:

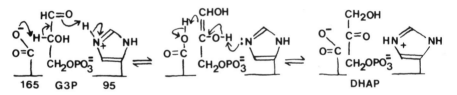

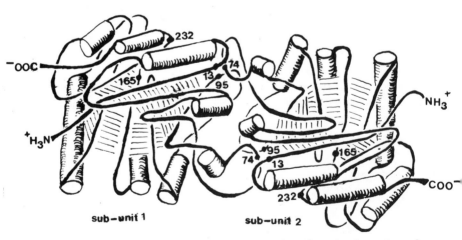

Fig. 11.2 — A simplified representation of the three-dimensional structure of chicken muscle triose phosphate isomerase as revealed by the x-ray diffraction studies of Phillips and colleagues (1975). (Conventions as for Fig. 2.8.)

11.4 METAL-ACTIVATED ENZYMES AND METALLOENZYMES

11.4.1 Introduction

More than a quarter of all known enzymes require the presence of metal atoms for full catalytic activity. Metal atoms usually exist as cations and often have more than one oxidation state, as with ferrous (Fe^{2+}) and ferric (Fe^{3+}) iron. We have noted that this positive change can stabilise transition-states by electrostatic interactions, giving one mechanism for catalysis by metals (Chapter 11.2.2). However, irrespective of the oxidation state and charge carried, a metal ion can bind a particular number of groups (**ligands**) by accepting free electron pairs to form co-ordinate bonds in specific orientations (Chapter 2.3.2).

Therefore, metal ions can be involved in enzyme catalysis in a variety of ways: they may accept or donate electrons to activate electrophiles or nucleophiles, even in neutral solution; they themselves may act as electrophiles or nucleophiles; they may mask nucleophiles to prevent unwanted side reactions; they may bring together enzyme and substrate by means of co-ordinate bonds, possibly causing strain to the substrate in the process; they may hold reacting groups in the required three-dimensional orientation; they may simply stabilise a catalytically-active conformation of the enzyme.

With **metalloenzymes**, the metal is tightly bound and retained by the enzyme on purification; with **metal-activated** enzymes, the binding is less tight and the purified enzymes may have to be activated by the addition of metal ions. There is no clear-cut division between the two groups.

Mildvan (1970) has pointed out that ternary complexes formed between an enzyme (E), metal ion (M) and substrate (S) may be enzyme bridge complexes ($M-E-S$), substrate bridge complexes ($E-S-M$) or metal bridge complexes ($E-M-S$ or $E\diagup\overset{M}{\underset{S}{\diagdown}}$). Metalloenzymes cannot form substrate bridge complexes since the purified enzyme exists as $E-M$.

The involvement of metal ions in enzymes may be investigated by nmr, esr and proton relaxation rate (PRR) enhancement techniques; some examples are given below.

11.4.2 Activation by Alkali Metal Cations (Na^+ and K^+)

Alkali metal cations bind only weakly to form complexes with enzymes, but K^+, the most abundant intracellular cation, is known to activate a great many enzymes, particularly those catalysing phosphoryl transfer or elimination reactions. It appears that the role of K^+ is largely to bind to negatively charged groups on an inactive form of the enzyme and thus cause a change in conformation to a more active form. However, in some cases it may also aid substrate binding. For example, muscle **pyruvate kinase**, a tetrameric enzyme which catalyses the reaction:

$$CH_2 = C.CO_2^- + H^+ + ADP \rightleftharpoons CH_3C.CO_2^- + ATP$$
$$| \qquad\qquad\qquad\qquad\qquad\quad ||$$
$$OPO_3^{2-} \qquad\qquad\qquad\qquad\quad O$$

phosphoenolpyruvate   pyruvate
(PEP)

has a requirement for alkali metal cations and for Mn^{2+} (or Mg^{2+}), all of which bind in the region of the active site. Various studies have indicated that the carboxyl group of PEP binds to the enzyme-bound K^+, whereupon a conformational change takes place which facilitates the progress of the reaction via an $E-Mn^{2+}-PEP$ complex.

11.4.3 Activation by Alkaline Earth Metal Cations (Ca^{2+} and Mg^{2+})

Oxygen atoms are often involved in the bonds of both alkali metal and alkaline earth metal cations, bonds of the latter being relatively stronger. The divalent cations, Ca^{2+} and Mg^{2+}, can form 6 co-ordinate bonds to produce octahedral complexes.

Mg^{2+} is accumulated by cells in exchange for transport of Ca^{2+} in the opposite direction. As might be expected, therefore, enzymes requiring Ca^{2+} for activation are mainly extracellular ones, e.g. the salivary and pancreatic α-amylases: the Ca^{2+} appears to play a role in maintaining the structure required for catalytic activity. A variety of intracellular enzymes require Mg^{2+} for activity, and in most cases this requirement can be replaced *in vitro* by one for Mn^{2+}. Mn^{2+} is paramagnetic, which enables the system to be more easily investigated.

It has been shown that all possible types of ternary bridge complexes involving divalent cations can exist. Most kinases form $E-S-M$ complexes, where S is the reacting nucleotide. Let us consider, as an example, the reaction catalysed by muscle **creatine kinase**:

$$creatine + MgATP \rightleftharpoons MgADP + phosphocreatine + H^+.$$

The true substrate is MgATP and the reaction proceeds via the formation of the complex $E \diagdown^{ATP-Mg^{2+}}_{creatine}$. Cohn and colleagues (1971) have shown that the divalent cation binds to the α- and β-phosphates of the nucleotide but not to the terminal (γ) phosphate which is transferred to creatine. The cation helps in the orientation of the complex and may also assist in the breaking of the pyrophosphate bond by withdrawing electrons from the β-phosphate. The overall reaction has a random-order rapid-equilibrium kinetic mechanism and dead-end complexes may be formed (see Chapter 9.3.2); identification of these, e.g. creatine-E-ADP-Mn^{2+}, has helped in the elucidation of the reaction mechanism.

In contrast, the reaction catalysed by **pyruvate kinase** (see Chapter 11.4.2) involves a cyclic metal bridge complex

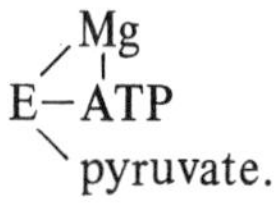

$$
\begin{array}{c}
\diagup \text{Mg} \\
\text{E--ATP} \\
\diagdown \\
\text{pyruvate.}
\end{array}
$$

Metal bridge complexes are found with many enzymes which use pyruvate or phosphoenolpyruvate as substrate. Another example is **enolase**, which catalyses the reaction:

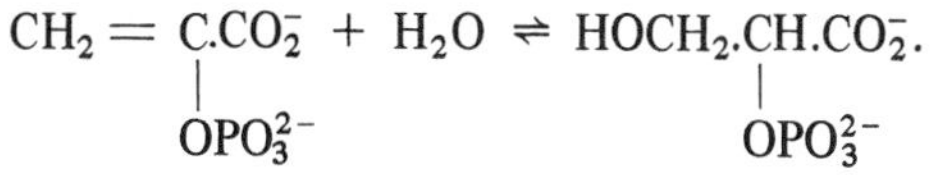

$$
\begin{array}{cc}
\text{CH}_2 = \underset{\underset{\text{OPO}_3^{2-}}{|}}{\text{C.CO}_2^-} + \text{H}_2\text{O} \rightleftharpoons \text{HOCH}_2.\underset{\underset{\text{OPO}_3^{2-}}{|}}{\text{CH.CO}_2^-}.
\end{array}
$$

phosphoenolpyruvate   2-phosphoglycerate

Enolase is a dimeric enzyme, two Mg^{2+} ions being required to stabilise the active dimer; a further two Mg^{2+} ions are required if each of the two active sites binds a substrate. Nowak, Mildvan and colleagues (1973) have demonstrated that the enzyme-bound cation probably binds to water, forming a co-ordinated hydroxyl group which can attack the phosphoenolpyruvate:

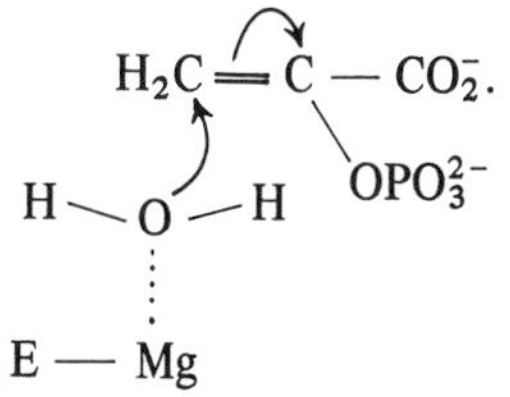

$$
\begin{array}{c}
\text{H}_2\text{C} = \text{C} - \text{CO}_2^-. \\
\diagdown \\
\text{OPO}_3^{2-} \\
\text{H} - \text{O} - \text{H} \\
\vdots \\
\text{E} - \text{Mg}
\end{array}
$$

A few enzymes, e.g. *E. coli* **glutamine synthetase**, have a mechanism involving an enzyme bridge complex. Here the divalent cation presumably has a purely structural role.

11.4.4 Activation by Transition Metal Cations (Cu, Zn, Mo, Fe and Co Cations)

Transition metal ions bind to enzymes much more strongly than the metal ions discussed above, and usually form metalloenzymes; this makes their involvement in enzyme-catalysis relatively easy to investigate. They are found in only trace amounts in living organisms, for larger amounts can be toxic.

The trace metals Mo and Fe are found in **nitric–oxide reductase**, the nitrogen-activating complex of nitrogen-fixing bacteria, and Fe is a component of haemoglobin, the oxygen-carrying haemoprotein of the erythrocytes of vertebrates. Another trace metal, Co, is found in vitamin B_{12} (see Chapter 11.5.10). We will now consider, in a little more detail, an example of a Cu– and a Zn-metalloenzyme.

Superoxide dismutase is a copper–metalloenzyme which catalyses the removal of the highly reactive O_2^- produced, for example, by oxidation of xanthine by molecular oxygen in the presence of **xanthine oxidase**. Fridovich and colleagues (1969) have demonstrated that the dismutase reaction is as follows:

$$2O_2^- + 2H^+ \rightleftharpoons H_2O_2 + O_2.$$

Bovine erythrocyte superoxide dismutase is a dimeric protein containing two Cu^{2+} ions and two Zn^{2+} ions. The Zn^{2+} ions appear to have a structural rather than a catalytic role, while the Cu^{2+} ions are involved in the reaction sequence:

$$E - Cu^{2+} + O_2^- \rightarrow E - Cu^+ + O_2$$

$$E - Cu^+ + O_2^- \xrightarrow{+2H^+} E - Cu^{2+} + H_2O_2.$$

In contrast to the above, Zn^{2+} has a catalytic role in the reaction catalysed by **carboxypeptidase A**, where the C-terminal amino-acid of a polypeptide is removed, provided it has a non-polar side chain. Carboxypeptidase A from bovine pancreas is a monomeric enzyme which contains one atom of zinc. X-ray crystallography studies (by Lipscomb and colleagues, 1967) and the determination of the complete amino-acid sequence (by Bradshaw and co-workers, 1969) have shown that the active site contains the co-ordinated Zn^{2+} ion bound to histidine-69, glutamate-72, histidine-196 and H_2O, as well as a groove for the polypeptide substrate and a hydrophobic pocket for binding the side chain of the C-terminal amino-acid (Fig. 11.3); the terminal carboxyl group of the substrate forms an electrostatic interaction with arginine-145.

During the reaction, the carbonyl oxygen of the peptide bond being hydrolysed replaces the water molecule bound to Zn^{2+}; the metal ion probably facilitates cleavage of the peptide bond by withdrawing electrons from this carbonyl group. Vallee (1964) proposed a general mechanism for the reaction which involved acid and base groups on the enzyme; x-ray diffraction studies have shown that tyrosine-248 is located in such a position in the enzyme-substrate complex that it could donate a proton to the nitrogen of the peptide

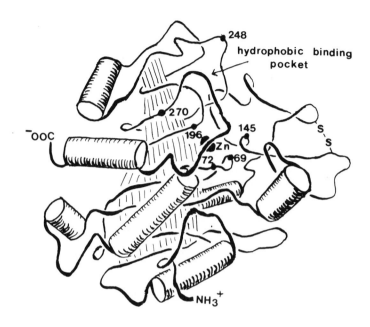

Fig. 11.3 A simplified representation of the three-dimensional structure of carboxypeptidase A, as revealed by the x-ray diffraction studies of Lipscomb and colleagues (1967). (Conventions as for Fig. 2.8.)

bond being hydrolysed and there is also evidence that the carboxyl group of glutamate-270 acts as a general base catalyst to make the attacking water molecule nucleophilic. Thus the reaction appears to involve:

where R″ is the side chain of the C-terminal amino-acid, and R′ represents the rest of the peptide molecule.

X-ray diffraction studies have also shown that substrate-binding results in arginine-145, glutamate-270 and especially tyrosine-248 moving to new positions close to the substrate, forcing water molecules out of the active site and thus creating a hydrophilic environment; the substrate is so tightly enclosed that it would not have been able to get into this position but for the mechanism of the conformational change.

11.5 THE INVOLVEMENT OF COENZYMES IN ENZYME-CATALYSED REACTIONS

11.5.1 Introduction
Coenzymes are organic compounds required by many enzymes for catalytic activity; they are often vitamins, or derivatives of vitamins. Sometimes they can act as catalysts in the absence of enzymes, but not so effectively as in conjunction with an enzyme.

As with metal-enzyme linkages, there is a range of bond strengths for coenzyme-enzyme links, the point of distinction between tightly-bound cofactor (prosthetic group) and loosely-bound cofactor being arbitrary. Coenzymes which are prosthetic groups form an integral part of the active site of an enzyme and undergo no net change as a result of acting as a catalyst; loosely-bound coenzymes can be regarded as **co-substrates** since they often bind to the enzyme-protein together with the other substrates at the start of a reaction and are released in an altered form at the end of it. They are regarded as coenzymes since they usually bind to the enzyme before the other substrates are bound, since they participate in many reactions, and since they may be reconverted to their original form by many enzymes present within cells.

Some important coenzymes are discussed below.

11.5.2 Nicotinamide nucleotides (NAD⁺ and NADP⁺)
These are derived from the vitamin **niacin**, which is nicotinamide or nicotinic acid. The structure of nicotinamide adenine dinucleotide (NAD^+) in its oxidised and reduced forms is given below:

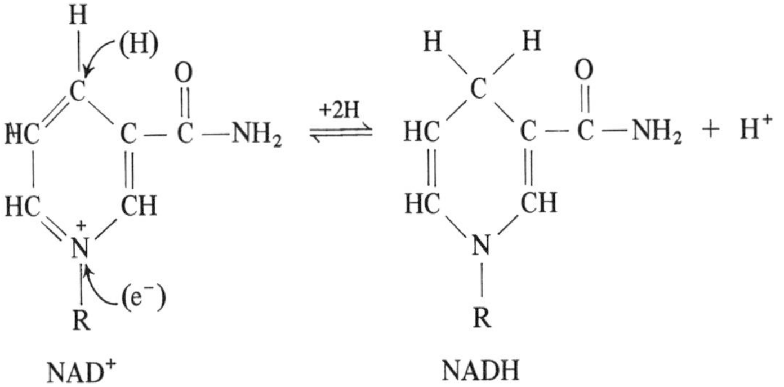

$$NAD^+ \quad\quad\quad\quad NADH$$

where $-R$ represents -D-ribose-phosphate-phosphate-D-ribose-adenine.

It can be seen that the reduction of NAD^+ to NADH requires two reducing equivalents per molecule; one electron (e^-) and one hydrogen atom ($H = H^+ + e^-$) add to the pyridine ring of nicotinamide as shown. The pyridine ring is conjugated, so the positive charge may be delocalised, making several points vulnerable to nucleophilic attack. However, the exact mechanism of the reaction is not known.

Nicotinamide adenine dinucleotide phosphate ($NADP^+$) is identical to NAD^+, except that the $2'$-position of the D-ribose unit attached to adenine is phosphorylated. This does not affect the oxidation/reduction properties, but results in NAD^+ and $NADP^+$ acting as coenzymes for different enzymes: enzymes utilising NAD^+ usually have a catabolic function, the NADH produced being an energy source for the cell; anabolic enzymes, in contrast, frequently involve NADPH as coenzyme.

The names and abbreviations given above are those currently recommended by the International Union of Biochemistry: NAD^+ has also been known as diphosphopyridine nucleotide (DPN^+) or Coenzyme I; $NADP^+$ as triphosphopyridine nucleotide (TPN^+) or Coenzyme II.

Needless to say, NAD^+ and $NADP^+$ act as coenzymes for oxidation/reduction reactions; they are loosely bound, and leave the enzyme in a changed form at the end of the reaction.

The kinetic mechanism of horse liver **alcohol dehydrogenase**, an enzyme which involves NAD^+ as coenzyme in the catalytic oxidation of primary or secondary alcohols, was discussed in Chapter 9.4. It was noted that NAD^+ is the first substrate to be bound and NADH the last product to leave, the dissociation of NADH from the enzyme being the rate-limiting step of the overall reaction; this is one reason why NAD^+ is regarded as a coenzyme rather than simply as a substrate. Horse liver alcohol dehydrogenase is a dimer, each sub-unit containing one binding site for NAD^+ and two sites for Zn^{2+}; only one of these zinc ions is directly involved in catalysis. From results of x-ray diffraction studies, Bränden and colleagues (1975) have deduced that the ternary complex may be of the form:

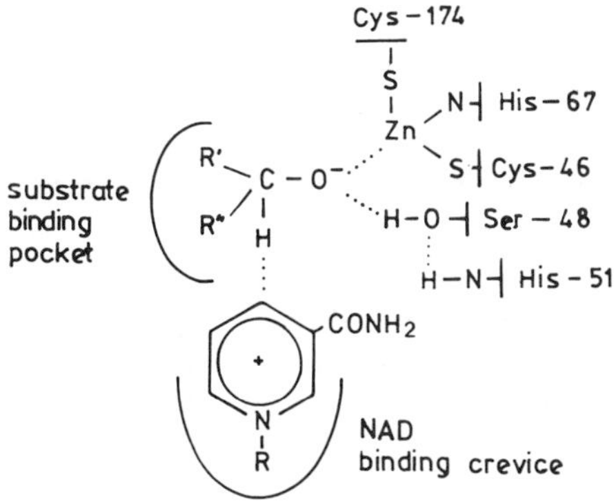

and the reaction mechanism may involve:

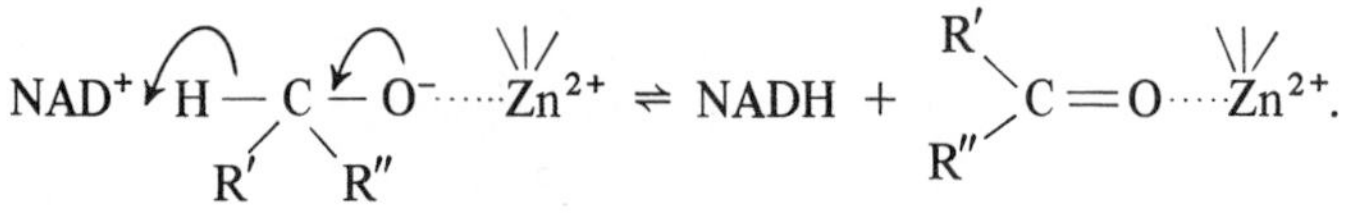

$$\text{NAD}^+ \overset{\curvearrowright}{\text{H}} - \text{C} \overset{\curvearrowleft}{-} \text{O}^- \cdots \overset{\backslash/}{\text{Zn}^{2+}} \rightleftharpoons \text{NADH} + \underset{\text{R}''}{\overset{\text{R}'}{\diagdown}} \text{C}=\text{O} \cdots \overset{\backslash/}{\text{Zn}^{2+}}.$$

Dogfish muscle **lactate dehydrogenase**, which catalyses the reaction:

$$\text{CH}_3.\underset{\underset{\text{OH}}{|}}{\text{CH}}.\text{CO}_2^- + \text{NAD}^+ \rightleftharpoons \text{CH}_3.\underset{\underset{\text{O}}{||}}{\text{C}}.\text{CO}_2^- + \text{NADH} + \text{H}^+$$

L–lactate   pyruvate

has also been investigated in detail. It is a tetrameric enzyme, each sub-unit having a binding site for NAD^+; no metal ions are bound. Like alcohol dehydrogenase, the reaction has a compulsory-order mechanism, the coenzyme binding first and bringing about a conformational change in the enzyme which enables the substrate to bind. From the studies of Adams and co-workers (1973) it seems that the reaction mechanism involves:

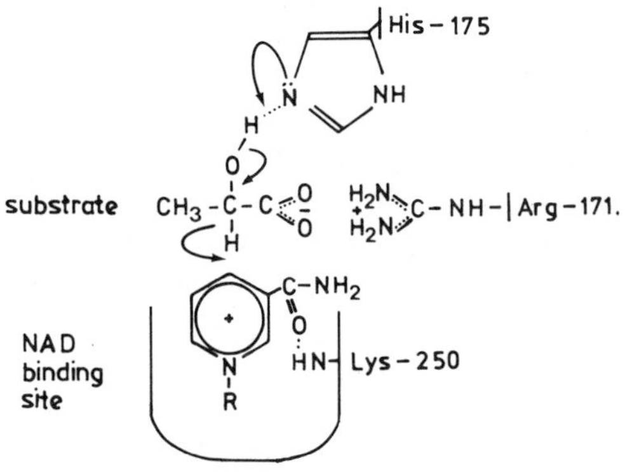

Arginine–171 binds the carboxyl group of the substrate by electrostatic inter-action and histidine–195 acts as an acid-base catalyst, removing a proton from lactate during its oxidation (see also Fig. 12.7).

11.5.3   Flavin Nucleotides (FMN and FAD)

Flavin nucleotides are derived from riboflavin, vitamin B_2; like the nicotinamide nucleotides, they function in oxidation/reduction reactions, the reducing equivalents being carried by the fused three-ringed system of flavin as shown below:

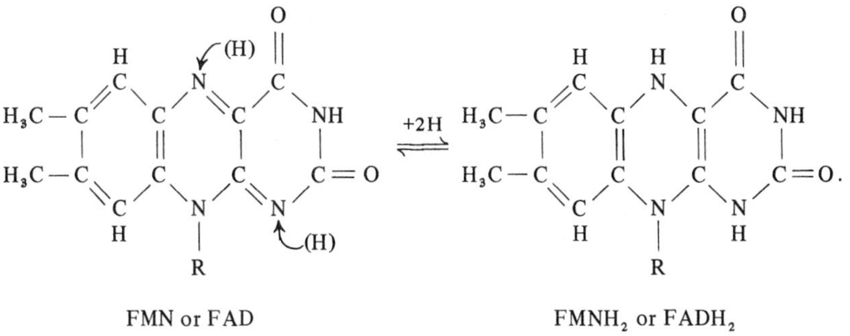

FMN or FAD   FMNH$_2$ or FADH$_2$

For flavin mononucleotide (FMN), $-R$ = -ribitol-phosphate; for flavin adenine dinucleotide (FAD), $-R$ = -ribitol-phosphate-phosphate-D-ribose–adenine.

In contrast to NAD$^+$ and NADP$^+$, FMN and FAD are prosthetic groups and cannot be separated from the protein without denaturing it: the protein–flavin nucleotide complex is termed a **flavoprotein**.

Because the flavin nucleotide does not have an independent existence, reactions catalysed by flavoproteins usually involve the transfer of reducing equivalents from a donor via the flavin to some specific external acceptor. For example, **glucose oxidase**, which catalyses the reaction

$$\text{D-glucose} + O_2 \rightleftharpoons \text{D-glucono-}\delta\text{-lactone} + H_2O_2$$

utilises FAD as prosthetic group and O_2 as hydrogen acceptor. Bright and Gibson (1967) have shown that this is a two-stage reaction:

$$\text{E-FAD} + \text{D-glucose} \rightleftharpoons \text{E-FADH}_2 + \text{D-glucono-}\delta\text{-lactone}$$

$$\text{E-FADH}_2 + O_2 \rightleftharpoons \text{E-FAD} + H_2O_2.$$

With some flavoproteins, the reduction of the flavin has been shown to be a two-step process, involving an unstable free radical **semiquinone** as intermediate:

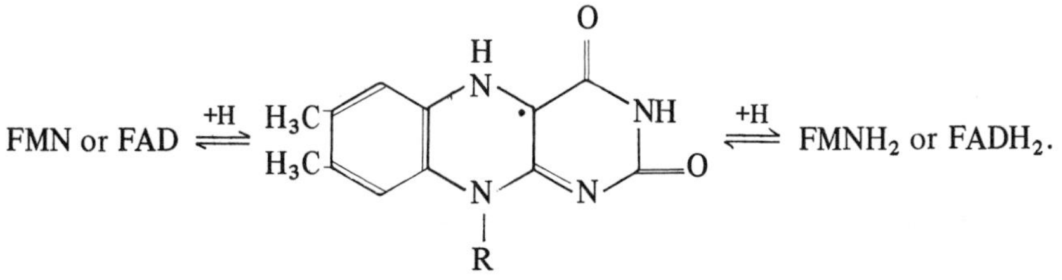

flavosemiquinone
(FMNH· or FADH·)

The unpaired electron, which can be delocalised about the ring system, has been revealed by esr studies. Many flavoproteins are also metalloproteins and one of the roles of the metal ion could be to stabilise this semiquinone.

However, it seems likely that not all reactions involving flavin nucleotides as coenzymes proceed via an identical mechanism: the reaction catalysed by **NADH dehydrogenase** has been shown to involve semiquinone formation, but that catalysed by **glucose oxidase** (discussed above) apparently does not. Similarly, Massey and his colleagues (1973) have pointed out that the reoxidation of E-FADH$_2$ can proceed by a variety of mechanisms: where molecular oxygen is the acceptor, the products may be H_2O_2 (with oxidases), H_2O and hydroxylated products (with hydroxylases), or the superoxide anion (O_2^-.) and flavin semiquinone.

11.5.4 Adenosine Phosphates (ATP, ADP and AMP)

The nucleoside phosphates ATP, ADP and AMP are involved in phosphate transfer reactions.

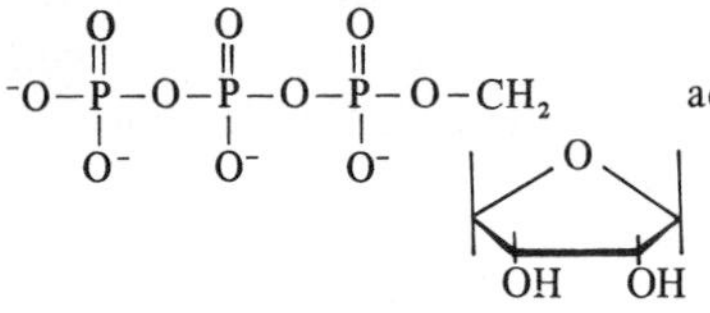

ATP 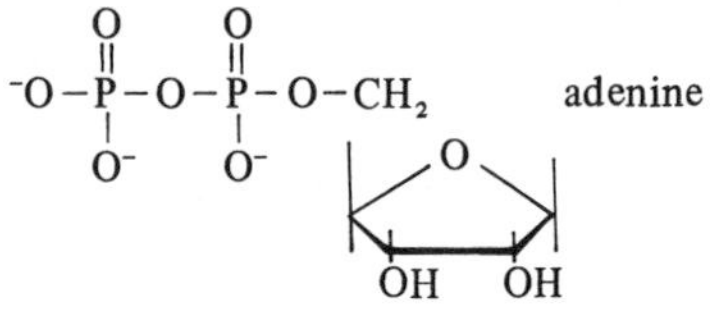   ADP

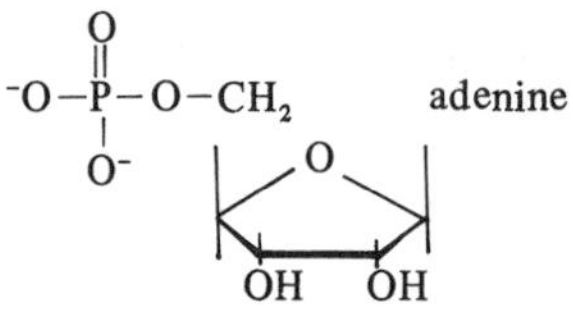

AMP

ATP and ADP may be interconverted by the reaction

$$ATP + H_2O \rightleftharpoons ADP + P_i.$$

This tends to go strongly in the forward direction as written ($\Delta G^{\ominus\prime} = -31.0$ kJ.mol^{-1}) because the four negative charges which are in close proximity on ATP make it an unstable molecule (although the product, ADP, itself has three negative charges close together), and because the reverse reaction requires negatively charged ADP to react with negatively charged P_i. The reaction can be coupled to others, so that phosphate may be transferred between ATP and other organic compounds without ever being present as free P_i (Chapter 6.2.5).

The importance of ATP in energy metabolism is that, by comparison to other organic phosphates, it is only moderately unstable. Hence it may be synthesised by the transfer of phosphate to ADP from a more unstable organic phosphate (e.g. phosphoenolpyruvate) by substrate level phosphorylation, or by oxidative phosphorylation; however ATP is sufficiently unstable to be able to force the transfer of the phosphate to a whole variety of other compounds, thus driving such processes as biosynthesis, active transport and muscular contraction.

In the cell, adenosine phosphates are stabilised by binding to Mg^{2+} ions, and their metabolism is strictly mediated by enzymes (see Chapter 11.4.3); in some instances two phosphate groups, rather than one, may be removed from ATP to liberate inorganic pyrophosphate:

$$ATP + H_2O \rightleftharpoons AMP + (PP)_i.$$

Adenosine phosphates, like the nicotinamide nucleotides, are loosely bound by enzymes and may be regarded both as coenzymes and as co-substrates/co-products of the reactions in which they participate.

11.5.5 Coenzyme A (CoA.SH)

Coenzyme A has the structure:

$$HS.CH_2CH_2NH\text{-pantothenic acid-}OPO_3^-.PO_3^-\text{-ribose-3-P-adenine.}$$

With carboxylic acids it can form thioesters:

$$RCO_2H + HS.CoA \rightleftharpoons R.COSCoA + H_2O.$$
$$\text{acyl-CoA}$$

These thioesters are of great importance in biochemical metabolism since they can be attacked by electrophiles (including other acyl-CoA molecules and CO_2) to form addition compounds, and by nucleophiles (including water) to displace the $-SCoA$ group:

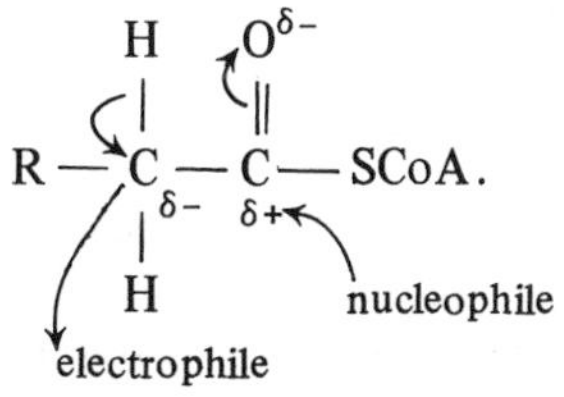

$$R-\overset{\overset{\displaystyle H}{|}}{\underset{\underset{\displaystyle H}{|}}{C}}-\overset{\overset{\displaystyle O^{\delta-}}{\|}}{C}-SCoA.$$

electrophile   nucleophile

Some examples are given in Chapters 11.5.6, 11.5.8 and 11.5.10.

11.5.6 Thiamine Pyrophosphate (TPP)

Thiamine pyrophosphate is derived from vitamin B_1, thiamine, and has the structure:

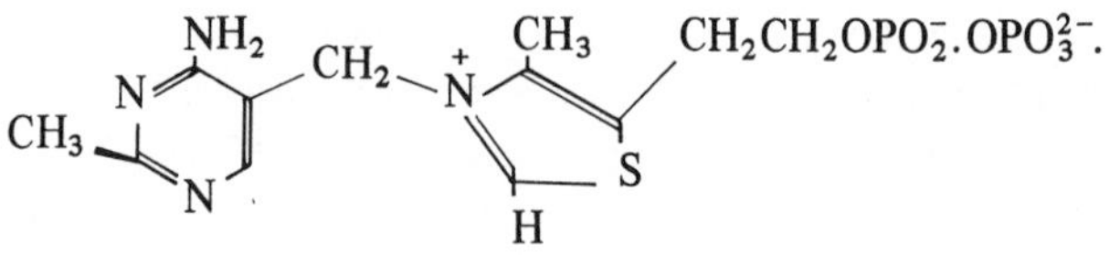

The **thiazole ring** can lose a proton to produce a negatively charged carbon atom:

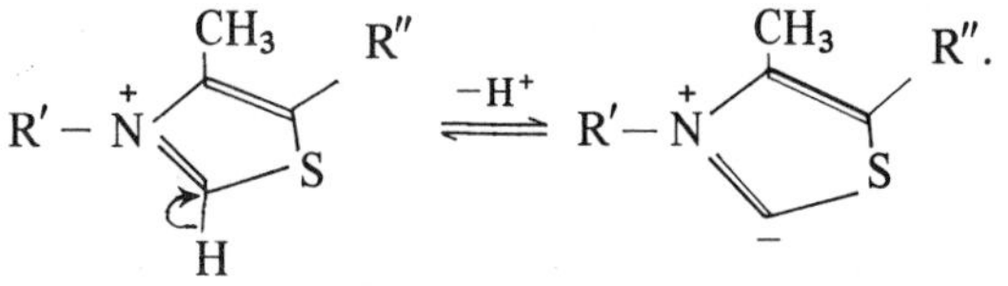

This is a potent nucleophile and can participate in covalent catalysis, particularly with α-keto (oxo) acid decarboxylase, α-keto acid oxidase, transketolase and phosphoketolase enzymes.

For example, **pyruvate decarboxylase**, found in yeast and some other micro-organisms, utilises TPP to catalyse the production of acetaldehyde from pyruvate. Breslow (1957) proposed the following reaction mechanism:

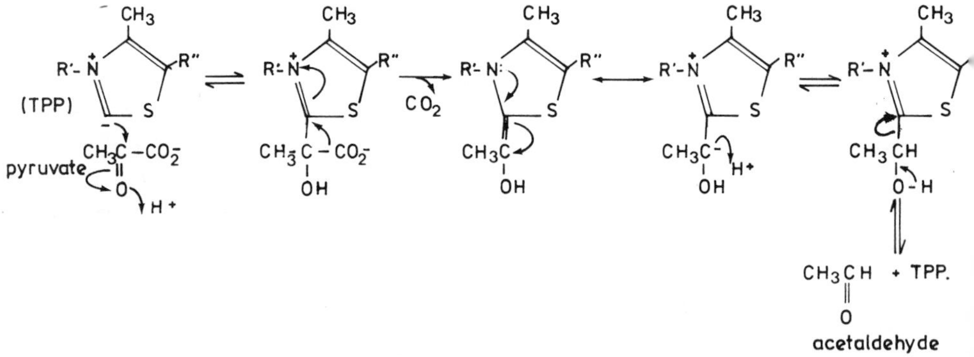

The actual decarboxylation step is facilitated by electrophilic catalysis as the thiazole ring withdraws electrons. The reaction will proceed in the absence of enzyme, but the acetaldehyde formed tends to react with the $TPP-C^--OH$ complex to produce acetoin as the final product. Juni (1961) has suggested that the enzyme stabilises the TPP-acetaldehyde complex and prevents this condensation occurring.

The multienzyme complex known as **pyruvate dehydrogenase** (see Chapter 5.2.5) also catalyses the decarboxylation of pyruvate, but it utilises a second coenzyme, **lipoic acid**, to introduce an oxidation step and a third coenzyme, **coenzyme A (CoA.SH)** to react with the acetyl–lipoamide complex, giving acetyl-CoA as the final product. The TPP–C^-— OH complex is formed as above,

$$\text{CH}_3$$

and then the reaction is thought to proceed as follows:

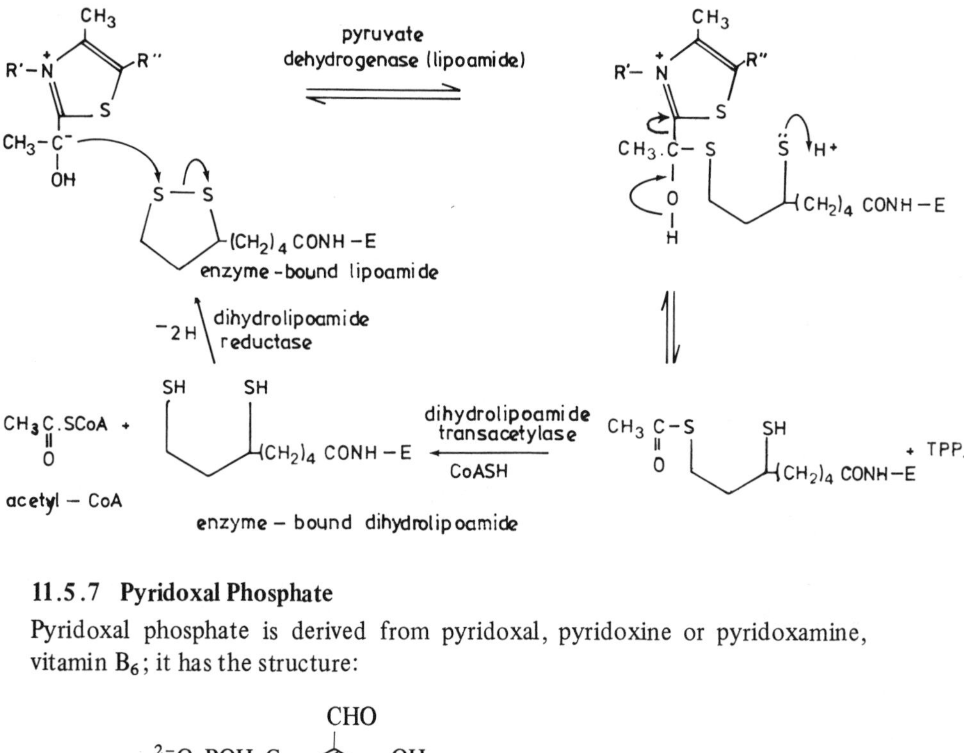

11.5.7 Pyridoxal Phosphate

Pyridoxal phosphate is derived from pyridoxal, pyridoxine or pyridoxamine, vitamin B_6; it has the structure:

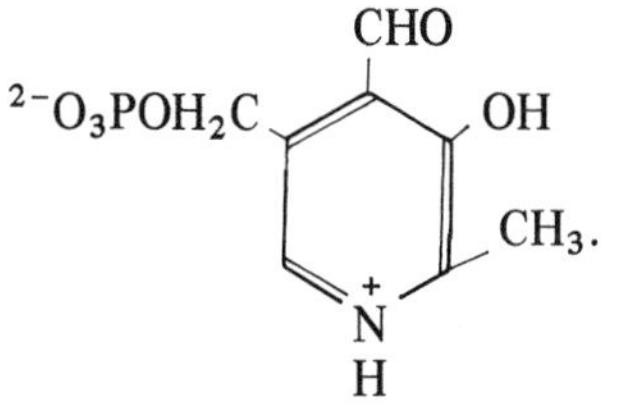

The coenzyme is important in amino-acid metabolism, being involved in aminotransferase (transaminase), decarboxylase and racemase reactions. There is much evidence to suggest that in each case a **Schiff's base** linkage ($-N=CH-$) is formed, involving the aldehyde group of the coenzyme (as suggested by Braunstein and Snell, 1953). The phenolic group of the coenzyme is also important, as it may help to stabilise the Schiff's base intermediate:

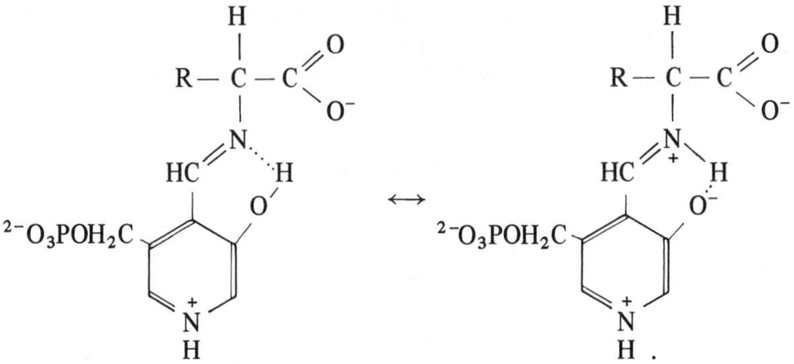

A conjugated double bond system links the pyridine ring with the substrate, and the positively-charged nitrogen atoms tend to withdraw electrons from the α-carbon of the amino-acid, weakening its bonding to R, H and CO_2^-. Any of these three bonds may thus be cleaved as a result of this electrophilic catalysis to form an anion which is stabilised by the conjugated system.

The simplest mechanism to consider is that of the **amino-acid racemase** enzymes, where a racemic mixture of D- and L-stereoisomers is produced from a single stereoisomeric form. The coenzyme binds to the enzyme by a Schiff's base linkage to a side chain amino group of a lysine residue and also by an electrostatic linkage between the coenzyme-phosphate and a positively-charged group on the enzyme. When the amino-acid substrate binds to the enzyme, the carbon of the Schiff's base undergoes electrophilic attack by the α-amino group of the substrate and a new Schiff's base linkage is formed, this time between coenzyme and substrate. The weak C—H bond in the complex is then broken to release a proton:

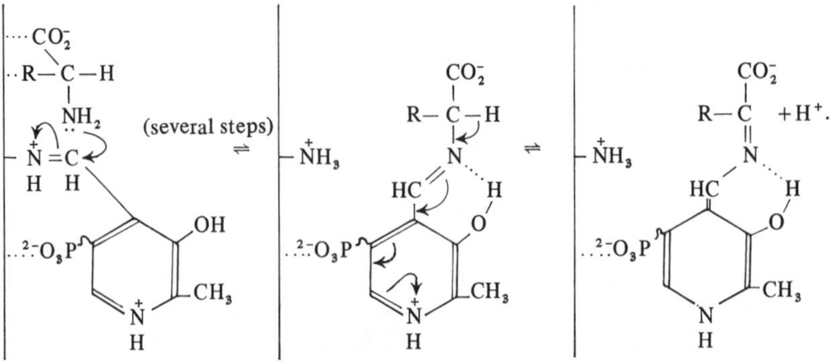

The whole process is reversible, so a proton could add on to the complex and re-form the free amino-acid. However the initial loss of the proton also results in the loss of asymmetry of the α-carbon atom, so the amino-acid produced is not necessarily the same stereoisomer as that present at the start.

Aminotransferase (transaminase) reactions proceed as above, with loss of a proton from the Schiff's base. However, the resulting complex is then attacked in a different place by a proton and undergoes hydrolysis to form pyridoxamine phosphate, with the liberation of an oxo acid:

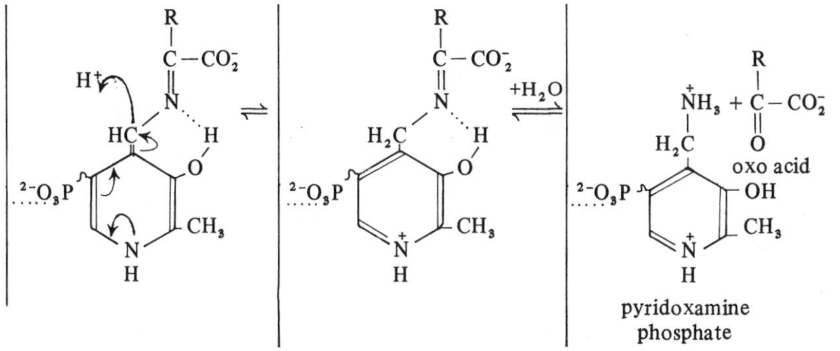

pyridoxamine phosphate

The reverse sequence may then take place, initiated by the attack on pyridoxamine phosphate by a different oxo acid, and so producing a different amino-acid from that initially present. A mechanism of this type has been established for **aspartate aminotransferase** by Jenkins and co-workers (1966). The two-stage reaction sequence is:

L-glutamate + E-pyridoxal-P $\rightleftharpoons$ 2-oxoglutarate + E-pyridoxamine-P

oxaloacetate + E-pyridoxamine-P $\rightleftharpoons$ L-aspartate + E-pyridoxal-P.

Kinetic studies have indicated a ping-pong bi-bi mechanism, which is consistent with the above chemical mechanism.

Decarboxylase reactions involving amino-acid substrates commence in much the same way, but the Schiff's base complex loses the α-carboxyl group rather than the hydrogen atom; the resulting complex is then hydrolysed to reform the coenzyme and give an amine (RCH_2NH_2) as product.

11.5.8 Biotin

Biotin, like lipoic acid, is always found firmly bound to a side chain amino group of one of the lysine residues of a protein.

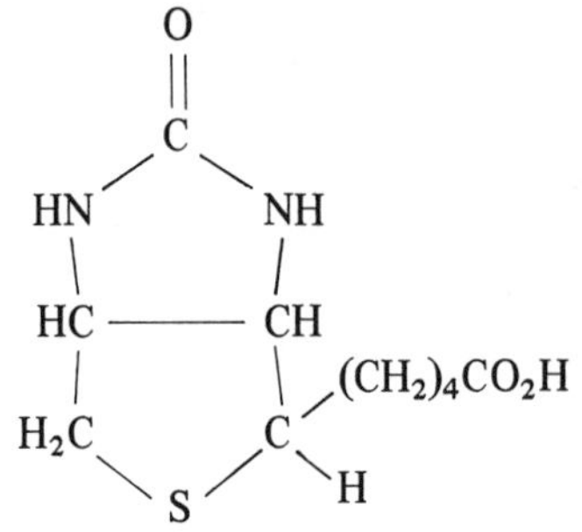

free biotin 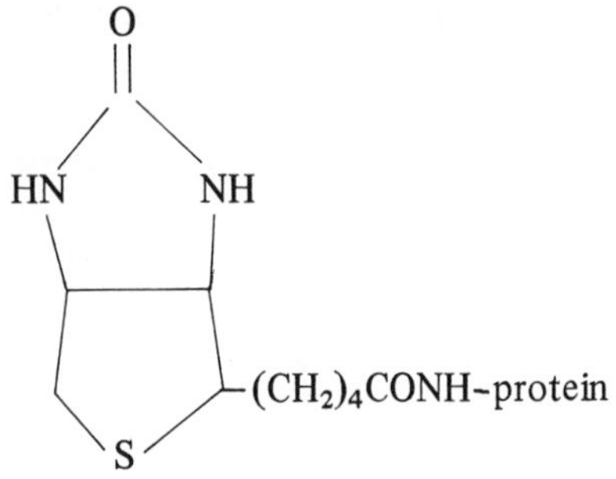   protein-bound biotin

Protein-bound biotin can link with CO_2 to form a biotin carboxyl carrier protein (BCCP), which is important in carboxylation reactions, e.g. that catalysed by acetyl-CoA carboxylase.

Acetyl-CoA carboxylase from *E. coli* has been shown, by Lane and co-workers (1971) and Vagelos and colleagues (1973), to dissociate into three distinct sub-units: one is the BCCP, another is biotin carboxylase, which catalyses the reaction:

$$ATP + HCO_3^- + BCCP \rightleftharpoons BCCP - CO_2^- + ADP + P_i$$

and the third is carboxyltransferase, which mediates the reaction:

$$BCCP - CO_2^- + acetyl\text{-}CoA \rightleftharpoons malonyl\text{-}CoA + BCCP.$$

The BCCP appears to act as a flexible arm, transporting the CO_2^- from the active site of biotin carboxylase to that of carboxyltransferase, where it is presented to the acetyl-CoA. The chemical mechanism has not been established beyond doubt, but it may involve:

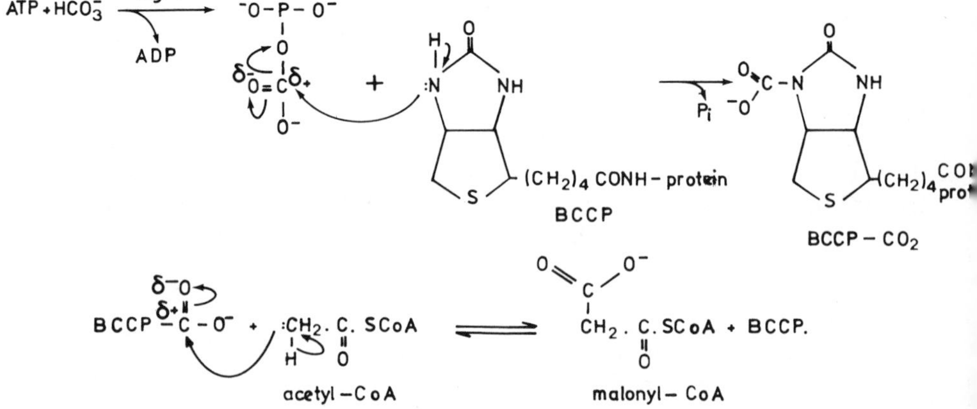

11.5.9 Tetrahydrofolate

Tetrahydrofolate (FH_4) is derived from the vitamin folic acid and has the structure:

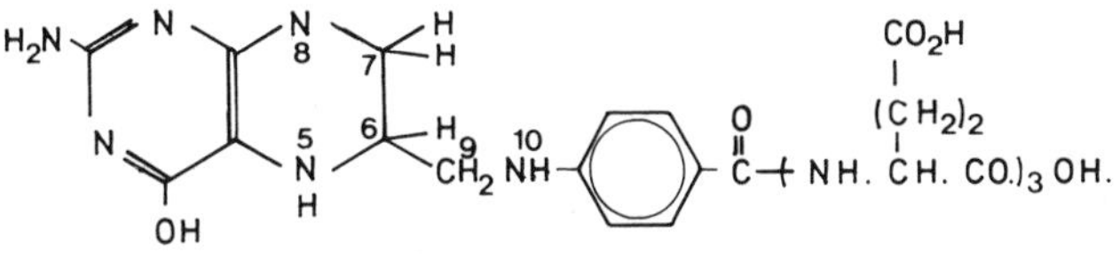

Atoms N5 and N10 can carry **one-carbon units** for transfer to a suitable acceptor. **Formyltetrahydrofolate synthetase** catalyses the addition of formate to FH_4 by the reaction:

$$ATP + formate + FH_4 \xrightleftharpoons{Mg^{2+}} 10\text{-formyl-}FH_4 + ADP + P_i.$$

The mechanism for this is likely to resemble that for the carboxylation of biotin (Chapter 11.5.8), except that it involves addition of a formyl group and not a carboxyl group. 10-Formyl-FH_4 can then react further to produce a variety of one-carbon groups carried by the coenzyme:

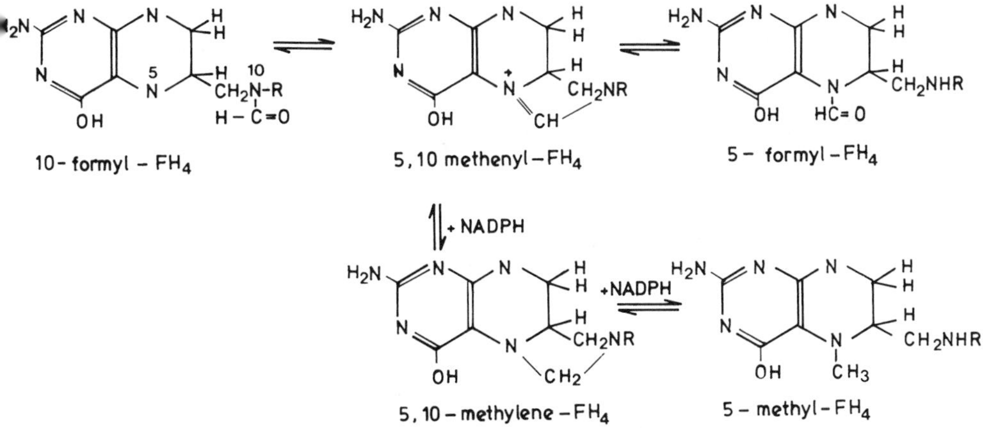

These, and other, one-carbon units may then be transferred to an acceptor. Consider, for example, the reaction catalysed by **serine transhydroxymethylase**:

$$\underset{\text{glycine}}{H_2N.CH_2CO_2H} + 5,10\text{-methylene-}FH_4 \underset{+H_2O}{\rightleftharpoons} \underset{\text{L-serine}}{H_2N\overset{\overset{\displaystyle CH_2OH}{|}}{C}H.CO_2H} + FH_4.$$

Pyridoxal phosphate is also involved and, as is usual with reactions involving this coenzyme (see Chapter 11.5.7), the substrate (glycine) binds by a Schiff's base linkage and then goes on to lose a proton from the α-carbon atom; the methylene-FH_4 complex then presents the one-carbon unit in a suitable orientation and the reaction is completed.

11.5.10 Coenzyme B_{12}

Hodgkin (1956) showed that **cobalamin (vitamin B_{12})** has a central core of a cobalt ion surrounded by a **corrin ring** (Fig. 11.4): the metal ion is linked to the four nitrogen atoms of the ring by one covalent and three co-ordinate bonds,

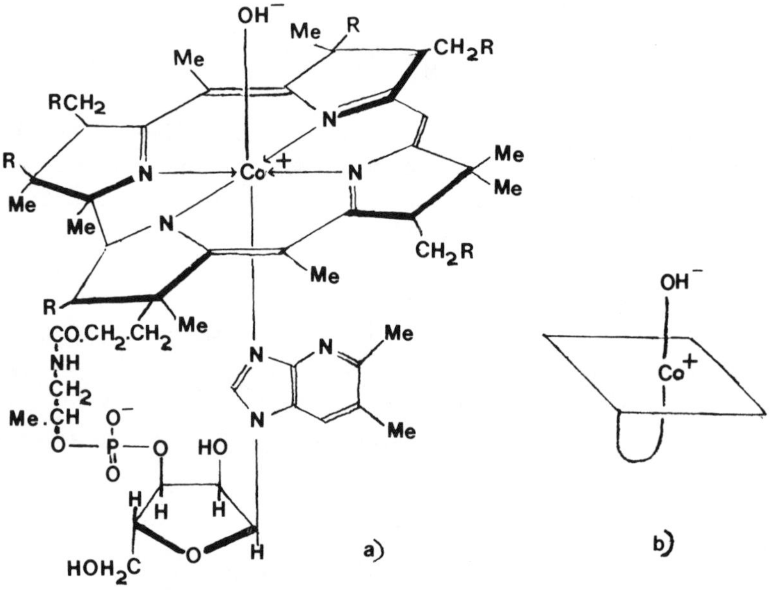

Fig. 11.4 Vitamin B_{12} (hydroxocobalamin): (a) complete structure ($-R = -CH_2CONH_2$, $-Me = -CH_3$); (b) simplified representation.

but, because of resonance, the four bonds are almost equivalent. One of the pyrrole components of the corrin ring is joined to dimethylbenzimidazole, which co-ordinates Co from "below" the ring; the sixth co-ordination site is believed to be occupied *in vivo* by OH^-. In this naturally occurring compound, **hydroxocobalamin (B_{12a})**, the metal ion has an oxidation state of +3, but possesses only a single positive charge because of its interactions with various groups; this is indicated as $Co(III)^+$.

The coenzyme forms are obtained after activation of the vitamin by NADH-linked reducing systems:

$$Co(III)^+ \xrightarrow[-OH]{e} \dot{C}o(II) \xrightarrow{e} \ddot{C}o(I).$$

$$B_{12a} \qquad\qquad B_{12r} \qquad\quad B_{12s}$$

The Co(I) in B_{12s} is strongly nucleophilic and can be alkylated, a bond of largely covalent character being formed. Reaction with ATP, mediated by an adenosyltransferase, results in the formation of **5-deoxyadenosylcoenzyme B_{12}** (Fig. 11.5(a)), also called **5′-deoxyadenosylcobalamin**. Alternatively, methylation may take place, with the formation of **methylcobalamin** (Fig. 11.5(b)).

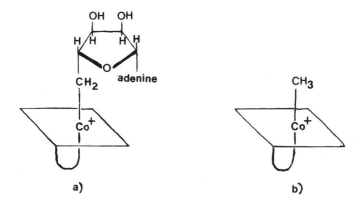

Fig. 11.5   Simplified representation of the B_{12} coenzymes: (a) 5′-deoxyadenosylcobalamin; (b) methylcobalamin (for detailed structure, see Fig. 11.4).

5′-Deoxyadenosylcobalamin acts as a coenzyme for a variety of transfer reactions, usually those of an intramolecular kind, in animals and bacteria. An example is the reaction catalysed by **methylmalonyl-CoA mutase**:

$$H_2\overset{\frown H}{C} - CH.CH_2CO_2H \rightarrow H_3C.CH.CH_2CO_2H.$$

propionyl-CoA   methylmalonyl-CoA

There is evidence, as with all reactions of this type, that the hydrogen atom is transferred via the coenzyme.

Methylcobalamin acts as a coenzyme in several methyl transfer reactions, e.g. in the conversion of homcysteine to methionine in some bacteria and mammals:

$$\text{N5-methyl-FH}_4 + \overset{\overset{+}{N}H_3}{\underset{\underset{\text{homocysteine}}{|}}{\text{HS.CH}_2\text{CH}_2.\text{CHCO}_2^-}} \rightarrow \text{FH}_4 + \overset{\overset{+}{N}H_3}{\underset{\underset{\text{methionine}}{|}}{\text{H}_3\text{C.S.CH}_2\text{CH}_2\text{CHCO}_2^-}}.$$

The methyl group is transferred via the cobalt ion of B_{12}: in other bacteria, as well as in fungi and higher plants, this reaction does not involve B_{12}.

The detailed mechanisms of action of 5'-deoxyadenosyl- and methyl-cobalamin are not known for certain; they may well involve further changes in the oxidation state of cobalt.

SUMMARY OF CHAPTER 11

Enzyme-catalysis can include, within a single reaction mechanism, acid-base, electrostatic and covalent catalysis as well as proximity effects, orbital steering and stress/strain factors. In this way the total free energy of activation required for the reaction to proceed is reduced and also spread out over several stages. The free energy requirement for each stage may be provided in some way from the energy made available by collisions between the enzyme–substrate complex and solute molecules.

Some enzymes can catalyse reactions without requiring cofactors. Other enzymes require the presence of metal ions for full catalytic activity: these ions can play a structural role or act as catalysts in a variety of ways. Coenzymes are also required by many enzymes, in some cases in addition to metal ions: coenzymes usually play a catalytic role.

FURTHER READING

As for Chapter 10, plus:

Boyer, P. D. (1970), *The Enzymes,* 2, 3rd edition, Academic Press.

Muirhead, H. (1983), Triose phosphate isomerase, pyruvate kinase and other α/β-barrel enzymes, *Trends in Biochemical Sciences,* 8 (pages 326–330).

Sykes, P. (1981), *A Guidebook to Mechanism in Organic Chemistry,* 5th edition, Longman.

Walsh, C. T. (1979), *Enzymatic Reaction Mechanisms,* Freeman.

Walsh, C. T. (1984), Enzyme mechanisms, *Trends in Biochemical Sciences,* 9 (pages 159–162).

CHAPTER 12

The Binding of Ligands to Proteins

12.1 INTRODUCTION

In this chapter we will discuss the ·binding of ligands to monomeric and oligomeric proteins. Anything which binds to an enzyme or other protein is a ligand, regardless of whether or not it is a substrate and undergoes a subsequent reaction. Here, in general, we will be considering binding processes where no subsequent reaction is taking place, e.g. the binding to a protein of a non-substrate, or of a substrate for a two-substrate reaction in the absence of the second substrate. However, we will briefly consider what effects the binding characteristics might have on the kinetics of a subsequent reaction. We will also take into consideration the possibility of interaction between binding sites, particularly in the case of oligomeric proteins where there are several identical binding sites for the same ligand (i.e. one on each identical sub-unit).

12.2 THE BINDING OF A LIGAND TO A PROTEIN HAVING A SINGLE LIGAND-BINDING SITE

Consider the binding of a ligand (S) to a protein (E), according to the following reaction:

$$E + S \rightleftharpoons ES.$$

The **binding constant**, $K_b = \dfrac{[ES]}{[E][S]}$ (note that $K_b = \dfrac{1}{K_s}$).

The **fractional saturation**, $\overline{Y}$, of the protein is given by:

$$\overline{Y} = \frac{[ES]}{[E_o]} = \frac{[ES]}{[ES] + [E]} = \frac{K_b[E][S]}{K_b[E][S] + [E]} = \frac{K_b[S]}{K_b[S] + 1}.$$

From this, it can be seen that a plot of $\overline{Y}$ against [S], at constant $[E_o]$, will be hyperbolic (Fig. 12.1).

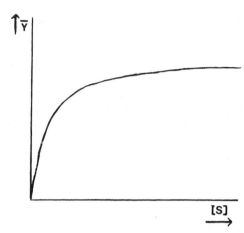

Fig. 12.1 Graph of fractional saturation ($\bar{Y}$) against ligand concentration ([S]) at fixed concentration of a protein having a single binding site for S.

Let us now consider the situation where the binding of S to E is the first step in a process whereby a product P is formed. If the reaction proceeds under steady-state conditions, where $[S_o] \gg [E_o]$ and $[S] \simeq [S_o]$, then [ES] does not vary with time and, in the most straightforward system, v_o is proportional to [ES]. Under these conditions, $\dfrac{V_o}{V_{max}} = \dfrac{[ES]}{[E_o]} = \bar{Y}$, so a graph of v_o against $[S_o]$ will be the same shape as that of $\bar{Y}$ against [S], i.e. hyperbolic. This hyperbolic relationship between v_o and $[S_o]$ under steady-state conditions is, of course, predicted by the Michaelis–Menten equation (see Chapters 7.1.1 and 7.1.2).

If, on the other hand, the reaction proceeds in a way which is not consistent with all of the assumptions made in the derivation of the Michaelis–Menten equation, then the kinetic characteristics of the reaction will not usually run parallel to the binding characteristics.

12.3 COOPERATIVITY

If more than one ligand-binding site is present on a protein, there is a possibility of interaction between the binding sites during the binding process. This is termed **cooperativity**.

Positive cooperativity is said to occur where the binding of one molecule of a substrate or ligand *increases* the affinity of the protein for other molecules of the same or different substrate or ligand.

Negative cooperativity occurs when the binding of one molecule of a substrate of ligand *decreases* the affinity of the protein for other molecules of the same or different substrate or ligand.

Homotropic cooperativity occurs when the binding of one molecule of a substrate or ligand affects the binding to the protein of subsequent molecules of the *same* substrate or ligand (i.e. the binding of one molecule of A affects the binding of further molecules of A).

Heterotropic cooperativity occurs when the binding of one molecule of a substrate or ligand affects the binding to the protein of molecules of a *different* substrate or ligand (i.e. the binding of one molecule of A affects the binding of B).

Cooperative effects may be positive and homotropic, positive and heterotropic, negative and homotropic or negative and heterotropic. **Allosteric inhibition** (Chapter 8.2.7) is an example of negative heterotropic cooperativity and **allosteric activation** an example of positive heterotropic cooperativity.

12.4 POSITIVE HOMOTROPIC COOPERATIVITY AND THE HILL EQUATION

Let us consider the simplest case of positive homotropic cooperativity in a dimeric protein: there are two identical ligand-binding sites, and when the ligand binds to one, it increases the affinity of the protein for the ligand at the other site, so the reaction sequence is:

$$M_2 + S \xrightarrow{\text{slow}} M_2S$$

$$M_2S + S \xrightarrow{\text{rapid}} M_2S_2$$

(where M is the monomeric sub-unit, termed a **protomer**, and M_2 is the dimeric protein).

If the increase in affinity is sufficiently large, M_2S will react with S almost immediately it is formed; under these conditions, $[M_2S_2] \gg [M_2S]$ and $\overline{Y} = \dfrac{[M_2S_2]}{[(M_2)_0]}$, where $[(M_2)_0]$ is the total concentration of dimer present. Also, a graph of $\overline{Y}$ against $[S]$ will be sigmoidal (S-shaped) rather then hyperbolic (Fig. 12.2).

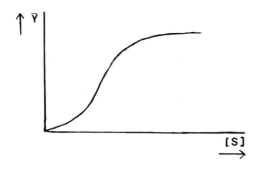

Fig. 12.2   Graph of $\overline{Y}$ against $[S]$, at fixed protein concentration, where the binding shows positive homotropic cooperativity.

For **complete cooperativity**, where each protein molecule must be either free of ligand or completely saturated, the reaction may be written

$$M_2 + 2S \rightleftharpoons M_2S_2.$$

The binding constant of this reaction is given by the expression

$$K_b = \frac{[M_2S_2]}{[M_2][S]^2}$$

from which

$$\overline{Y} = \frac{K_b[S]^2}{1 + K_b[S]^2}.$$

Alternatively, taking logs,

$$\log K_b + 2\log[S] = \log\left(\frac{[M_2S_2]}{[M_2]}\right) = \log\left(\frac{[M_2S_2]}{[(M_2)_o] - [M_2S_2]}\right).$$

In the general case of complete positive homotropic cooperativity of a protein with n identical binding sites, this becomes

$$\log K_b + n\log[S] = \log\left(\frac{[M_nS_n]}{[(M_n)_o] - [M_nS_n]}\right) = \log\left(\frac{\overline{Y}}{1 - \overline{Y}}\right).$$

This is called the **Hill equation**, after its deriver. If it is obeyed, a graph of $\log(\frac{\overline{Y}}{1-\overline{Y}})$ against $\log[S]$ will be linear with slope $= n$ and intercept $= \log K_b$. Such a graph is called a **Hill plot**, and its experimentally determined slope is known as the **Hill coefficient** and given the general symbol h (Fig. 12.3).

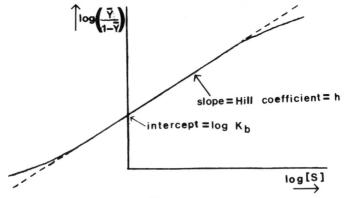

Fig. 12.3 The Hill plot of log $\dfrac{\overline{Y}}{1-\overline{Y}}$ against log[S], at fixed protein concentration, where the binding shows positive homotropic cooperativity.

At values of $\overline{Y}$ below 0.1 and above 0.9, the slopes of Hill plots tend to a value of 1, indicating an absence of cooperativity. This is because at very low ligand concentrations there is not enough ligand present to fill more than one site on most protein molecules, regardless of affinity; similarly at high ligand concentrations there are extremely few protein molecules present with more than one binding site remaining to be filled.

The Hill coefficient is therefore taken to be the slope of the linear, central portion of the graph, where the cooperative effect is expressed to its greatest extent (Fig. 12.3). For systems where cooperativity is complete, the Hill coefficient (h) is equal to the number of binding sites (n). Proteins which exhibit only a partial degree of positive cooperativity may still give a Hill plot with a linear central section, but in such cases h will be less than n, and the linear section is likely to be shorter than that for a system where cooperativity is more nearly complete.

In the case where S is a substrate and the reaction proceeds to yield products in such a way that the Michaelis–Menten equilibrium assumption is valid, then initial velocity is proportional to the concentration of enzyme-bound substrate, i.e. $v_o \propto [MS]$, and $\dfrac{v_o}{V_{max}} = \dfrac{[MS]}{[M_o]} = \overline{Y}$ (where [MS] is the number of substrate-bound sub-units present per unit volume, and $[M_o]$ is the total number of sub-units per unit volume, i.e. $[M_o] = [M] + [MS]$).

Under these conditions, $\dfrac{\overline{Y}}{1-\overline{Y}} = \dfrac{v_o}{V_{max}-v_o}$, so a Hill plot of $\log(\dfrac{v_o}{V_{max}-v_o})$ against $\log[S_o]$ may be substituted for the one shown in Fig. 12.3. The slopes of the two graphs will have the same value and meaning. Note that although the relationship $\overline{Y} = \dfrac{v_o}{V_{max}}$ may be assumed valid for systems involving monomeric enzymes under general steady-state conditions, the same is not true for the more complicated systems involving oligomeric enzymes; in the latter case, $\overline{Y} = \dfrac{v_o}{V_{max}}$ only if the binding process is at or very near equilibrium.

One of the main problems in constructing a Hill plot from kinetic data is to obtain an accurate estimate of V_{max}: this is particularly true for cooperative systems, since the primary plots (Chapters 7.1.4 and 7.1.5) are not linear. Nevertheless, an estimate of V_{max} can be obtained from an Eadie–Holstee or other plot, enabling a Hill plot to be constructed and a Hill coefficient (h) determined. The primary plot can then be redrawn substituting $[S]^h$ for $[S]$, which should give more linear results and a more accurate estimate of V_{max}. If this differs markedly from the initial estimate of V_{max}, the Hill plot should then be redrawn, incorporating the new (and better) estimate of V_{max}.

12.5 THE ADAIR EQUATION FOR THE BINDING OF A LIGAND TO A PROTEIN HAVING TWO BINDING SITES FOR THAT LIGAND

12.5.1 General Considerations

Let us now investigate the binding of a ligand to a protein having a number of identical binding sites for that ligand, making no assumptions at all about cooperativity. The **intrinsic** (or **microscopic**) **binding constant**, K_b, for each site is defined as the binding constant which would be measured if all the other sites on the protein were absent. Since all the sites are identical in the example we are considering, each will have the same K_b. However the **actual**, or **apparent**, **binding constant** for each step of the reaction will not be the same. In the case of a dimeric protein (M_2) having two identical binding sites for a ligand (S):

$$M_2 + S \rightleftharpoons M_2S \quad \text{apparent binding constant} = K_{b1}$$

$$M_2S + S \rightleftharpoons M_2S_2 \quad \text{apparent binding constant} = K_{b2}.$$

Note that K_{b1} and K_{b2} depend solely on the position in the reaction sequence and do not refer to any particular binding site.

Fractional saturation,

$$\overline{Y} = \frac{\text{number of protomers per unit volume which are bound to ligand}}{\text{total number of protomers per unit volume}}$$

$$= \frac{[MS]}{[M_o]} = \frac{[MS]}{[MS + M]}.$$

However, there are no isolated protomers present: they are part of the dimeric protein. Hence it is necessary to express $\overline{Y}$ in terms of the various protein–ligand complexes which are actually present.

The species M_2 consists of 2 protomers, both unbound;

the species M_2S consists of 1 bound and 1 unbound protomer; and

the species M_2S_2 consists of 2 protomers, both bound.

Therefore, the total concentration of ligand-bound protomers present ([MS]) is given by $[MS] = [M_2S] + 2[M_2S_2]$. Similarly, the total concentration of unbound protomers present ([M]) is given by $[M] = 2[M_2] + [M_2S]$.

Also,
$$[MS] + [M] = [M_2S] + 2[M_2S_2] + 2[M_2] + [M_2S]$$

$$= 2([M_2] + [M_2S] + [M_2S_2]).$$

$$\therefore \overline{Y} = \frac{[MS]}{[MS] + [M]} = \frac{[M_2S] + 2[M_2S_2]}{2([M_2] + [M_2S] + [M_2S_2])}.$$

By definition,

$$K_{b1} = \frac{[M_2S]}{[M_2][S]}, \quad \text{so} \quad [M_2S] = K_{b1}[M_2][S].$$

Also, $\quad K_{b2} = \frac{[M_2S_2]}{[M_2S][S]}, \quad \text{so} \quad [M_2S_2] = K_{b2}[M_2S][S] = K_{b1}K_{b2}[M_2][S]^2.$

Substituting for $[M_2S]$ and $[M_2S_2]$ in the expression for $\bar{Y}$ obtained above,

$$\bar{Y} = \frac{K_{b1}[M_2][S] + 2K_{b1}K_{b2}[M_2][S]^2}{2([M_2] + K_{b1}[M_2][S] + K_{b1}K_{b2}[M_2][S]^2)}$$

$$= \frac{K_{b1}[S] + 2K_{b1}K_{b2}[S]^2}{2(1 + K_{b1}[S] + K_{b1}K_{b2}[S]^2)}.$$

This is the **Adair equation** (see Chapter 12.9) for the binding of a ligand to a dimeric protein.

12.5.2 Where There is No Interaction Between the Binding Sites

Let us now look at the relationship between the intrinsic and apparent constants where there is no interaction between the binding sites. We will compare the reaction for the dimer with that for the hypothetical isolated protomer under identical conditions of molar concentration, assuming that each binding site behaves in an identical manner, regardless of its surroundings.

The **first step** in the reaction involving the dimer is

$$M_2 + S \rightleftharpoons M_2S \text{ (binding constant } K_{b1})$$

whereas the reaction for the protomer is

$$M + S \rightleftharpoons MS \text{ (binding constant } K_b).$$

In diagramatic form, these reactions can be written:

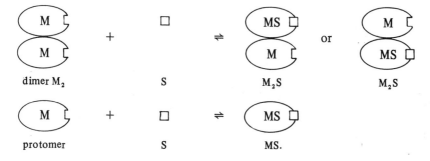

In the forward direction, the dimer has two free binding sites whereas the isolated protomer has only one. Therefore, the ligand is two times more likely to bind to a molecule of the dimer than to a molecule of the isolated protomer. In the reverse direction in both cases there is only one site from which S can dissociate, i.e. that to which it is attached. Hence there is no difference between the rates of dissociation of the dimer and the isolated protomer. Taking the forward and back reactions together, we see that $K_{b1} = 2K_b$.

The **second step** in the reaction involving the dimer is

$$M_2S + S \rightleftharpoons M_2S_2 \quad \text{(binding constant } K_{b2}).$$

In diagrammatic form, this is:

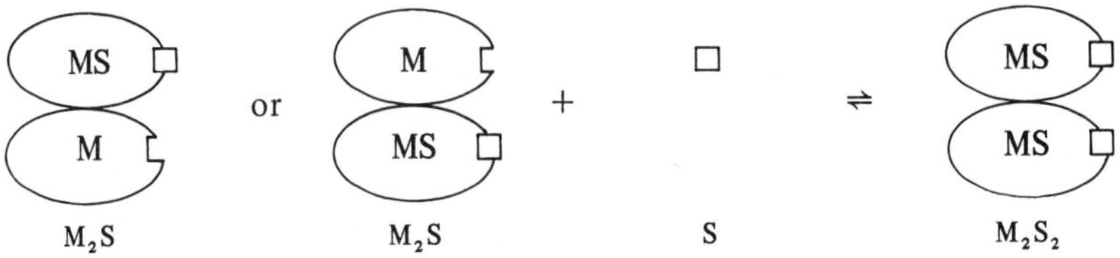

$$M_2S \qquad\qquad M_2S \qquad\qquad S \qquad\qquad M_2S_2$$

while for the isolated protomer we again have

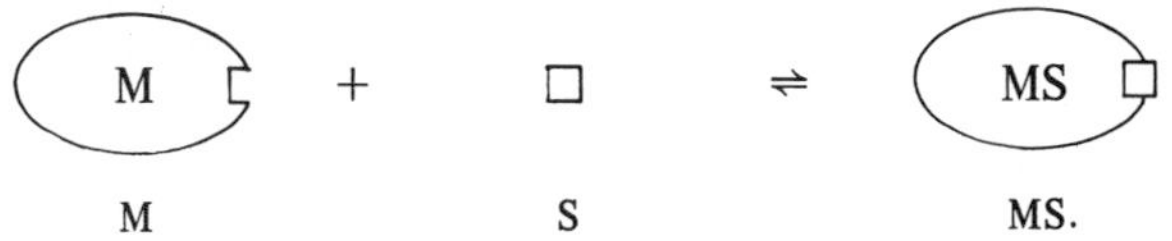

$$M \qquad\qquad S \qquad\qquad MS.$$

In the forward direction, both the dimer and the hypothetical isolated protomer have one free binding site and so the ligand is equally likely to bind to either. In the reverse direction, there are two sites in the dimer from which S can dissociate, but only one on the isolated protomer. Hence a molecule of ligand is twice as likely to dissociate from a molecule of dimer M_2S_2 than from a molecule of protomer MS. Therefore, for the overall reaction, $K_{b2} = \frac{1}{2}K_b$.

If we substitute these relationships in the general equation for $\bar{Y}$ (Chapter 12.5.1).

$$\bar{Y} = \frac{2K_b[S] + 2.2K_b.\frac{1}{2}K_b[S]^2}{2(1 + 2K_b[S] + 2K_b[S]^2)}$$

$$= \frac{K_b[S] + K_b^2[S]^2}{(1 + 2K_b[S] + K_b^2[S]^2)} = \frac{K_b[S](1 + K_b[S])}{(1 + K_b[S])^2}$$

$$= \frac{K_b[S]}{1 + K_b[S]}.$$

This is identical to the expression obtained for a protein with a single ligand-binding site, which gives a hyperbolic plot of $\overline{Y}$ against [S] (Chapter 12.2). In general, for the binding of a ligand (S) to a protein having several identical binding sites for the ligand, a hyperbolic plot of $\overline{Y}$ against [S] will be obtained provided there is no interaction between the binding sites. If this binding is the first step in a process by which S is converted to a product in such a way that the equilibrium assumption is valid and $v_0 \propto [MS]$, then a plot of v_0 against $[S_0]$ will also be hyperbolic. This conclusion has already been stated (in Chapter 7.1.3); here we have seen the justification for that statement.

One further relationship can be obtained for the reaction involving the binding of a ligand to a dimeric protein with no interaction between the binding sites. From the above discussion, $K_{b1} = 2K_b$ and $K_{b2} = \frac{1}{2}K_b$. Hence $K_{b1} = 4K_{b2}$.

12.5.3 Where There is Positive Homotropic Cooperativity

If the binding of the first molecule of the ligand increases the affinity of the protein for the ligand, the second step of the binding process will be faster than it is in the situation where there is no interaction between the binding sites, i.e. where $K_{b1} = 4K_{b2}$.

Hence for positive homotropic cooperativity, $K_{b1} < 4K_{b2}$. According to the Adair equation, this relationship results in a sigmoidal plot of $\overline{Y}$ against [S] being obtained (see Fig. 12.4(a)); the sigmoidal character of the curve is more marked the greater the degree of cooperativity.

When cooperativity is complete, $K_b = \dfrac{K_b[S]^2}{1 + K_b[S]^2}$, where K_b in this case is the binding constant for the overall process $M_2 + 2S \rightleftharpoons M_2S_2$ (see Chapter 12.4).

12.5.4 Where There is Negative Homotropic Cooperativity

Negative cooperativity results in the second step of the binding process being slower than it would be if there were no interaction between the binding sites.

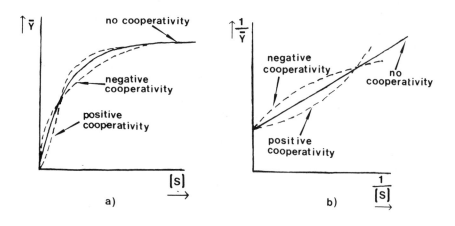

a)

b)

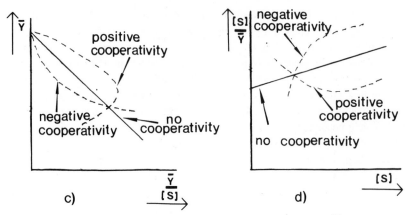

Fig. 12.4 Plots of (a) $\bar{Y}$ against [S], (b) $\dfrac{1}{\bar{Y}}$ against $\dfrac{1}{[S]}$, (c) $\bar{Y}$ against $\dfrac{\bar{Y}}{[S]}$ and (d) $\dfrac{[S]}{\bar{Y}}$ against [S], all at constant $[E_0]$, showing the effects of positive and negative homotropic cooperativity.

Hence, for negative homotropic cooperativity, $K_{b1} > 4K_{b2}$. In this case, a plot of $\bar{Y}$ against [S] is neither sigmoidal nor a true rectangular hyperbola (see Fig. 12.4 (a)).

12.6 THE ADAIR EQUATION FOR THE BINDING OF A LIGAND TO A PROTEIN HAVING THREE BINDING SITES FOR THAT LIGAND

For a trimeric protein (M_3) having three identical binding sites for a ligand (S), there are three steps in the binding process:

$$M_3 + S \rightleftharpoons M_3S \quad \text{(apparent binding constant } K_{b1})$$
$$M_3S + S \rightleftharpoons M_3S_2 \quad \text{(apparent binding constant } K_{b2})$$
$$M_3S_2 + S \rightleftharpoons M_3S_3 \quad \text{(apparent binding constant } K_{b3}).$$

Using reasoning exactly as for the dimeric protein in Chapter 12.5,

$$\bar{Y} = \frac{K_{b1}[S] + 2K_{b1}K_{b2}[S]^2 + 3K_{b1}K_{b2}K_{b3}[S]^3}{3(1 + K_{b1}[S] + K_{b1}K_{b2}[S]^2 + K_{b1}K_{b2}K_{b3}[S]^3)}.$$

This is the **Adair equation** for a trimeric protein.

If there is no interaction between the binding sites,

$$K_{b1} = 3K_b, \quad K_{b2} = K_b \quad \text{and} \quad K_{b3} = \tfrac{1}{3}K_b.$$

Hence $K_{b1} = 3K_{b2}$ and $K_{b2} = 3K_{b3}$, and the Adair equation reduces, as before, to $\overline{Y} = \dfrac{K_b[S]}{1 + K_b[S]}$.

If there is positive homotropic cooperativity,

$$K_{b1} < 3K_{b2} \quad \text{and} \quad K_{b2} < 3K_{b3}$$

and, if cooperativity is complete, the Hill coefficient $(h) = 3$.

If there is negative homotropic cooperativity,

$$K_{b1} > 3K_{b2} \quad \text{and} \quad K_{b2} > 3K_{b3}.$$

12.7 THE ADAIR EQUATION FOR THE BINDING OF A LIGAND TO A PROTEIN HAVING FOUR BINDING SITES FOR THAT LIGAND

A tetrameric protein (M_4) having four identical binding sites for a ligand (S) will have four steps in the binding process, with apparent binding constants K_{b1}, K_{b2}, K_{b3} and K_{b4}.

The Adair equation for a tetrameric protein is found to be

$$\overline{Y} = \frac{K_{b1}[S] + 2K_{b1}K_{b2}[S]^2 + 3K_{b1}K_{b2}K_{b3}[S]^3 + 4K_{b1}K_{b2}K_{b3}K_{b4}[S]^4}{4(1 + K_{b1}[S] + K_{b1}K_{b2}[S]^2 + K_{b1}K_{b2}K_{b3}[S]^3 + K_{b1}K_{b2}K_{b3}K_{b4}[S]^4)}.$$

If there is no interaction between the binding sites,

$$K_{b1} = 4K_b, \ K_{b2} = \tfrac{3}{2}K_b, \ K_{b3} = \tfrac{2}{3}K_b, \quad \text{and} \quad K_{b4} = \tfrac{1}{4}K_b.$$

Under these conditions, $K_{b1} = \tfrac{8}{3}K_{b2}$, $K_{b2} = \tfrac{9}{4}K_{b3}$, $K_{b3} = \tfrac{8}{3}K_{b4}$ and the Adair equation again reduces to $\overline{Y} = \dfrac{K_b[S]}{1 + K_b[S]}$.

If there is positive homotropic cooperativity,

$$K_{b1} < \tfrac{8}{3}K_{b2}, \ K_{b2} < \tfrac{9}{4}K_{b3} \quad \text{and} \quad K_{b3} < \tfrac{8}{3}K_{b4},$$

and if the cooperativity is complete, the Hill coefficient $(h) = 4$.

If there is negative homotropic cooperativity,

$$K_{b1} > \tfrac{8}{3}K_{b3}, \ K_{b2} > \tfrac{9}{4}K_{b3} \quad \text{and} \quad K_{b3} > \tfrac{8}{3}K_{b4}.$$

12.8 INVESTIGATION OF COOPERATIVE EFFECTS

12.8.1 Measurement of the Relationship between $\overline{Y}$ and [S]

If there is some measurable difference between a ligand in its free and protein-bound forms, or between the free protein and the protein–ligand complex, then the relationship between fractional saturation ($\overline{Y}$) and free ligand concentration ([S]) is relatively easy to determine. For example, as mentioned in Chapter 9.4.2, there is a difference in absorbance at 350 nm between free NADH and NADH bound to alcohol dehydrogenase; hence it is possible to investigate the binding of NADH to this enzyme at different NADH concentrations in the absence of all other substrates. Other methods for the investigation of ligand-binding to protein include the observation of changes in the fluorescence or nmr spectra, or the measurement by ion-selective electrodes of the loss of free ligand as binding takes place.

In general, for an oligomeric protein (E, or M_n) having n identical and non-interacting binding sites for a ligand (S),

$$\overline{Y} = \frac{K_b[S]}{1 + K_b[S]}.$$

This is valid where M_n and S are at or near equilibrium, regardless of whether or not a product is being formed, since in either case, [S] and [MS] may be assumed constant (Chapter 12.5.2). If possible, it is best to investigate under conditions where equilibrium can be ensured, e.g. to determine the binding characteristics for one substrate of a multi-substrate reaction in the absence of the other substrates: this minimises the assumptions being made and excludes possible heterotropic effects.

This relationship between $\overline{Y}$ and [S] in the absence of cooperativity is the equation of a rectangular hyperbola, like the Michaelis–Menten equation derived in Chapter 7.1. As with the Michaelis–Menten equation, it is possible to manipulate the binding equation to obtain linear relationships between variables: if the equation is obeyed, linear plots are obtained of $1/\overline{Y}$ against $1/[S]$, $\overline{Y}$ against $\overline{Y}/[S]$ and $[S]/\overline{Y}$ against [S] (exactly analogous to the Lineweaver–Burk, Eadie-Hofstee and Hanes plots of Chapter 7.1). These are shown in Fig. 12.4.

Where positive homotropic cooperativity occurs, a sigmoidal plot of $\overline{Y}$ against [S] is obtained; the other plots are non-linear, as shown in Fig. 12.4. In general, it is considered that departures from linearity are more obvious on Eadie-Hofstee and Hanes-type plots than on those of the Lineweaver-Burk type.

Where negative homotropic cooperativity occurs, the plot of $\overline{Y}$ against [S] is neither sigmoidal nor a rectangular hyperbola, although it could easily be mistaken for the latter. For this reason, it is essential to investigate the other relationships, the plots for negative cooperativity being non-linear and of the opposite curvature to those for positive cooperativity (Fig. 12.4).

12.8.2 Measurement of the Relationship Between v_0 and $[S_0]$

If S is a substrate and reacts to form products in such a way that the binding process remains at or near equilibrium, then $[MS]$ is constant, v_0 is proportional to $[MS]$ and $\overline{Y} = \dfrac{v_0}{V_{max}}$. Under these conditions and provided $[S_0] \gg [E_0]$, kinetic data may be used to plot the graphs shown in Fig. 12.4, with v_0 replacing $\overline{Y}$ and $[S_0]$ replacing $[S]$. The conclusions would be unchanged.

This gives a more versatile way of investigating cooperative effects, for only a limited number of binding processes can be monitored directly by the use of spectroscopy or ion-selective electrodes. However, more assumptions are involved, and complexities in the kinetic mechanism could give misleading results (see Chapter 13.5).

12.8.3 The Scatchard Plot and Equilibrium Dialysis Techniques

For systems where a single ligand (S) binds to an oligomeric protein (E, or M_n) having n identical and non-interacting binding sites for that ligand,

$$\overline{Y} = \frac{[MS]}{[MS] + [M]} = \frac{K_b[S]}{1 + K_b[S]} \quad \text{(see Chapter 12.5).}$$

$$\therefore \ [MS] + [MS]K_b[S] = K_b[S][MS] + K_b[S][M].$$

$$\therefore \ K_b = \frac{[MS]}{[M][S]} = \frac{[MS]}{([M_0] - [MS])[S]}.$$

$$\therefore \ \frac{[MS]}{[S]} = K_b[M_0] - K_b[MS].$$

Scatchard (1949) pointed out that, under these conditions, a graph of $\dfrac{[MS]}{[S]}$ against $[MS]$ will be linear, with characteristics as shown in Fig. 12.5. Note that, since $\overline{Y} = \dfrac{[MS]}{[M_0]}$, this is basically a plot of the Eadie-Hofstee type (Fig. 12.4(c)), with the axes reversed.

The Scatchard plot may be used to determine the type of cooperativity, and also the number of binding sites, from the results of **equilibrium dialysis** studies: a solution of protein of known concentration ($[E_0] = [(M_n)_0]$) is dialysed against a solution of ligand of known concentration ($[S_0]$) and allowed to come to equilibrium. (Note that this limits the use of such investigations to systems where binding is not a prelude to product formation, and to systems where both protein and ligand are stable for several hours.) The ligand will be able to pass freely through the dialysis membrane, but the protein will be trapped within its compartment (e.g. dialysis bag). The concentration of free ligand outside the

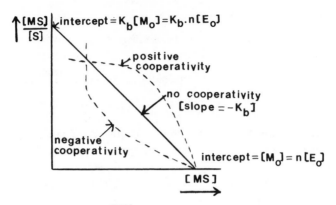

Fig. 12.5   Scatchard plot of $\dfrac{[MS]}{[S]}$ against [MS], at fixed $[E_o]$, showing the effects of positive and negative cooperativity.

protein compartment can be easily determined at any time, and at equilibrium it should be equal to the free ligand concentration within the protein compartment (= [S]) (Fig. 12.6). Radioactive-labelled ligands are often used for equilibrium dialysis experiments, since they result in greater sensitivity being obtained.

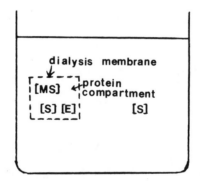

Fig. 12.6   Diagrammatic representation of an equilibrium dialysis experiment, showing the concentrations present in the two compartments at equilibrium.

If the volume of ligand within the protein compartment is negligible compared to the total volume of liquid present, then $[S] = [S_o] - [MS]$, from which [MS] may be calculated. Alternatively, and without making this assumption, the total ligand concentration within the protein compartment (= [MS] + [S]) can be determined, and [MS] calculated as the difference between this and the total ligand concentration outside the protein compartment (= [S]).

For example, the binding of NADH to **lactate dehydrogenase** has been investigated by these techniques: it is found that there is no interaction between the binding sites on the four sub-units. The location of this binding site is shown in Fig. 12.7, the adenine component binding to tyrosine-85 and nicotinamide to lysine-250.

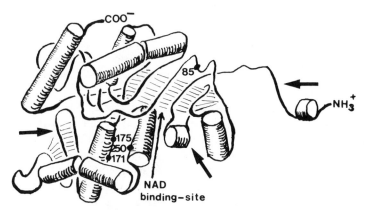

Fig. 12.7   A simplified representation of the three-dimensional structure of one of the four identical sub-units of dogfish muscle lactate dehydrogenase as revealed by the x-ray diffraction studies of Adams, Rossmann and colleagues (1972). (Conventions as for Fig. 2.8.) Areas of contact with the three other sub-units are indicated by arrows.

Similar results to those obtained in equilibrium dialysis experiments may be obtained by the use of ultracentrifugation or gel filtration techniques, both of which involve moving an initially ligand-free protein through a solution of ligand and observing the changes which take place as it binds ligand.

12.9 THE BINDING OF OXYGEN TO HAEMOGLOBIN

The stimulus for much of the work described in Chapter 12 was experimental evidence regarding the binding of oxygen to haemoglobin. In 1904, Bohr and co-workers showed that if the fractional saturation of haemoglobin with oxygen was plotted against the partial pressure of oxygen gas (equivalent to the concentration), a curve was obtained which was clearly sigmoidal.

Hill (1909) explained this in the basis of interaction between binding sites causing positive cooperativity. At that time it was known that each haem (iron protoporphyrin) group bound one oxygen molecule, and Hill correctly suggested that each haemoglobin sub-unit contained one haem group, but it was not known how many sub-units made up the oligomeric protein. Hill assumed that cooperativity was complete, so if there were n sub-units in the haemoglobin molecule, the overall reaction was

$$Hb + n\, O_2 \rightleftharpoons Hb\,(O_2)_n.$$

On this basis he derived the Hill equation (Chapter 12.4) and found the Hill coefficient (h) to be about 2.8.

It was subsequently shown that there were four binding sites to each haemoglobin molecule, so cooperativity was far from complete. Adair, (1925) then developed the theory of ligand binding to protein which was described in general terms in Chapter 12.5: he saw that oxygen molecules could bind to a haemoglobin molecule in four separate steps, each with a different apparent binding constant, and derived the Adair equation for a tetrameric protein; he also showed what the relationship between the apparent binding constants must be to explain positive cooperativity.

Results from x-ray diffraction studies, reported by Perutz and co-workers in 1960, showed that the four binding sites are in very similar environments, so the assumption that they behave identically is a reasonable one. However, these studies also showed that the four haem groups are completely spatially separate in the molecule, so direct interaction between the binding sites is impossible. It seems likely, therefore, that the mechanism of cooperativity involves interactions between sub-units at places other than the binding sites.

All four C-terminal amino-acid residues, and possibly some others, form electrostatic linkages with groups on other sub-units in the oxygen-free molecule (deoxyhaemoglobin), but not in the fully-oxygenated molecule (oxyhaemoglobin). Conformational changes also take place as the oxygen binds to the haemoglobin molecule, the binding site on each sub-unit being a Fe(II) atom attached to a histidine residue and to the four pyrrole groups of a protoporphyrin ring. In the unbound form, the Fe atom is too large to fit into the hole in the centre of the porphyrin ring, so lies about 0.75 Å out of the plane of this ring; when oxygen fills the vacant sixth coordination position of the Fe atom it decreases the atomic radius, enabling the metal atom to move into the plane of the porphyrin ring. This it proceeds to do, pulling the histidine residue after it and so altering the tertiary structure to the sub-unit. The tyrosine adjacent to the C-terminus is forced out of a pocket between two helical regions, where in deoxyhaemoglobin it plays a role in stabilising the tertiary structure, and with it moves the C-terminal amino-acid. As a result, the electrostatic linkages with other sub-units are broken and a less constrained (or more relaxed) conformational state is assumed.

Although it is still not entirely clear how this facilitates oxygen binding to other sub-units, one relevant factor is that the breaking of some electrostatic interactions between sub-units when the first molecule of oxygen binds means that there are fewer such interactions remaining to be broken when subsequent molecules bind, so these processes are energetically more favourable than the first.

SUMMARY OF CHAPTER 12

If there are several ligand-binding sites on a protein, it is possible that there could be interaction between them: the binding of one ligand might increase or decrease the affinity of another site on the protein for the same or a different ligand. Such interaction between binding sites is called a **cooperative effect**: positive cooperative effects increase affinity, while negative effects decrease it; homotropic effects concern identical ligands, whereas heterotropic effects concern different ligands.

If the ligand is a substrate and goes on to give a product in such a way that the Michaelis-Menten equilibrium assumption is valid, then initial velocity is proportional to the concentration of enzyme-bound substrate and cooperative effects are reflected in the kinetics of the overall reaction; in the presence of cooperativity, Michaelis-Menten plots will not be rectangular hyperbolae, and other primary plots, e.g. those of Lineweaver–Burk and Eadie–Hofstee, will not be linear.

From initial studies on the binding of oxygen to **haemoglobin**, Hill derived an equation relating fractional saturation to ligand concentration; this is strictly valid only where positive homotropic cooperativity is total. Adair formulated an equation which is of more general application; it is valid for any oligomeric protein which has several identical binding sites for a particular ligand, since it makes no assumptions about cooperativity.

Cooperative effects can be investigated by the use of spectroscopy to determine fractional saturation, by equilibrium dialysis experiments in association with the **Scatchard plot**, or by kinetic studies under steady-state conditions.

FURTHER READING

Ferdinand, W. (1976), *The Enzyme Molecule* (pages 85-99), Wiley.

Gutfreund, H. (1972), *Enzymes: Physical Principles* (Chapter 4), Wiley.

Holbrook, J. J., Liljas, A., Steindel, S. J. and Rossmann, M. G. (1975), Lactate dehydrogenase, in Boyer, P. D. (ed.), *The Enzymes,* **11**, 3rd edition, Academic Press.

Newsholme, E. A. and Start, C. (1973), *Regulation in Metabolism* (Chapter 2), Wiley.

Perutz, M. F. (1970), Stereochemistry of cooperative effects of haemoglobin, *Nature,* **228** (pages 726-739).

Wharton, C. W. and Eisenthal, R. (1981), *Molecular Enzymology* (Chapter 7), Blackie.

PROBLEMS

12.1 A single-substrate enzyme-catalysed reaction was investigated at fixed total enzyme concentration and the following results were obtained:

$[S_0]$ (mmol.l^{-1})	1.0	1.67	2.0	2.5	3.3	5.0	10.0
v_0 (μmol.min^{-1})	1.10	1.43	1.54	1.75	2.00	2.56	4.00

Draw Michaelis-Menten, Lineweaver–Burk, Eadie–Hofstee and Hanes plots of this data. Assuming the reaction was proceeding under steady-state conditions in each case, what type of cooperative effect is indicated?

12.2 The following results were obtained during an investigation of the binding of a ligand to a protein at fixed total protein concentration:

ligand concentration $(mmol.l^{-1})$	1.0	1.67	2.0	2.5	3.3	5.0	10.0
fractional saturation			0.06	0.14	0.19	0.24	0.35	0.53	0.80.

What can you conclude about the binding of the ligand? Draw a Hill plot from this data and determine the Hill coefficient.

12.3 An enzyme was dialysed against one of its substrates at a series of different initial substrate concentrations. The system was allowed to come to equilibrium in each case and the total concentration of substrate inside and outside the dialysis bag was measured. The following results were obtained at equilibrium:

Total enzyme concentration $(mmol.l^{-1})$		Total substrate concentration $(mmol.l^{-1})$	
inside dialysis bag	outside bag	inside dialysis bag	outside bag
2.0	0	2.40	0.80
2.0	0	3.33	1.28
2.0	0	5.25	2.34
2.0	0	8.55	4.55
2.0	0	11.60	6.78
2.0	0	17.90	12.10
2.0	0	34.5	27.6 .

What can you deduce from this data about the binding of the substrate?

Sigmoidal Kinetics and Allosteric Enzymes

13.1 INTRODUCTION

In Chapter 12 we discussed how interaction between the ligand-binding sites of oligomeric proteins could give rise to cooperative binding, and how this would be reflected in departures from linearity of Lineweaver-Burk and similar plots if the ligand was a substrate. This is an important consideration, for many enzymes are oligomeric proteins made up of several identical sub-units or protomers. As we shall see later (Chapter 13.5), similar departures from linearity may be seen in the absence of cooperative binding if the kinetic mechanism of the reaction is not straightforward. However, first we must consider in a little more detail how the cooperative binding of ligand to protein may occur. How do the binding sites interact?

It appears that with most proteins, as with haemoglobin (Chapter 12.9), binding sites are clearly separated and so cannot interact directly. Hence it seems that the mechanism of cooperative binding must involve more general interactions between sub-units and the occurrence of conformational changes. The simplest treatment considers that each protomer can exist in two conformational forms: the **T-form** is that which predominates in the unliganded protein, whereas the **R-form** predominates in the protein–ligand complexes. On the basis of the findings with haemoglobin, the T-form may be taken to represent a tensed (or constrained) sub-unit and the R-form a more relaxed one, but this is not necessarily always the case.

From this starting point, Monod, Wyman and Changeux (1965) and Koshland, Némethy and Filmer (1966) have put forward models to account for cooperative binding. These models do not give a detailed chemical explanation for cooperativity, but they provide a framework within which the factors involved may be discussed.

13.2 THE MONOD-WYMAN-CHANGEUX (MWC) MODEL

13.2.1 The MWC Equation

The MWC model is sometimes called the **symmetrical model** because it is based on the assumption that, in a particular protein molecule, all of the protomers must be in the same conformational state: all must be in the R-form or all in the T-form, no hybrids being found because of supposed unfavourable interactions between sub-units in different conformational states. The two conformational forms of the protein are in equilibrium in the absence of ligand, and the equilibrium is disturbed by the binding of the ligand; this alone can be the explanation for cooperative effects.

Let us consider a dimeric protein having two identical binding sites for a substrate or ligand (S). In the absence of ligand, there will be equilibrium between the two conformational forms of the dimer $(R_2 \rightleftharpoons T_2)$, the equilibrium constant being termed the **allosteric constant** and given the symbol L. The hybrid RT is held to be unstable and ignored.

The ligand can bind to either of the sites on the R_2 molecule, each having an intrinsic dissociation constant K_R. In the simplest form of the hypothesis, it is assumed that S does not bind to T to any appreciable extent. Therefore, the only processes which need to be considered (apart from any subsequent reaction to form products) are:

$$R_2 \rightleftharpoons T_2 \qquad \text{(equilibrium constant L),}$$

$$R_2 + S \rightleftharpoons R_2S \qquad \text{(intrinsic dissociation constant } K_R\text{),}$$

$$\text{and} \quad R_2S + S \rightleftharpoons R_2S_2 \qquad \text{(intrinsic dissociation constant } K_R\text{).}$$

In diagrammatic form this may be written:

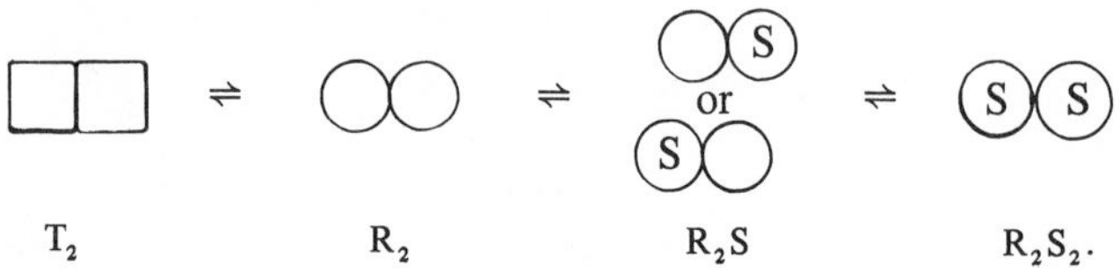

$$T_2 \qquad\qquad R_2 \qquad\qquad R_2S \qquad\qquad R_2S_2.$$

Let us apply exactly the same logic to this sequence of reactions as we applied in Chapter 12.5: again we are considering the binding of a ligand to a dimeric protein, but on this occasion we have the extra complication of two conformational forms. We will assume that the binding of one molecule of S to R_2 does not alter the affinity of the other binding site for S.

The concentration of bound sub-units present $= [R_2S] + 2[R_2S_2]$. The total concentration of sub-units present $= 2[R_2] + 2[R_2S] + 2[R_2S_2] + 2[T_2]$.

$\therefore$ Fractional saturation $\overline{Y} =$

$$\frac{[R_2S] + 2[R_2S_2]}{2([R_2] + [R_2S] + [R_2S_2] + [T_2])} = \frac{[R_2S] + 2[R_2S_2]}{2([R_2] + [R_2S] + [R_2S_2] + L[R_2])}.$$

For the **first step** of the binding process, $R_2 + S \rightleftharpoons R_2S$, the apparent binding constant $K_{b1} = \dfrac{[R_2S]}{[R_2][S]}$.

$$\therefore \ [R_2S] = K_{b1}[R_2][S].$$

Since there are two unbound sites which may be filled in the forward reaction but only one bound ligand to dissociate in the reverse reaction,

$$K_{b1} = 2 \times \text{intrinsic binding constant} = \frac{2}{\text{intrinsic dissociation constant}} = \frac{2}{K_R}.$$

Hence substituting for K_{b1} in the expression for $[R_2S]$ above,

$$[R_2S] = \frac{2}{K_R}[R_2][S].$$

For the **second step** of the binding process, $R_2S + S \rightleftharpoons R_2S_2$, the apparent binding constant $K_{b2} = \dfrac{[R_2S_2]}{[R_2S][S]}$.

$$\therefore \ [R_2S_2] = K_{b2}[R_2S][S] = K_{b1}K_{b2}[R_2][S]^2.$$

Since there is only one unbound site which may be filled in the forward reaction but two bound ligand molecules to dissociate in the reverse reaction,

$$K_{b2} = \tfrac{1}{2} \times \text{intrinsic binding constant} = \frac{1}{2 \times \text{intrinsic dissociation constant}} = \frac{1}{2K_R}.$$

Hence, substituting for K_{b1} and K_{b2} in the expression for $[R_2S_2]$ above,

$$[R_2S_2] = \frac{2}{K_R} \cdot \frac{1}{2K_R}[R_2][S]^2 = \frac{[R_2][S]^2}{(K_R)^2}.$$

Now substituting for R_2S and R_2S_2 in the expression for $\overline{Y}$ above,

$$\overline{Y} = \frac{\dfrac{2[R_2][S]}{K_R} + \dfrac{2[R_2][S]^2}{(K_R)^2}}{2\left([R_2] + \dfrac{2[R_2][S]}{K_R} + \dfrac{[R_2][S]^2}{(K_R)^2} + L[R_2]\right)}$$

$$= \frac{\dfrac{[S]}{K_R}\left(1 + \dfrac{[S]}{K_R}\right)}{L + \left(1 + \dfrac{[S]}{K_R}\right)^2}.$$

This is the **Monod–Wyman–Changeux equation** for a dimeric protein. It may similarly be shown that for a protein consisting of n protomers, each with a binding site for the substrate or ligand (S), the MWC equation is

$$\bar{Y} = \frac{\dfrac{[S]}{K_R}\left(1 + \dfrac{[S]}{K_R}\right)^{n-1}}{L + \left(1 + \dfrac{[S]}{K_R}\right)^n}.$$

According to this equation, the greater the value of L, the more sigmoidal a plot of $\bar{Y}$ against $[S]$. If $L = 0$, a hyperbolic curve is obtained; a hyperbolic curve is also obtained, as would be expected, for a monomeric protein, i.e. where $n = 1$, and for the situation where the substrate can bind equally well to the R and the T conformational forms.

13.2.2   How the MWC Model Accounts for Cooperative Effects

The MWC equation is consistent with a sigmoidal binding curve, even though its derivation assumes that the binding of one molecule of ligand does not affect the affinity for the ligand of other binding sites on the molecule. The explanation for the cooperative effects lies in the R_n/T_n equilibrium.

When L is large, this equilibrium is in favour of the T_n form in the absence of ligand. If ligand is introduced, but at vary low concentrations, there will not be enough present to react significantly with the small amounts of R_n present, so very little formation of R_nS, R_nS_2 and the other liganded species of protein will take place. At higher ligand concentrations, however, there will be enough ligand present to force formation of significant amounts of R_nS, R_nS_2 etc.; thus, some free R_n will be removed from the system, thereby disturbing the R_n/T_n equilibrium and causing more R_n to be formed from T_n. This freshly formed R_n can also react with ligand, resulting in yet more formation of R_nS, R_nS_2 and the other liganded forms. Hence the T_n species can be regarded as a reservoir of R_n which only becomes available when the ligand concentration is high enough to cause the formation of appreciable amounts of protein–ligand complex; there will be a surge in the binding curve in the region of the critical ligand concentration.

At still higher ligand concentrations, more of the reservoir of protein will be utilised, and this process will continue until a ligand concentration is reached which is high enough to force conversion of all T_n to R_n; at this point the protein will be fully saturated with ligand.

Thus the overall binding curve will be **sigmoidal**, a characteristic of positive homotropic cooperativity; it will be apparent from the above that the MWC model *cannot* explain negative homotropic cooperativity.

13.2.3 The MWC Model and Allosteric Regulation

One of the main reasons for the introduction of the MWC model was an attempt to explain the phenomena of allosteric inhibition and activation: Umbarger (1956) first found that isoleucine could inhibit **threonine dehydratase**, an enzyme involved in its biosynthesis in bacteria; other similar examples of **end-product inhibition**, and also of **allosteric activation**, were soon reported. In 1963, Monod, Changeux and Jacob put forward the allosteric theory of regulation: they pointed out that these naturally occurring metabolic **regulators** (also called **effectors** and **modifiers**) generally do not resemble the substrate in structure, so are likely to bind to the enzyme at a separate site and affect the binding of the substrate by heterotropic cooperativity. The word **allosteric** was originally used to stress the difference in shape between regulator and substrate (*allo* meaning *other*); since then it has been used loosely to describe any kind of cooperative effect, homotropic as well as heterotropic.

According to the MWC model, allosteric inhibitors bind to the T-form of the enzyme, stabilising it and thus increasing the value of L. Allosteric activators have the opposite effect, binding to and stabilising the R-form and decreasing the value of L. In either case, the binding of the modifier to one of the forms of the enzyme will disturb the R/T equilibrium and therefore show some degree of sigmoidal character if investigated in the absence of substrate. Enzymes subject to allosteric control may fit into either of two categories: they may be **K-series** or **V-series** enzymes.

K-series enzymes are those where the presence of the modifier changes the binding characteristics of the enzyme for the substrate but does not affect the V_{max} of the reaction. The term K_m has no real meaning for an allosteric enzyme, particularly if the binding rather than the kinetic properties are being considered: a more appropriate term is $S_{0.5}$, which is the ligand concentration required to produce 50% saturation of the protein. For a K-series enzyme, $(S_{0.5})_{substrate}$, i.e. the substrate concentration required to half-saturate the enzyme, varies with the concentration of modifier. The MWC hypothesis is that the substrates of such enzymes bind preferentially to the R-form, giving a sigmoidal binding curve as discussed in Chapter 13.2.1; the subsequent reaction is straightforward, so the shape of the Michaelis–Menten plot is determined simply by that of the binding curve. Allosteric inhibitors, by increasing the value of L, increase the sigmoidal nature of the binding curve for substrate; thus they decrease the fractional saturation of an enzyme with its substrate at low and moderate substrate concentrations, decreasing the value of v_o under these conditions (Fig. 13.1). Allosteric activators, on the other hand, tend to increase the hyperbolic nature of the substrate binding curve. In each case, the degree of allosteric effect depends on the concentration of modifier, but the value of V_{max} is not affected.

V-series enzymes are those where the presence of a modifier results in a change in the V_{max} but not in the value of the apparent K_m (or $S_{0.5}$) for the

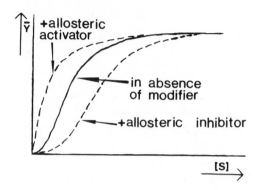

Fig. 13.1 Effect of allosteric activators and inhibitors on the binding of a substrate to a K-series enzyme, at fixed concentrations of modifier and enzyme.

substrate; the binding curve (and Michaelis–Menten plot) for the substrate at constant modifier concentration is a rectangular hyperbola, but the binding curve for the modifier itself is sigmoidal. This can be explained, according to the MWC model, if the substrate can bind equally well to the R- and T-forms of the enzyme, but the reaction catalysed by the R-form is faster than that catalysed by the T-form. V-series enzymes are much less common than K-series enzymes, but Tipton and colleagues (1974) have shown that possible examples include **fructose 1,6-bisphosphatase**, of which AMP is an allosteric inhibitor, and **pyruvate carboxylase**, activated by acetyl–CoA.

Enzymes are also likely to exist in which the R- and T-forms have different affinities for the substrate and also catalyse the reaction at different rates; in this case, allosteric modifiers would affect both the V_{max} and apparent K_m values.

13.2.4 The MWC Model and the Hill Equation

For the MWC model where the substrate binds only to the R-form of the enzyme, the fractional saturation, as we saw in Chapter 13.2.1, is given by the expression

$$\bar{Y} = \frac{\dfrac{[S]}{K_R}\left(1 + \dfrac{[S]}{K_R}\right)^{n-1}}{L + \left(1 + \dfrac{[S]}{K_R}\right)^{n}}.$$

If L is very large, most of the enzyme will usually be in a form (T) which will not bind S, keeping free [S] relatively high. Also, if it is the R-form that binds S, K_R will be relatively low. Hence $[S]/K_R$ will tend to be large, so $(1 + \dfrac{[S]}{K_R}) \simeq \dfrac{[S]}{K_R}$.

Under these conditions,

$$\bar{Y} = \frac{\left(\dfrac{[S]}{K_R}\right)^n}{L + \left(\dfrac{[S]}{K_R}\right)^n} = \frac{\dfrac{[S]^n}{K_R^n . L}}{1 + \dfrac{[S]^n}{K_R^n . L}}.$$

However, K_R, L and n are all constant, so $\dfrac{1}{K_R^n . L} = $ constant $(= K')$.

$$\therefore \quad \bar{Y} = \frac{K'[S]^n}{1 + K'[S]^n}.$$

This is a form of the Hill equation (see Chapter 12.4) and implies that, If L is sufficiently large, the only enzyme species present are T_n, R_n and $R_n S_n$.

Hence, if a Hill plot of $\log(\dfrac{\bar{Y}}{1 - \bar{Y}})$ against $\log[S]$, or of $\log(\dfrac{v_o}{V_{max} - v_o})$ against $\log[S_o]$, is drawn from experimental data and the Hill coefficient (h) is found to be equal to the number of binding sites (n), as determined by an independent experiment, then this series of assumptions must be valid for the system under investigation. A value of $h = n$ will therefore imply that the MWC model is operating in this instance, that S does not bind to the T-form of the enzyme and that L is very large; no other model has been proposed which is consistent with the Hill equation.

In the simple system discussed in Chapter 12.4, a value of $h < n$ was taken to imply that cooperativity was not complete; in the slightly more complicated system being considered here, a value of $h < n$ would indicate that one (or more) of the assumptions made above was not valid for the enzyme under study. It would not in itself exclude the possibility that the MWC was operating because, for example, the value of L might not be large enough to enable a Hill-type equation to be obtained. Since allosteric inhibitors are assumed to increase the value of L, it is common for the Hill coefficient to be determined in the presence of an allosteric inhibitor, so that the best indication as to whether or not the MWC model is operating may be obtained. For example, Scarano and co-workers (1967) showed that, for the reaction catalysed by donkey spleen **deoxycytidine monophosphate deaminase**,

$$dCMP + H_2O \rightleftharpoons dUMP + NH_3$$

the Hill coefficient in the presence of the allosteric inhibitor dTTP is 4. From ·other evidence it was known that there are 4 binding sites for dCMP, so it was concluded that the MWC model operates for this reaction.

According to this model, the limiting value of h is n, this being obtained when the substrate binds only to the R-form of the enzyme and where L is very large; a value of h > n should never be obtained.

13.3 THE KOSHLAND-NÉMETHY-FILMER (KNF) MODEL

13.3.1 The KNF Model for a Dimeric Protein

The KNF model differs from the MWC one in that it does not exclude hybrids between the two conformational forms of the protein. Therefore, for a dimeric protein where each protomer can exist in R- and T-forms, the species R_2, T_2, R_2S, R_2S_2, R.TS, RS.TS, T_2S and T_2S_2 can all exist. However, in order to explain cooperative effects, some restrictions have to be made.

In the KNF linear sequential model, the only protein species present to any appreciable extent at (or near) equilibrium are T_2, T.RS and R_2S_2. The reaction sequence may therefore be written:

$$T_2 + S \rightleftharpoons T.RS \quad \text{(apparent binding constant } K_{b1})$$

$$T.RS + S \rightleftharpoons R_2S_2 \quad \text{(apparent binding constant } K_{b2}).$$

There is no fundamental difference between this reaction sequence and that used in Chapter 12.5 to derive the Adair equation for a dimeric protein existing in one conformational form. Hence, in both cases, fractional saturation

$$\overline{Y} = \frac{K_{b1}[S] + 2K_{b1}K_{b2}[S]^2}{2(1 + K_{b1}[S] + K_{b1}K_{b2}[S]^2)}.$$

If $K_{b1} = 4K_{b2}$, there is no cooperativity.

If $K_{b1} < 4K_{b2}$, there is positive homotropic cooperativity.

If $K_{b1} > 4K_{b2}$, there is negative homotropic cooperativity.

The KNF linear sequential model was developed from the induced-fit theory of Koshland (see Chapter 4.4) and implies that the substrate or ligand induces a conformational change to take place (T → R) as it binds to the T-form of the protein:

$$T_2 + S \rightarrow T.TS \rightarrow T.RS.$$

However the same results could be obtained by an alternative pathway, in which there is an R/T equilibrium which strongly favours the T-form, but where S can only bind to the R-form and so disturbs the equilibrium:

$$T_2 + S \rightleftharpoons T.R + S \rightarrow T.RS.$$

In both cases there are negligible amounts of T.TS, T.R and similar species present at (or near) equilibrium; the KNF linear sequential model may therefore be analysed in terms of either of these alternative pathways, and the one chosen was that where the substrate can only bind to the R-form.

The following constants are introduced:

(a) K_t, an **equilibrium constant** for the conformational change $T \rightleftharpoons R$, so that

$$K_t = \frac{[R]}{[T]}$$

(b) K_b, a **binding constant** for the reaction $R + S \rightleftharpoons RS$, so that

$$K_b = \frac{[RS]}{[R][S]}$$

(c) K_{RT}, K_{RR} and K_{TT}, **interaction constants** indicating the relative stabilities of the various conformational forms of the oligomeric protein, such that

$$K_{RT} = \frac{[RT]}{[R][T]}, \quad K_{RR} = \frac{[RR]}{[R][R]} \quad \text{and} \quad K_{TT} = \frac{[TT]}{[T][T]}.$$

Since we are only interested in comparing the stabilities of these species, K_{TT} is arbitrarily given the value of 1. On this basis if, for example, K_{RT} has a value greater than 1, then RT will be more stable than TT, which will facilitate binding of S; on the other hand, if $K_{RT} < 1$, RT will be less stable than TT and binding of S will be difficult.

Let us now analyse the step $T_2 + S \rightleftharpoons T.RS$ in terms of these constants:

$$K_{b1} = 2.K_t.K_b.\frac{K_{RT}}{K_{TT}}$$

(Note that the factor 2 is introduced because there are two equally possible binding sites in the forward direction.)

Similarly, for the step $T.RS + S \rightleftharpoons RS.RS$,

$$K_{b2} = \tfrac{1}{2}.K_t.K_b.\frac{K_{RR}}{K_{RT}}.$$

If $K_{TT} \simeq K_{RT} \simeq K_{RR}$, then $K_{b1} = 4K_{b2}$ and there is no cooperativity, all interactions between the protomers being identical.

If $K_{TT} \simeq K_{RT} \ll K_{RR}$, then $K_{b1} < 4K_{b2}$ and positive homotropic cooperativity results. The binding of S to one protomer traps it in the R-form; this results in the other protomer staying mainly in the ligand-binding R-form, since R–R interactions are more favourable than R–T interactions.

If $K_{TT} \simeq K_{RT} \gg K_{RR}$, then $K_{b1} > 4K_{b2}$ and negative homotropic cooperativity results. The binding of S to one protomer again traps it in the R-form; in this instance, this results in the other protomer staying mainly in the T-form which cannot bind ligands, since R-T interactions are more favourable than R-R interactions.

Note that if $K_{TT} > K_{RR} \gg K_{RT}$ we have the conditions assumed for the MWC model, interactions between RT hybrids being very unfavourable. Since $K_{RT} \ll K_{RR}$ we can confirm that positive homotropic cooperativity, but not negative homotropic cooperativity, is possible for the MWC model.

As pointed out by Eigen (1967), the KNF linear sequential model and the MWC model are special cases of a general model in which all combinations are possible (Fig. 13.2). The MWC model may be termed the **concerted** form of the general model.

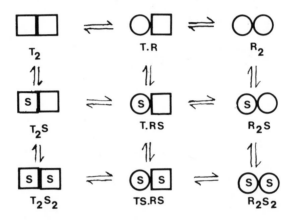

Fig. 13.2   General scheme for the binding of a ligand (S) to a dimeric protein where each protomer can exist in two conformational forms. In the KNF linear sequential model, the only protein species present at (or near) equilibrium are T_2, T.RS and R_2S_2; in the simplest form of the MWC (concerted) model, the only protein species present at (or near) equilibrium are T_2, R_2, R_2S and R_2S_2.

13.3.2   The KNF Model for Any Oligomeric Enzyme

A similar treatment to that discussed in Chapter 13.3.1 can be applied to any oligomeric protein with a number of identical binding sites for a ligand, using the appropriate form of the Adair equation. The only extra complication is in deciding which of the protomers can interact, and thus which interaction constants have to be considered.

In the case of a tetramer, for example, it is possible that each of the protomers can interact with the other three; this arrangement is called a **tetrahedral model** (Fig. 13.3). Alternatively, each protomer may only be able to interact with two other protomers, forming a **square model**. A further possibility is a **linear model**, where two of the protomers can interact with one other protomer and two can interact with two other protomers. These models refer only to possible interactions and do not necessarily describe the arrangements of the sub-units in space.

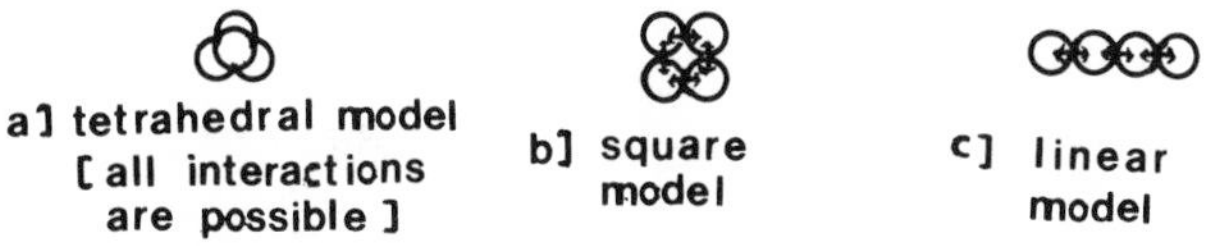

a] tetrahedral model
[all interactions
are possible]

b] square
model

c] linear
model

Fig. 13.3 Possible interactions between sub-units in a tetrameric protein.

Also with any oligomeric protein there exists the possibility that the concerted form of the general model, corresponding to the MWC model, may operate.

13.3.3 The KNF Model and Allosteric Regulation

In contrast to the MWC model, where the explanation for allosteric regulation is relatively straightforward, the KNF model allows for the possibility that allosteric modifiers may act in a variety of ways. For example, the modifier could bind to the same form of the enzyme as the substrate and cause either the same or a different conformational change to take place; also, the binding of the modifier might or might not prevent the subsequent binding of the substrate to the same sub-unit. The overall effect will depend on factors of this type and also on the various interaction constants involved.

13.4 DIFFERENTIATION BETWEEN MODELS FOR COOPERATIVE BINDING IN PROTEINS

There are a variety of ways of investigating whether the cooperative binding of a ligand to a protein results from a mechanism based on the concerted (MWC) model or on some other model. As we discussed earlier (Chapter 13.2.4), if positive homotropic cooperativity is observed and the **Hill coefficient** is found to be equal to the number of binding sites, then it is likely that the MWC model is operating. On the other hand, if **negative homotropic cooperativity** is found, then the MWC model is excluded. For example, Koshland and colleagues (1968) have reported that the binding of NAD^+ to rabbit muscle **glyceraldehyde 3-phosphate dehydrogenase** shows negative homotropic cooperativity, so the MWC model cannot be operating in this instance. The reaction catalysed by this enzyme is:

D-glyceraldehyde 3-phosphate + NAD^+ + P_i $\rightleftharpoons$

3-phospho-D-glyceroyl phosphate + NADH + H^+.

Relaxation studies (see Chapter 7.2.2) are ideally suited for the investigation of processes involving conformational changes, since these are likely to be extremely rapid and difficult to follow by any other technique. Kirschner, Eigen and co-workers (1966) have used the **temperature-jump method** to investigate the binding of NAD^+ to yeast glyceraldehyde-3-phosphate dehydrogenase, which has 4 binding sites for this coenzyme; rate constants for three different processes could be identified from the results, the slowest process being independent of NAD^+ concentration. It was concluded that this supported the MWC model, the two fastest processes being the binding of NAD^+ to the R- and T-forms of the enzyme, and the slowest being the $R_4 \rightleftharpoons T_4$ transformation: four processes dependent on NAD^+ concentration would have been expected if the KNF model was operating. Thus it would appear that the binding of NAD^+ to yeast glyceraldhyde 3-phosphate dehydrogenase proceeds in a different way from the binding of NAD^+ to the same enzyme from rabbit muscle. However, it will be realised that if two processes have very similar rate constants, it is likely that they would appear to be a single process in relaxation studies. Hence, in general, findings from such studies must be supported by independent evidence before a firm conclusion can be reached.

The binding curves predicted by the MWC and KNF models are not exactly the same, and on this basis computers may be able to help determine the most probable model in a particular instance if supplied with suitable experimental data. Needless to say, the degree of accuracy and reliability required of such data is extremely high.

With some proteins it may be possible to investigate the fractional saturation of each conformational form at different ligand concentrations by the use of optical rotation or spectroscopic techniques. Again this may help to distinguish between possible binding models.

13.5 SIGMOIDAL KINETICS IN THE ABSENCE OF COOPERATIVE BINDING

13.5.1 Ligand-Binding Evidence versus Kinetic Evidence

Kinetic studies are often performed to investigate possible cooperative binding of a substrate to an enzyme since they are often easier to carry out than direct binding studies; cooperative binding effects are reflected in non-hyperbolic Michaelis-Menten plots and in departures from linearity of Lineweaver-Burk and similar plots derived from the Michaelis-Menten equation. However, such kinetic findings cannot be said to prove the existence of cooperative binding unless there is corroborative evidence: the Michaelis-Menten plot only follows exactly the characteristics of the binding plot if the reaction is straightforward and proceeds at too slow a rate to significantly affect the equilibria of the binding processes (Chapter 12.4). Hence, if direct binding studies show that substrate-binding is not a cooperative process, but corresponding kinetic studies show non-hyperbolic Michaelis-Menten plots and departures from linearity in other primary plots, then it must be concluded that the kinetic mechanism of the reaction is not consistent with all of the Michaelis–Menten assumptions (Chapter 7.1.1); some situations where this might be found are discussed below.

It should also be mentioned that if an enzyme preparation contains a mixture of isoenzymes having different K_m values, then both binding and kinetic plots may show irregularities which are not due to cooperative binding.

13.5.2 The Ferdinand Mechanism

Ferdinand (1966) showed that a random-order ternary-complex mechanism for a two-substrate enzyme-catalysed reaction can lead to sigmoidal kinetics being observed in the absence if cooperatinve binding:

It is assumed that one of the pathways, let us say the one via E. AX, is kinetically preferred and will proceed faster than the other, even though both pathways are possible. Also, the affinity of E for B is less than that of $E.AX$ for B. We will investigate this system at constant $[E_o]$ and $[B_o]$ but variable $[AX_o]$.

At low $[AX_o]$, E will react mainly with B and so the reaction will proceed via the slower pathway $E \rightleftharpoons E.B \rightleftharpoons E.AX.B \rightarrow$ products. At higher $[AX_o]$ there will be a switch-over to the faster pathway $E \rightleftharpoons E.AX \rightleftharpoons E.AX.B \rightarrow$ products. E will be depleted as a result of its rapid reaction with AX, so the EB which is formed will tend to dissociate back to $E + B$ rather than to proceed to $E.AX.B$; this will provide yet more E to react with AX and ensure maximum utilisation of the faster pathway.

Therefore, a graph of v_o against $[AX_o]$ at constant $[B_o]$ will be sigmoidal, even where there is no possibility of cooperative binding; the surge in the curve is explained by the switch-over from the slow to the rapid pathway.

Jensen and Trentini (1970) have shown that the reaction catalysed by **phospho-2-keto-3-deoxyheptonate aldolase** from *Rhodomicrobium vannielli* may be of this type. The enzyme catalyses the reaction

phosphoenolpyruvate + erythrose 4-phosphate $\rightleftharpoons$

7–phospho–2–keto–3–deoxyarabinoheptonate + P_i

and the preferred pathway is that where phosphoenolpyruvate binds first.

13.5.3 The Rabin and Mnemonical Mechanisms

Rabin (1967) has shown that even a single-substrate reaction catalysed by an enzyme with a single binding site for the substrate can show sigmoidal kinetics, provided the enzyme can exist in more than one conformational form. Consider the sequence for the forward reaction in (I) below.

E is assumed to be thermodynamically more stable than the other conformational form of the enzyme, E'. The rate-limiting step of the whole sequence is $ES \rightarrow E'S$, $E' \rightarrow E$ also being slow. The formation of $E'S$ from $E' + S$ will be appreciable provided free E' is present.

At low substrate concentrations, the overall rate of the reaction $ES \rightarrow E'S \rightarrow E' + P$ will be very slow compared to the rate of the reaction $E' \rightarrow E$; therefore, the amount of free E' present will be low, and the supplementary pathway $E' + S \rightarrow E'S$ will not be used.

At higher substrate concentrations, the rate of the reaction $ES \rightarrow E'S \rightarrow E' + P$ will be increased, so E' will be formed faster than it can be converted back to E; therefore, appreciable amounts of E' will be present. This will result in the formation of more $E'S$ via the supplementary route $E' + S \rightarrow E'S$ and so a further increase in the rate of product formation.

Hence, as soon as the substrate concentration is high enough to produce E' appreciably faster than it can be converted back to E, the supplementary pathway $E' + S \rightarrow E'S$ comes into operation and the overall rate of reaction escalates. A plot of v_o against $[S_o]$ at fixed total enzyme concentration will therefore be sigmoidal.

Allosteric modifiers could act on such a system by increasing or decreasing the rates of the isomerisation steps $E' \rightarrow E$ and/or $ES \rightarrow E'S$.

A variation of the Rabin mechanism which has been proposed for some two-substrate reactions, e.g. that catalysed by rat liver glucokinase, is the mnemonical mechanism shown in (II) above. Glucokinase is a monomeric enzyme, so co-operativity is out of the question, but the reaction exhibits sigmoidal kinetics with respect to variable glucose (G) concentration at high concentrations of MgATP, possibly because the latter prevents the $E/E'/E'G$ system coming to equilibrium.

SUMMARY OF CHAPTER 13

Cooperative binding in oligomeric enzymes can be explained by the Monod-Wyman-Changeux (MWC) or the Koshland-Némethy-Filmer (KNF) hypothesis; these are seen as special cases of a more general hypothesis. Only the MWC model can explain binding characteristics consistent with the Hill equation, but this model cannot explain negative homotropic cooperativity.

Allosteric inhibitors usually increase, and allosteric activators usually decrease, the degree of positive homotropic cooperativity in the binding of a substrate. The MWC model explains this on the basis of allosteric inhibitors stabilising a conformational form of the enzyme which does not bind to the substrate, and allosteric activators stabilising one which does. In the KNF model, allosteric modifiers may act in a variety of ways.

Sigmoidal kinetics can be seen in the absence of cooperative binding and may be a consequence of the kinetic mechanism of the reaction.

FURTHER READING

Cornish-Bowden, A. (1979), *Fundamentals of Enzyme Kinetics* (Chapter 8), Butterworth.

Engel, P. C. (1977), *Enzyme Kinetics* (Chapter 7), Chapman and Hall.

Ferdinand, W. (1976), *The Enzyme Molecule* (pages 186-207), Wiley.

Kurganov, B. I. (1982), *Allosteric Enzymes,* Wiley.

Newsholme, E. A. and Start, C. (1973), *Regulation in Metabolism* (Chapter 2), Wiley.

Tipton, K. F. (1979), Kinetic properties of allosteric and cooperative enzymes, in Bull, A. T., Lagnado, J. R., Thomas, J. D. and Tipton, K. F. (eds.), *Companion to Biochemistry,* Volume 2, Longman.

PROBLEMS

13.1 Deoxycytidine monophosphate deaminase catalyses the reaction:

$$dCMP + H_2O \rightleftharpoons dUMP + NH_3.$$

The reaction was investigated in the absence and presence of a fixed concentration (1 mmol.l^{-1}) dTTP, the same concentration of enzyme being present in every case. The following results were obtained:

Conc^n of dCMP (mmol.l^{-1})	Initial velocity of reaction (μmol NH$_3$ produced min^{-1}.mg protein^{-1})	
	In absence of dTTP	In presence of dTTP
0.65	0.56	0.43
0.77	0.78	0.50
0.89	1.28	0.57
1.00	1.74	0.73
1.13	2.25	1.08
1.35	3.06	1.74
1.62	3.68	2.64
1.90	4.03	3.24
2.46	4.26	3.95
3.09	4.31	4.26.

The enzyme was found to dissociate into a number of identical sub-units and ultra-centrifuge studies were performed on enzyme and sub-units. The sedimentation coefficient (s) of the sub-unit was 4.6×10^{-13}s and of the enzyme 1.6×10^{-12}s. The corresponding diffusion coefficients (D) were 5.96×10^{-7} cm^2.s^{-1} and 5.31×10^{-7} cm^2.s^{-1} (all values corrected for water at 20°C). The partial specific volume ($\bar{v}$) in each case is 0.736, and the density of the water (ρ) at 20°C is 0.998. What can you conclude from the data? (Note that according to Svedberg's equation, molecular weight $= \dfrac{RTs}{D(1 - \bar{v}\rho)}$ where $R = 8.314 \times 10^7$ erg.K^{-1}.mol^{-1} and T = temperature (°K).

13.2 Alcohol dehydrogenase catalyses the reaction:

$$\text{acetaldehyde} + \text{NADH} \rightleftharpoons \text{ethanol} + \text{NAD}^+.$$

An alcohol dehydrogenase enzyme was investigated as follows.

(a) A series of equilibrium dialysis experiments was performed. In each case 2.5 mmol.l^{-1} enzyme was present in the dialysis bag, and dialysed against different initial concentrations of NADH in a suitable buffer. No ethanol, acetaldehyde or NAD$^+$ was present. The volume of liquid within the bag was very small compared to the total volume of liquid present, and remained constant throughout each experiment. Each system was allowed to come to equilibrium, then a sample of the solution surrounding the bag was removed, diluted 1 in 100 in buffer, and the absorbance at 340 nm determined in 1 cm cells against a distilled water blank. The following results were obtained:

Initial NADH conc^n
(nmol.l^{-1})   4   5   6   7   8   10   12

Absorbance   0.044 0.059 0.081 0.103 0.131 0.193 0.274

                        14   16   18   20

                        0.367 0.473 0.582 0.694   .

(Molar extinction coefficient of NADH at 340 nm $= 6.22 \times 10^3$.)

(b) The rate of the forward reaction (as written above) at different initial
concentrations of NADH and NAD$^+$ was investigated, and the results given
below. The enzyme concentration was the same in each case. The initial
conc^n of acetaldehyde was always 3 mmol.l^{-1}, and the initial conc^n of
ethanol zero.

Initial conc^n NADH (mmol.l^{-1})	Initial conc^n NAD (mmol.l^{-1})	Absorbance (340 nm) at time t (min) $=$					
		0.5	1.0	1.5	2.0	2.5	3.0
1.25	0	0.470	0.455	0.440	0.427	0.413	0.400
1.25	1	0.472	0.460	0.447	0.434	0.423	0.410
1.25	2	0.475	0.463	0.452	0.440	0.430	0.423
1.5	0	0.467	0.447	0.430	0.410	0.395	0.378
1.5	1	0.468	0.453	0.437	0.420	0.405	0.390
1.5	2	0.470	0.458	0.445	0.435	0.418	0.410
2.0	0	0.460	0.437	0.413	0.390	0.370	0.350
2.0	1	0.463	0.443	0.422	0.400	0.380	0.365
2.0	2	0.467	0.449	0.430	0.412	0.398	0.383
2.5	0	0.457	0.430	0.403	0.377	0.353	0.330
2.5	1	0.460	0.437	0.413	0.388	0.365	0.346
2.5	2	0.464	0.442	0.420	0.400	0.380	0.362
3.33	0	0.454	0.423	0.393	0.362	0.333	0.307
3.33	1	0.457	0.430	0.400	0.373	0.348	0.325
3.33	2	0.460	0.437	0.412	0.386	0.365	0.345
5.0	0	0.450	0.413	0.378	0.342	0.310	0.280
5.0	1	0.452	0.420	0.385	0.353	0.320	0.302
5.0	2	0.455	0.425	0.395	0.364	0.338	0.314
10.0	0	0.442	0.400	0.360	0.317	0.278	0.242
10.0	1	0.445	0.405	0.364	0.325	0.290	0.260
10.0	2	0.445	0.407	0.370	0.332	0.303	0.277

(c) Equilibrium isotope exchange studies showed an increased rate of transfer of label from NADH to NAD^+ with increased concentration of acetaldehyde (maintaining a constant [ethanol] : [acetaldehyde] ratio) at all concentrations of acetaldehyde tested.

What can you conclude about the enzyme mechanism from this data?

13.3 A single-substrate enzyme-catalysed reaction gave the following results at fixed enzyme concentration in the presence or absence of a fixed concentration (10 mmol.l^{-1}) of an inhibitor A.

Initial substrate conc^n (mmol.l^{-1})		Product conc^n (μmol.l^{-1}.min^{-1}) at time t (min)					
		$t = $ 0.5	1.0	1.5	2.0	2.5	3.0
0.5	uninhibited	4.0	7.5	11.0	15.0	19.0	22.0
0.5	inhibited	0.15	0.3	0.45	0.6	0.75	0.9
1.0	uninhibited	9.5	19.0	28.5	38.0	47.5	57.0
1.0	inhibited	1.3	2.5	3.7	5.0	6.3	7.5
2.0	uninhibited	35.0	69.0	104	138	172	206
2.0	inhibited	10.0	20.0	30.0	40.0	50.0	60.0
5.0	uninhibited	165	330	495	660	825	990
5.0	inhibited	119	238	357	476	595	714
10.0	uninhibited	339	677	1020	1350	1690	2030
10.0	inhibited	357	714	1070	1430	1790	2140
20.0	uninhibited	438	876	1310	1750	2190	2630
20.0	inhibited	476	952	1430	1900	2380	2860
50.0	uninhibited	480	959	1440	1920	2400	2880
50.0	inhibited	499	997	1500	1990	2490	2990
100.0	uninhibited	490	980	1470	1960	2450	2940
100.0	inhibited	500	1000	1500	2000	2500	3000

Gel filtration experiments gave a single band for the enzyme corresponding to a molecular weight of about 211,000 daltons. SDS-electrophoresis experiments also gave a single band for the enzyme, corresponding to a molecular weight of 70,500 daltons.

What can be concluded from these results about the mechanism of the reaction and the nature of the inhibition by A?

CHAPTER 14

The Significance of Sigmoidal Behaviour

14.1 THE PHYSIOLOGICAL IMPORTANCE OF COOPERATIVE OXYGEN-BINDING BY HAEMOGLOBIN

Since many proteins show evidence of cooperative ligand-binding, it is reasonable to ask if there is a physiological advantage in them doing so. In the case of haemoglobin, the advantages of its cooperative oxygen-binding are easy to see.

Haemoglobin is a tetrameric protein which can bind four molecules of oxygen in a sigmoidal manner (Chapter 12.9). It is found in the blood of vertebrates, where it transports oxygen from the alveoli of the lungs to the capillaries of muscle and other tissue; there the oxygen is released to diffuse into the cells. These cells possess no haemoglobin, but they do contain its monomeric relative, **myoglobin**, which has only one binding site for oxygen and so must bind in a hyperbolic fashion. The myoglobin can store the oxygen and, when required, facilitate its transport to **cytochrome oxidase** in the inner mitochondrial membrane of the cell, where the gas completes its physiological role and accepts electrons which have passed down the respiratory pathway.

If we consider the binding curves for haemoglobin and myoglobin (Fig.14.1) we see that the partial pressure of oxygen in the capillaries of the lung is sufficient to cause saturation of both haemoglobin and myoglobin. The oxygen tension in the tissues is much lower, but it is still sufficient for the almost complete saturation of myoglobin; however, haemoglobin, because of its different binding characteristics, is only about 40% saturated at this lower pO_2.

Thus we see that the sigmoidal oxygen-binding of haemoglobin provides a mechanism by which 60% of the oxygen taken up in the lungs can be released in the capillaries of the tissues; if myoglobin was to act as an oxygen carrier in blood, a much greater differential in oxygen tension between lungs and tissues would be required before the required amount of oxygen could be released.

In general, fractional saturation is far more sensitive to changes in ligand concentration if the binding mechanism is sigmoidal rather than hyperbolic: according to the simple binding equation for a non-cooperative system (Chapter 12.2), an 80-fold increase in ligand concentration is required to change the

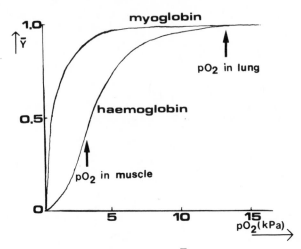

Fig. 14.1   Graphs of fractional saturation ($\overline{Y}$) against pO_2 for the binding of oxygen to myoglobin and haemoglobin.

fractional saturation from 0.1 to 0.9; in contrast, according to the MWC equation for the situation where a high degree of positive homotropic cooperativity is present (Chapter 13.2.4), this identical change in fractional saturation can be brought about by a 4-fold increase in ligand concentration (also see Fig. 14.2).

In the case of haemoglobin, the degree of positive homotropic cooperativity, and hence of sigmoidal binding, is influenced by heterotropic factors: in isolation, its $(S_{0.5})_{oxygen}$ is about 0.1 kPa, which is much the same to that of myoglobin; however, the presence of **2,3-bisphosphoglycerate (BPG)** in red blood cells increases the $(S_{0.5})_{oxygen}$ value of haemoglobin *in vivo* to about 3.5 kPa (which is the value shown in Fig. 14.1). BPG fits into the central cavity of deoxyhaemoglobin and forms, through its phosphate groups, electrostatic interactions with three positively charged groups in each of the β-chains; the central cavity of oxyhaemoglobin is too small to contain BPG, so the binding of oxygen results in the ejection of the modifier. BPG may be considered to have some of the characteristics of a MWC allosteric inhibitor, since it stabilises the conformational form of the protein which has least affinity for oxygen.

The sigmoidal nature of oxygen-binding is also made more marked by increasing the concentration of H^+ and CO_2, and the binding of these in turn is affected by the partial pressure of oxygen. Three protons are taken up by haemoglobin as oxygen is released because two terminal amino groups and one histidine residue are in more negatively charged environments in deoxyhaemoglobin and can more readily accept a proton; carbon dioxide may bind to any of the four terminal amino groups of haemoglobin to form a carbamate group

$$RNH_2 + CO_2 \rightleftharpoons RNHCO_2^- + H^+$$

and again this takes place more readily when the haemoglobin is in the deoxy form.

Muscle produces a great deal of CO_2 and H^+ as metabolic end-products, and the presence of these in the muscle capillaries facilitates the release of oxygen from haemoglobin. This in turn makes the binding of protons and CO_2 to the haemoglobin more favourable, enabling them to be carried through the venous system back to the alveolar capillaries of the lungs. There the high oxygen tension results in the binding of oxygen to haemoglobin and the concomitant release of protons and CO_2; this enables excess CO_2 to be removed from the body in expired air. The interrelation between the binding to haemoglobin of oxygen, H^+ and CO_2 is termed the **Bohr effect**.

14.2 ALLOSTERIC ENZYMES AND METABOLIC REGULATION

14.2.1 Introduction

Before we go on to discuss the possible roles of allosteric enzymes in metabolic regulation, we must first briefly consider the environment in which this regulation takes place. In previous chapters we have usually restricted our discussions about the properties of enzymes to those observed under simple *in vitro* conditions, e.g. at fixed concentration of enzyme in a dilute solution in the absence of product and of other enzymes. In contrast, one of the essential features of living cells is that conditions such as concentrations of substrate, product and enzyme are, to a greater or lesser degree, constantly changing; also concentrations of enzymes *in vivo* are usually considerably greater than those used *in vitro* for steady-state investigations; furthermore, the reaction catalysed by one enzyme is always linked to reaction catalysed by other enzymes, forming a **metabolic pathway** in a highly organised environment.

Hence it cannot be blindly assumed that findings made *in vitro* are applicable to the situation *in vivo*: for example, although it has been shown *in vitro* that the H_4 isoenzyme of LDH is inhibited by high concentrations of pyruvate (pyruvate and NAD^+ forming a dead-end ternary complex with the enzyme), it is not certain that pyruvate concentrations *in vivo* could ever reach high enough levels for this to be of significance in metabolic regulation; also, the characteristics of the LDH isoenzymes may vary according to whether they are freely soluble or membrane-bound, as some may be *in vivo*. For these and other reasons (see Chapter 5.2.2), extensive *in vitro* investigations have so far failed to establish beyond doubt the physiological roles of the isoenzymes of LDH.

The total concentration of each enzyme present in a cell is determined by the rate of its **synthesis** and the rate of its **breakdown**. The former is influenced by such factors as **induction** and **repression** (mainly in prokaryotic cells) and by the presence of **hormones** (agents of translational or transcriptional control in eukaryotic cells), thus ensuring that each cell synthesises only the enzymes required at that time (see Chapter 3.1.5). Similarly, the breakdown of enzymes (catalysed by proteolytic enzymes) is subject to some degree of control: in general, enzymes are broken down more rapidly when they represent the only source of energy available to the cell (e.g. in starvation) or when some change in function is taking place which requires the synthesis of different proteins (e.g. in germinating seeds); in addition to this, large enzymes and those involved in metabolic control tend to be broken down more rapidly than others. On the other hand, many enzymes are more resistant to breakdown if their substrates are present in high concentrations.

Control mechanisms which affect the rates of synthesis of degradation can only serve as **coarse** (or **long-term**) **agents of metabolic regulation**, because at least several minutes are likely to elapse before they can bring about a significant change in the total concentration of the enzyme in question; also, mechanisms of this type, and those hormonal mechanisms which control metabolism by regulating the entry of substrates into cells, are likely to affect a group of cells rather than a single one. Hence, in order to meet the needs of the moment in each individual cell, **fine** (or **acute**) **mechanisms of metabolic regulation** exist in which the **activity** of an enzyme, rather than its total **concentration**, is controlled. It is this subject of metabolic regulation by the control of enzyme activity that particularly concerns us in the present chapter.

14.2.2 Characteristics of Steady-State Metabolic Pathways

Metabolic pathways often contain branch points, at which metabolites may enter or leave by alternative routes, as in the following example:

$$P \rightleftharpoons \rightleftharpoons Q \rightleftharpoons \rightleftharpoons R \rightleftharpoons \rightleftharpoons S \rightleftharpoons \rightleftharpoons T.$$
$$\begin{array}{ccc} \updownarrow & \updownarrow & \updownarrow \\ \updownarrow & \updownarrow & \updownarrow \\ X & Y & Z \end{array}$$

The sequences between successive branch points may be regarded as separate units, for each usually contains a regulated step. Let us consider one such unbranched sequence of steps, as follows:

$$A \overset{E_1}{\rightleftharpoons} B \overset{E_2}{\rightleftharpoons} C \overset{E_3}{\rightleftharpoons} D \rightleftharpoons E \rightleftharpoons F.$$

Such a system is not likely to be found at equilibrium *in vivo*, because it would then be unable to provide any free energy for the organism (Chapter 6.1.5). If, instead, it is assumed to be at steady-state, then there will be a net flux through the system in one particular direction (let us say A to F) and the concentrations of all intermediates will be constant. This implies that A is being fed into the system (by metabolic routes or by transport from outside the cell or cellular compartment) at a constant rate and F is being dissipated (by further metabolic routes or by transport) at the same rate; this rate must also be the net flux through the system, i.e. the difference between the rates of the forward and back reactions for each step and for the overall process.

Although this steady-state assumption must obviously be a simplification of the situation *in vivo*, it is nevertheless a reasonable one: concentrations of metabolic intermediates are usually maintained within quite a narrow range within the living cell. In general, the level of each intermediate is found at a concentration slightly below the $(S_{0.5})$ value for the enzyme which utilises it as substrate, i.e. the next one in the sequence. Therefore, in the example being considered, the concentration of B is usually slightly below the $(S_{0.5})_B$ value of the enzyme E_2.

This general non-saturation of enzymes is an important condition for the setting up and maintenance of a steady-state: if, for example, enzyme E_2 was usually found to be very nearly saturated with B, this would mean that there was just enough E_2 present to handle the B being produced from A at its normal rate, but without there being any margin of safety; if the rate of production of B from A should increase, the reaction $B \to C$ could not be speeded up by any significant factor in response to this and so the concentration of B would rise, possibly catastrophically. Although metabolic regulation might limit the rate of production of B from A, it is clearly essential that sufficient E_2 is present to cope with the maximum rate at which B is likely to be produced, and that is what is found. Also, since all reactions are reversible to some degree, the maintenance of a steady-state in a particular direction requires that the concentrations of substrate and product for each enzyme bear such a relationship to the characteristics of the enzyme that the forward reaction is favoured, e.g. if $[B] = (S_{0.5})_B$ for enzyme E_2, then $[C] < (S_{0.5})_C$ for the same enzyme.

Since the system is at steady-state, none of the individual steps can be at equilibrium. An indication of the disequilibrium of each step can be obtained from the **mass action ratio** (Γ) as compared to the **apparent equilibrium constant** (K'_{eq}). For the step $B \rightleftharpoons C$, $\Gamma = \dfrac{[C]}{[B]}$. The **disequilibrium ratio** (ρ) is defined such that $\rho = \dfrac{\Gamma}{K'_{eq}}$ so, for the step $B \rightleftharpoons C$, $\rho = \dfrac{[C]}{[B]K'_{eq}}$. If the step is near equilibrium, $\Gamma \cong K'_{eq}$ and $\rho \cong 1$; if the step is some way from equilibrium, $\Gamma \ll K'_{eq}$ and $\rho \ll 1$.

By extending the treatment of simple steady-state kinetics (Chapter 7.1.2) to a system in which the product concentration cannot be ignored, it may be shown that, for the enzyme-catalysed reaction $B \rightleftharpoons C$,

$$v_f = \frac{V_{max}^B K_m^C [B]}{K_m^B K_m^C + K_m^C [B] + K_m^B [C]}$$

and

$$v_b = \frac{V_{max}^C K_m^B [C]}{K_m^B K_m^C + K_m^C [B] + K_m^B [C]}$$

where v_f is the rate of the forward reaction and v_b the rate of the back reaction.

$$\therefore \frac{v_b}{v_f} = \frac{V_{max}^C K_m^B [C]}{V_{max}^B K_m^C [B]}.$$

But, from the Haldane relationship (Chapter 7.1.7),

$$K_{eq}' = \frac{V_{max}^B K_m^C}{V_{max}^C K_m^B}.$$

$$\therefore \frac{v_b}{v_f} = \frac{[C]}{[B]K_{eq}'} = \rho.$$

The disequilibrium ratio for the overall sequence $A \rightleftharpoons F$ is the product of the disequilibrium ratios of each of the individual steps, and must be less than 1 for a steady-state system in which there is net production of F from A. In fact, although in such a system $v_f - v_b$ must be the same for each step and must be positive, many steps in metabolic pathways are found to be quite near to equilibrium; this can be consistent with the steady-state assumption provided both v_f and v_b are large.

14.2.3 Regulation of Steady-State Metabolic Pathways by Control of Enzyme Activity

Let us now consider how the steady-state discussed in Chapter 14.2.2 will be affected by changing the activity of an enzyme. We will compare the results for an enzyme catalysing a step which is near equilibrium with those for one catalysing a step far from equilibrium; for simplicity, we will assume in each case that the characteristics of the enzymes are not altered, so that the change in activity is equivalent to a change in concentration of the enzyme.

First let us consider the situation where ρ for $B \rightleftharpoons C$ is 0.99, i.e. the reaction is almost at equilibrium. The overall rate of reaction, v, is given by

$$v = v_f - v_b$$

$$= v_f - 0.99 v_f = 0.01 v_f.$$

The steady-state expressions for v_f and v_b show that both of these terms are proportional to $[E_o]$ (since V_{max}^C and V_{max}^B must both be proportional to $[E_o]$), and so the overall rate, v, must also be proportional to $[E_o]$.

If the effective concentration of $[E_o]$ is halved, then the immediate effect is that the values of v, v_f and v_b will all be halved. Since the rate of formation of B from A and the rate of conversion of C to D are not immediately affected, the concentration of C will start to fall, which will reduce the rate v_b still further without significantly altering v_f. When [C], and thus v_b, falls to such a level that the overall rate of reaction has been restored to its original value, then a new steady-state will have been set up. In fact it may be shown that this is achieved when [C] has fallen by only 1%; at this new value of [C], the new rate for the forward reaction, v_f', is given by $v_f' = 0.5v_f$, and the new rate of the back reaction, v_b', by $v_b' = 0.5.0.99v_b$; the new disequilibrium ratio, ρ', is given by $\rho' = 0.99\rho$, since [C] is reduced to 99% of its original value and [B] and K_{eq}' are unchanged.

According to the general relationship derived in Chapter 14.2.2,

$$\frac{v_b'}{v_f'} = \rho'.$$

$$\therefore \; v_b' = \rho'v_f' = 0.99\rho v_f' = 0.99.0.99v_f'$$

$$= 0.99.0.99.0.5v_f$$

$$= 0.49v_f.$$

Therefore, under these new conditions, the overall rate of reaction, v', is given by

$$v' = v_f' - v_b'$$

$$= 0.50v_f - 0.49v_f$$

$$= 0.01v_f.$$

This is the same as the original value, and so the flux through the whole system is unchanged.

If, in contrast to the above, the value of ρ for $B \rightleftharpoons C$ is 1×10^{-4}, i.e. the reaction is far from equilibrium, the overall rate of reaction,

$$v = v_f - v_b$$

$$= v_f - 10^{-4}.v_f$$

$$\simeq v_f.$$

Therefore, for such a reaction, v_f is approximately equal to the net flux through the system. If the effective concentration of $[E_o]$ is halved, the values of v, v_f and v_b will all be halved, as before. However, in this instance, compensation cannot come from a further decrease in v_b, for this is negligible to start with, so the overall rate will always be less than its original value. This step will therefore be rate-limiting for the overall process, and a new steady-state will be set up with a lower net flux than before.

Thus it may be seen that an enzyme catalysing a step which is near equilibrium is not likely to be important in metabolic regulation, because a considerable change in enzyme activity may take place without affecting the flux through the system. In contrast, an enzyme catalysing a step which is far from equilibrium may well be important in metabolic regulation: such steps are often rate-limiting, and changes in enzyme activity are accompanied by changes in the flux through the system.

Although the above discussion was concerned with systems where changes in enzyme activity are not accompanied by changes in enzyme characteristics, the general conclusions are equally valid for systems where enzyme characteristics may be affected, as is the case with most allosteric enzymes. **Allosteric inhibitors** usually increase the sigmoidal nature of Michaelis–Menten plots, so the effect of the inhibition is most marked at low and moderate substrate concentrations, and possibly non-existent under conditions where the substrate is plentiful and need not be conserved; **allosteric activators**, on the other hand, usually increase the hyperbolic character of Michaelis-Menten plots. Regardless of this, most allosteric modifiers act on enzymes catalysing steps far from equilibrium, i.e. where $\rho < 1 \times 10^{-2}$.

In pathways where the net flux may be in either direction *in vivo*, each regulatory step is usually associated with two enzymes: one catalyses the reaction in largely irreversible fashion in one direction while the other is responsible for the flux in the opposite direction; in general, for the reaction $B \rightleftharpoons C$, the isoenzyme with the lower $(S_{0.5})_B$ value will catalyse the reaction $B \rightarrow C$, while the isoenzyme with the lower $(S_{0.5})_C$ value will catalyse $C \rightarrow B$. An example of such a step occurs in glycolysis/gluconeogenesis with the interconversion of fructose 6-phosphate and fructose 1,6-bisphosphate (see Chapter 14.2.5).

In general, steps subject to allosteric inhibition occur early in metabolic sequences. For example, in the metabolic sequence whose main direction of flux is as follows

$$\rightarrow A \overset{E_1}{\rightarrow} B \rightarrow C \rightarrow D \rightarrow E \overset{E_5}{\rightarrow} F \overset{\nearrow}{\searrow}$$
$$\downarrow$$

the enzyme E_1, which catalyses the first step **committed** to the synthesis of F, might be inhibited by F or by one or more of the subsequent metabolites; this arrangement ensures that if F and its metabolic products are plentiful, more A is made available to be metabolised by alternative routes and excessive build-up of intermediates such as B and C is prevented. In most cases, this **switching** between alternative pathways is most sensitive when controlled by inhibitors which affect $(S_{0.5})$ rather than V_{max} values, since the concentrations of most metabolites are found to be well below the levels required for enzyme saturation (Chapter 14.2.2). Examples of allosteric inhibitors are discussed in Chapters 13.2.3, 14.2.5 and 14.2.6.

In contrast to the above, steps subject to allosteric activation often occur late in a metabolic sequence: thus, enzyme E_5 in the sequence under consideration might be activated by A or by a metabolic precursor of A; this arrangement helps to prevent levels of metabolic intermediates falling excessively low. An example of allosteric activation is the effect of fructose 1,6-bisphosphate on **pyruvate kinase**, the final enzyme in the glycolytic pathway; other examples are given in Chapters 14.2.5 and 14.2.6.

14.2.4   Allosteric Enzymes and the Amplification of Metabolic Regulation

Most modifiers of allosteric enzymes act to change the degree of sigmoidal character of a Michaelis-Menten plot, so having greatest effect on metabolic flux when the substrate concentration is low or moderate, and least effect when the substrate concentration is high (Chapter 14.2.3); this applies regardless of whether the modifier primarily affects the substrate binding or the kinetic characteristics of the enzyme, although most known examples fit into the former category.

In many cases, the process by which the modifier itself binds to the enzyme has sigmoidal characteristics, and this makes a significant contribution to the regulatory effect. The degree of inhibition or activation produced must depend on the amount of modifier bound to an enzyme, so the sensitivity of a particular mechanism of metabolic control must depend on whether a given change in modifier concentration produces a large or a small change in the fractional saturation of the enzyme with modifier, all other factors being equal. If the modifier concentration *in vivo* is approximately equal to the $(S_{0.5})$ value for its binding to the enzyme, a not unreasonable assumption, then a system with a sigmoidal modifier-binding process will show far greater sensitivity to changes in modifier concentration than will one with a hyperbolic modifier-binding process (Fig. 14.2). Thus the sigmoidal binding of a modifier may be said to **amplify** the regulatory effect.

The sensitivity of fractional saturation to changes in ligand concentration for processes with different binding characteristics was discussed in general terms in Chapter 14.1.

14.2.5   Other Mechanisms of Metabolic Regulation

Although in the present chapter we are concerned chiefly with the role of allosteric enzymes in metabolic regulation, it is necessary to mention briefly some other regulatory mechanisms: in some instances these may act in association with allosteric regulation, but not in every case.

One important factor in the regulation of metabolism is the availability of coenzymes such as NAD^+, NADH, ATP, ADP and AMP. Often these are

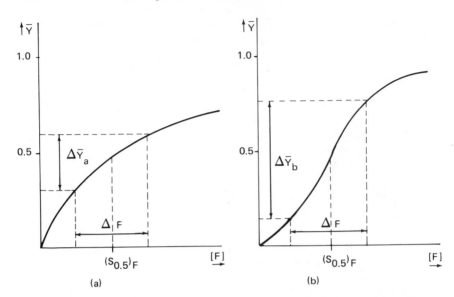

Fig. 14.2 Graphs showing the change of fractional saturation of a protein with ligand, F, with changes of [F], at constant protein concentration, for (a) a system where the binding of F is hyperbolic, and (b) a system where the binding of F is sigmoidal. Consider, for example, the situation where this represents the binding of a feedback inhibitor, F, to the enzyme E_1 in the metabolic sequence $A \xrightarrow{E_1} B \rightarrow C \rightarrow D \rightarrow E \rightarrow F$. If a given change in concentration of F, [ΔF], takes place near the $(S_{0.5})_F$ value in each case and produces a change in fractional saturation $\Delta \overline{Y}_a$ for the hyperbolic system and $\Delta \overline{Y}_b$ for the sigmoidal system, it can be seen that $\Delta \overline{Y}_b$ is much larger than $\Delta \overline{Y}_a$ and will cause a greater increase in the sigmoidal character of the binding curve for the substrate, A, all other factors being equal. In other words, a regulatory system where the modifier binds by a sigmoidal process will show far greater sensitivity to changes in modifier concentration in this range than will a system where modifier-binding is hyperbolic.

co-substrates or co-products of the reaction being regulated (Chapters 11.5.2 and 11.5.4), but in some instances they may bind to allosteric sites (e.g. **phosphofructokinase** appears to have two binding sites for ATP: one forms part of the active site, whilst the other lies elsewhere on the molecule and has a purely regulatory function).

In general, the affinities of enzymes for coenzymes of this type are strong enough to ensure that the binding sites are more or less saturated at all times, i.e. the $(S_{0.5})$ values are considerably smaller than the physiological concentrations of the coenzymes (n.b. this is one reason why they are regarded as coenzymes and not simply as co-substrates or co-products). Also, these coenzymes exist in alternative forms (e.g. NAD^+ and NADH) which resemble each other sufficiently for both to be able to bind to the same site on the enzyme, so there will be competition between them for binding. The total concentration of

$NAD^+ + NADH$, and of $ATP + ADP + AMP$, within a cell is usually approximately constant, so the regulatory effects of these coenzymes depends on concentration *ratios* (e.g. of $[NAD^+]$ to $[NADH]$) rather than on their individual concentrations. This applies regardless of whether the coenzyme functions simply as a co-substrate/co-product or whether it functions also (or instead) as an allosteric modifier; note, however, that equilibrium as well as non-equilibrium reactions may be regulated by co-substrate/co-product concentration ratios.

Another important factor in the control of metabolism within the cell, and one which could not be deduced from simple *in vitro* studies, is the phenomenon of **cellular compartmentation**. The different organelles present in eukaryotic cells (Fig. 14.3) possess different enzymes and metabolic pathways from each other and from the cytosol. These organelles are enclosed by membranes which, to a greater or lesser extent, prevent the passage of molecules between organelle and cytosol, except where this passage is under the control of a specific carrier transport system. Therefore, the control of the transport of substrates, products and cofactors across membranes provides another mechanism for the regulation of metabolism.

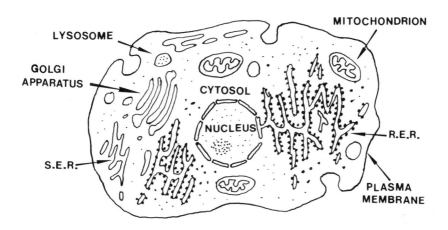

Fig. 14.3   Diagrammatic representation of a 'typical' animal cell, showing the presence of the various membrane-enclosed organelles. The part of the cell between the nucleus and the plasma membrane is termed the cytoplasm. (R.E.R. = rough endoplasmic reticulum, i.e. that to which ribosomes are attached; S.E.R. = smooth endoplasmic reticulum.)

For example, the β-oxidation of fatty acids proceeds in mitochondria, but the fatty acids must first be activated to fatty acyl-CoA in the cytosol; the fatty acid moiety can only enter a mitochondrion if coupled to the cofactor carnitine, so any factor affecting the formation of fatty acyl–carnitine, and thus the transport of activated fatty acids into mitochondria, will affect the overall rate of fatty acid oxidation.

The inner mitochondrial membrane is impermeable to NAD^+ and NADH, but reducing equivalents may enter and leave a mitochondrion in the form of malate (Fig. 14.4); in contrast, adenine nucleotides may cross the mitochondrial membranes, but only in exchange for other adenine nucleotides (regardless of their phosphate content). Thus there is clear evidence for compartmentation of nucleotide coenzymes in eukaryotic cells; there is also some evidence for this in prokaryotic cells, which is somewhat surprising in view of the lack of membrane-enclosed organelles in these cells.

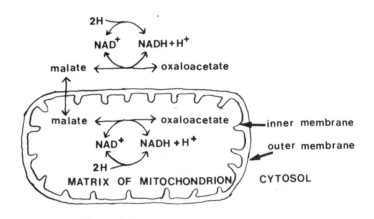

Fig. 14.4 Transport of reducing equivalents as malate between a mitochondrion and the cytosol of a cell. This process is termed the malate shuttle.

One reason for compartmentation within the cell is to provide a suitable environment for specialised enzymes: for example, the pH within lysosomes is low to facilitate the operation of the hydrolytic enzymes found there, most of which have a pH optimum of around 4-5.

Another reason for compartmentation is to prevent unnecessary conflict between metabolic pathways which happen to have a common intermediate but which are otherwise completely independent. For example, carbamoyl phosphate may be metabolised via the urea cycle as part of a mechanism for removing excess nitrogen (especially toxic ammonia) from the cell or it may be used for the synthesis of pyrimidine nucleotides (see Chapter 14.2.6). Both processes take place in the mammalian liver, but the former utilises carbamoyl phosphate synthesised in mitochondria from ammonium ions and bicarbonate:

$$NH_4^+ + HCO_3^- + 2ATP \rightarrow H_2N.CO.OPO_3^{2-} + 2ADP + P_i$$

whereas the latter proceeds from carbomoyl phosphate synthesised in the cytosol and using glutamine rather than ammonia as the nitrogen source; N-acetyl glutamate activates the **carbamoyl phosphate synthetase** enzyme present in mitochondria, but not that found in cytosol. Since there is normally little transport of carbamoyl phosphate across the mitrochondial membranes, this compartmentation enables the two different pathways to be regulated independently of each other.

Compartmentation may also take place in a chemical rather than a physical manner: a metabolite may exist in several isomeric forms, only one of them being able to act as substrate for a particular enzyme; alternatively, a metabolite may be able to bind to a metal ion, affecting metabolite-enzyme affinity (e.g. **aconitate hydratase** will convert isocitrate to citrate, but will not use Mg-isocitrate as substrate). Enzyme activity may also, of course, depend on metal ion concentration (Chapter 11.4). All of these factors may be able to play a role in metabolic regulation.

Enzymes which can exist in active and inactive forms usually have a very important regulatory function. In Chapter 5.1 we saw that many hydrolytic enzymes are synthesised as inactive zymogens and activated by the removal of peptide fragments as required; this is a crude mechanism, however, since the zymogens cannot be reformed from the active enzymes. Of more interest from the point of view of fine control of metabolism are systems where the enzyme is activated (or, in some cases, inactivated) by phosphorylation or adenylation: this is termed **covalent modification**.

Consider, for example, **glycogen phosphorylase** of animal muscle cells, which catalyses, in a largely irreversible fashion, the reaction

glycogen (n residues) + P_i → glucose 1-phosphate + glycogen (n−1 residues).

The enzyme exists in two forms: phosphorylase *b* is strongly inhibited *in vivo* by glucose 6-phosphate, and in any case is only active when AMP binds in an allosteric site, which rarely happens under physiological conditions because of competition by ATP for the same site; phosphorylase *a*, which has a phosphorylated serine residue on each of its two sub-units, is fully active under most physiological conditions. The largely inactive phosphorylase *b* and the active phosphorylase *a* may be interconverted by a system involving two enzymes, **phosphorylase kinase** and **phosphorylase phosphatase**.

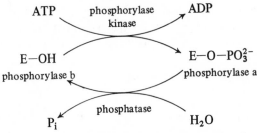

Phosphorylase kinase also requires phosphorylating for full activity: this is done by the action of another kinase (**protein kinase**), which is allosterically activated by cAMP. Hence an increased concentration of cAMP within a cell activates protein kinase, which in turn activates phosphorylase kinase, which in turn converts glycogen phosphorylase b to phosphorylase a, which in turn facilitates the breakdown of glycogen under physiological conditions. **Cascade systems** of this type provide another mechanism for the **amplification** of metabolic regulation, since small changes in concentration of a key substance (in this case cAMP) can have a considerable effect on metabolism; indeed, far greater amplification is possible from a cascade process than can be contributed by the sigmoidal binding of a modifier to a single allosteric enzyme.

Yet another amplification mechanism exists where there is an apparently **futile cycle** of intermediates. The best known example is the interconversion of fructose 6-phosphate (F—6—P) and fructose 1,6-bisphosphate (FBP), catalysed by the enzymes **phosphofructokinase (PFK)** and **fructose 1,6-bisphosphatase (FBPase)**:

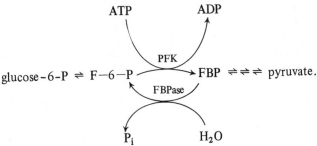

PFK and FBPase both catalyse more or less irreversible reactions, and these reactions both appear to be taking place under most conditions, irrespective of the direction of flux through the system. Since the reaction catalysed by PFK is accompanied by the breakdown of ATP, this set-up seems to be wasteful of energy (hence the term "futile cycle").

However Newsholme (1973) pointed out that this may be tolerated by the organism because the arrangement permits an extremely sensitive control mechanism to operate. A high [AMP]/[ATP] ratio results in inhibition of FBPase and activation of PFK; this favours glycolysis and the production of ATP at the expense of ADP and AMP (see Chapters 5.2.2 and 11.5.4). A low [AMP]/[ATP] ratio results in inhibition of PFK: this favours gluconeogenesis and the utilisation of ATP. Thus it may be seen that ATP controls its own production and utilisation.

The amplification of regulation which may be achieved by the existence of a futile cycle is far greater than that possible for a single enzyme, and can enable the direction of overall flux to be reversed (note that for a near-equilibrium reaction being catalysed by a single enzyme, an allosteric modifier will usually affect the forward and back reactions in a similar fashion, as discussed in Chapter 14.2.3; in contrast, a modifier of the enzyme involved in a futile cycle may affect the forward and back reaction in different ways, i.e. may activate one and inhibit the other, thus providing a much more sensitive mechanism for metabolic regulation).

It is of interest to note that, whereas glycogen breakdown is catalysed by **glycogen phosphorylase**, an enzyme activated by cAMP via a cascade system (as discussed above), glycogen synthesis is catalysed by a separate enzyme, **glycogen synthase**, which is *inactivated* by cAMP, a similar cascade system being involved. Also, the synthesis of cAMP is stimulated by the hormone **epinephrine** (**adrenaline**), via yet another cascade system.

14.2.6 Some Examples of Allosteric Enzymes Involved in Metabolic Regulation

There is no typical mechanism for the fine control of metabolism: already we have discussed examples of a variety of regulatory processes, most of them involving allosteric enzymes. In the previous section (Chapter 14.2.5) we considered how the involvement of PFK and FBPase in a futile cycle provides a very sensitive mechanism for metabolic regulation; we also saw how covalent modification of enzymes such as muscle glycogen phosphorylase can result in an amplification of regulation; many other allosteric enzymes act without being involved in such systems.

Aspartate transcarbamoylase (ATCase), also called aspartate carbamoyl transferase, from *E. coli* was the first enzyme in which the active and regulatory sites were shown to be clearly separated: indeed, not only are they found in different parts of the molecule, they are located on different sub-units (Fig. 14.5). Although this has facilitated the investigation of the characteristics of the enzyme, it is an unusual finding: most allosteric enzymes are oligomers consisting of *identical* sub-units.

ATCase catalyses the reaction:

$$
\begin{array}{l}
CO_2^- \\
| \\
CH_2 \\
| \quad {}^+ \\
HC-NH_3 \\
| \\
CO_2^-
\end{array}
\quad + \quad
\begin{array}{l}
O \\
\| \\
C-OPO_3^{2-} \\
| \\
NH_2
\end{array}
\quad \rightleftharpoons \quad
\begin{array}{l}
CO_2^- \\
| \\
CH_2 \\
| \\
HC-NH-C \\
| \\
CO_2^-
\end{array}
\begin{array}{l}
O \\
\| \\
 \\
NH_2
\end{array}
\quad + \quad P_i.
$$

aspartate   carbamoyl   N-carbamoyl
                phosphate   aspartate

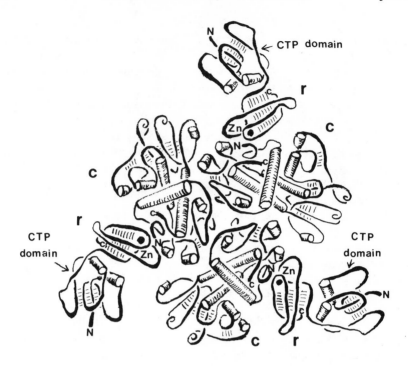

Fig. 14.5   A simplified representation of the three-dimensional structure of half
(c_3r_3) a molecule of *E. coli* ATCase as revealed by the x-ray diffraction studies of
Lipscomb and colleagues (1978). The active sites are at or near the interface
between adjacent c sub-units. The other c_3r_3 unit is behind the one shown, contact
being via the r sub-units. (Conventions as for Fig. 2.8.)

This is the first in a sequence of reactions leading to the biosynthesis of the
pyrimidine nucleotides UMP, UDP, UTP and, finally, CTP. Two important
control mechanisms regulate pyrimidine formation in *E. coli*: the synthesis of
ATCase is repressed in the presence of uracil, and the metabolic end-product,
CTP, controls its own synthesis by feedback inhibition of ATCase. Gerhart and
Pardee (1962) showed that CTP increases the sigmoidal nature of a Michaelis–
Menten plot for the reaction catalysed by ATCase, V_{max} being unchanged,
whereas the activator ATP (a purine nucleotide) decreases the sigmoidal nature
of the curve. From this, they and others developed the concept of allosteric
regulation.

Gerhart and Schachman (1965) dissociated the enzyme by the action
of p–mercuribenzoate and showed that one type of sub-unit (c) possesses
all the catalytic activity while another (r) contains the regulator sites. The
reaction as catalysed in the absence of the regulatory sub-units shows hyperbolic
characteristics and is not affected by ATP or CTP.

Sequence analysis by Weber (1968) and x-ray diffraction studies by Lipscomb and co-workers (1968) showed that the holoenzyme structure is $c_3(r_2)_3c_3$, two catalytic trimers (each of molecular weight 100,000 daltons) being separated by three regulatory dimers (each of molecular weight 33,000 daltons). The structural integrity of the holoenzyme is stabilised by Zn^{2+} ions.

Each regulatory dimer can bind two molecules of CTP, but although the two sub-units are apparently identical, the affinity for the first molecule of CTP is considerably greater than that for the second; the precise explanation for this negative homotropic cooperativity has yet to be found. Also, it is known that ATP and CTP compete for the same site, probably causing different conformational changes to take place when they bind.

The evidence currently available (e.g. from stopped-flow and relaxation measurements) suggests that the homotropic interactions (between catalytic sites) involve a concerted mechanism, whereas the heterotropic interactions (between catalytic and regulatory sites) involve a mechanism which is sequential.

The modifiers of most allosteric enzymes are relatively small molecules; however, **lactose synthase** (Chapter 5.2.3) is an example of an enzyme whose allosteric modifier (α-lactalbumin) is very large. Although α-lactalbumin may be regarded as a sub-unit of the enzyme, it binds only weakly to the other protein component (galactosyl transferase), but in doing so it completely changes the specificity of the enzyme, enabling glucose to act as a substrate. Morrison and Ebner (1971) concluded from their studies that the reaction proceeds by a compulsory-order mechanism, as follows:

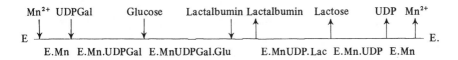

In the absence of α-lactalbumin, glucose dissociates very rapidly from the enzyme; in the presence of α-lactalbumin, glucose can bind to the enzyme long enough for a reaction to take place.

In the cell, galactosyl transferase is membrane-bound and apparently forms a multi-enzyme complex with other glycosyl transferases. Another multi-enzyme complex which is subject to allosteric regulation is pyruvate dehydrogenase (Chapter 5.2.5).

E. coli **pyruvate dehydrogenase** is activated by AMP and inhibited by ATP. The regulation of the mammalian system has similar features, but is more complicated than for *E. coli* pyruvate dehydrogenase and involves covalent modification: the mammalian complex actually contains **kinase** and **phosphatase** enzymes. Reed (1974) has shown that the kinase inactivates the complex by catalysing the phosphorylation of a serine residue in the decarboxylase-dehydrogenase component (E_1); this kinase is itself activated by acetyl–CoA and NADH, and inhibited by ADP. The phosphorylase, which catalyses the removal of this phosphate, is activated by Ca^{2+} and Mg^{2+} and inhibited by ATP and NADH. Thus, increased ratios of [ATP]/[ADP], [NADH]/[NAD^+] and [acetyl-CoA]/[CoASH] within mitochondria, where the pyruvate dehydrogenase complex is located, will tend to reduce its overall activity.

A situation where a metabolite is passed directly from one enzyme to another, whether in a soluble multienzyme complex or in an organised assembly of membrane-bound enzymes, would appear to be ideal for sensitive metabolic regulation.

SUMMARY OF CHAPTER 14

The fractional saturation of a protein with a ligand is far more sensitive to changes in ligand concentration if the binding is sigmoidal than if it is hyperbolic. This is important in the physiological roles of haemoglobin and of allosteric enzymes.

The **fine** (or **acute**) **control** of metabolic regulation frequently involves the activation or inhibition of allosteric enzymes, particularly of those catalysing reactions which are far from equilibrium. Allosteric inhibitors usually act on enzymes occurring early in metabolic sequences, whereas allosteric activators may act on enzymes lying towards the end of metabolic pathways: this helps to maintain the concentrations of most metabolic intermediates within reasonable limits.

The effects of allosteric modifications are most marked at low or moderate substrate concentrations, when it is particularly important that the substrate is utilised (or conserved) according to the precise needs of the cell. Sigmoidal binding of allosteric modifiers provides a mechanism for the **amplification** of regulation. Other amplification mechanisms include covalent modification and the establishment of "futile" cycles.

Metabolic regulation *in vivo* may be extremely complex: such factors as substrate or enzyme compartmentation and ratios of concentrations of pairs of coenzymes (e.g. [ATP]/[ADP]) may be involved, together with allosteric effects.

FURTHER READING

Atkinson, D. E. (1977), *Cellular Energy Metabolism and its Regulation* (Chapter 5), Academic Press.

Cohen, P. (ed.) (1980), *Recently Discovered Systems of Enzyme Regulation by Reversible Phosphorylation,* Elsevier.

Denton, R. M. and Pogson, C. I. (1976), *Metabolic Regulation,* Chapman and Hall.

Hammes, G. G. (1982), *Enzyme Catalysis and Regulation* (Chapters 8 and 9), Academic Press.

Holtzman, E. and Novikoff, A. B. (1984), *Cells and Organelles,* 3rd edition, Saunders.

Hoppe, J. and Wagner, K. G. (1979), cAMP-dependent protein kinase I, a unique allosteric enzyme, *Trends in Biochemical Sciences,* **4** (pages 282–285).

Kantrowitz, E. R., Pastra-Landis, S. C. and Lipscomb, W. N. (1980), *E. coli* aspartate transcarbamylase, *Trends in Biochemical Sciences,* **5** (pages 124–128 and 150–153).

Koshland, D. E. (1984), Control of enzyme activity and metabolic pathways, *Trends in Biochemical Sciences,* **9** (pages 155–159).

Scrimgeour, K. G. (1977), *Chemistry and Control of Enzyme Reactions* (Chapters 13 and 14), Academic Press.

Part 3

Application of Enzymology

CHAPTER 15

İnvestigation of Enzymes in Biological Preparations

15.1 CHOICE OF PREPARATION FOR THE INVESTIGATION OF ENZYME CHARACTERISTICS

It should be clear from previous chapters (particularly 7.1.3 and 14.2.1) that there is no ideal preparation in which to investigate the properties of an enzyme: different properties may have to be investigated in different preparations. **Highly-purified preparations** (see Chapter 16.2) are essential for the investigation of enzyme structure; they also allow enzyme characteristics (e.g. K_m and k_{cat}) to be determined and reaction mechanisms to be studied without complications arising from the presence of other enzymes and reactions. However, the removal of an enzyme from its natural environment may give a distorted impression of its characteristics *in vivo* and make the elucidation of its physiological role more difficult. In order to get the complete picture, therefore, it is necessary to investigate enzymes in a purified form and also in preparations where some cellular organisation has been retained.

The investigation of enzymes in single-cell micro-organisms is relatively straightforward, but the extra organisation of animals and plants presents further problems. **Whole animal** (or **plant**) **experiments** can give useful information about metabolic pathways (e.g. by studying the metabolic fate of ingested or injected radioactive isotopes), but reveal little about the function of individual enzymes.

Tissue slice techniques, as introduced by Warburg (1923), enable *in vitro* studies to be carried out on cells which retain a considerable degree of their organisation. A thin slice is cut from a tissue using a sharp implement such as a **microtome** (a device incorporating a specimen holder, a sharp blade and an advance mechanism, so that slices of reproducible thickness may be cut from the tissue). The slice is usually about 0.5 mm thick, because a thinner one would contain too large a portion of cells damaged during the slicing, and a thicker one would present problems relating to diffusion (see later). After preparation, the sliced tissue is incubated in a suitable medium at controlled pH and temperature, and the uptake and release of substances of interest may be measured. Therefore, in contrast to *in vivo* studies, the metabolism being investigated is that of one particular type of tissue, e.g. liver. However, it is still difficult to relate the results to any particular enzyme; also, in the absence of an intact vascular system, ingredients from the medium can only reach the innermost cells by diffusion through the others.

An alternative approach to the study of a tissue, and one which avoids causing damage to any cells, is the use of **perfusion techniques**. An intact tissue is removed from an animal and placed in a fluid environment which provides sufficient oxygen to keep the cells functioning while they are investigated. If the tissue is small and thin enough for each cell to be near the exterior, it may be incubated directly in the oxygenating medium; otherwise, it must be oxygenated by perfusion through blood vessels, when uptake and release of substances via these vascular routes may also be investigated. In either case, the main difference from the situation *in vivo* is that the isolated tissue is not linked to a nervous system.

Problems that result from different cells of a tissue not being equally accessible to the medium may be overcome by separating the cells from each other (i.e. by preparing a **tissue culture**): this is often done by treating the isolated tissue with collagenase or other proteolytic enzyme to break down the matrix and enable the individual cells to be dispersed; the conditions of proteolysis should be sufficiently mild so that each isolated cell retains most of its *in vivo* characteristics, but it is very difficult to prevent some damage occurring to surface proteins. Under favourable conditions, isolated cells in culture may grow and divide, exactly like micro-organisms, and thus increase the number of cells available for investigation. The main technical problem in long-term culture is the prevention of contamination, particularly by micro-organisms, but there is also a tendency for cells to lose their precise function (i.e. to de-differentiate) as time passes. So far, the most successful cultures of normal (as opposed to tumour) cells have come from skin biopsies, while cultures from liver samples have proved very difficult to grow.

All investigations of enzymes in intact cells must be complicated by transport through the **plasma membrane** (i.e. the membrane enclosing each cell). A simple way to overcome this problem is to prepare a **cell-free system**. This may be done by the **homogenisation** of a tissue in a suitable isotonic medium (e.g. 0.25 M sucrose), the aim usually being to disrupt the cells and release the contents without damaging the sub-cellular organelles; alternatively, a cell-free system may be obtained by **reconstitution** from previously isolated organelles and cytoplasm. Homogenates of soft tissues, such as liver, are often prepared by forcing the tissue and medium through the narrow gap (about 0.3 mm) between the walls of a glass tube and a close-fitting pestle. This pestle, which is made of teflon or glass, is moved up and down the tube like a piston and also rotated, either by hand (as in the TenBroeck apparatus) or by an electric motor (as in the Potter–Elvehjam homogeniser). Plants, whose cell walls contain a considerable amount of structural material, and tough, fibrous animal tissues, such as heart, may be homogenised by the use of a Waring blender (which resembles a domestic kitchen blender): the rapidly rotating blades successfully disrupt the cells, but unfortunately tend to disrupt the sub-cellular organelles as well. Cell-free preparations are frequently produced from micro-organisms by subjecting them to high-frequency sound waves (**sonication**); this procedure too causes considerable damage to cell membranes. It is important to maintain low temperatures throughout the production of *in vitro* preparations, for reasons discussed in Chapter 15.2.1.

Cell-free preparations provide a convenient and reproducible means of investigating enzymes and metabolism in the presence of some important cellular components, and interesting deviations from findings in pure enzyme preparations are sometimes revealed: for example, different estimates for the K_m of glucokinase have been obtained in liver homogenates and in pure solution. However, regardless of the method by which cell-free systems are prepared, a considerable degree of cellular organisation is lost in the process; hence the results obtained still cannot be taken to be an exact representation of what occurs in the intact organism. For the same reason, there should be caution in the interpretation of findings made with organelles which have been isolated from other cellular components by density-gradient centrifugation of some other method (see Chapter 15.3.2).

In summary, therefore, it is only possible to investigate the characteristics of individual enzymes in preparations where cellular organisation is totally or partially absent, so the characteristics observed are not necessarily identical to those existing *in vivo*; the differences between the preparation and the whole organism must always be borne in mind during the interpretation of results.

15.2 ENZYME ASSAY

15.2.1 Introduction

The purpose of enzyme assay is to determine how much of a given enzyme, of known characteristics, is present in a tissue homogenate, fluid or partially purified preparation. It is important that the specimens be treated carefully prior to assay, if the results are to have any meaning. The maintenance of cellular organisation and integrity is a characteristic of life, and cellular destruction by natural processes (**autolysis**), accompanied by changes in enzymes and breakdown of cofactors, commences on the death of an organism or the isolation of a tissue. Autolysis is minimised if a tissue is kept cold, so specimens for enzyme assay are usually stored at temperatures below $4°C$, both before and after the preparation of homogenates. Many enzymes are easily **denatured** at high or even moderate temperatures, so this is a further reason why all types of specimens or preparations should be kept cold prior to enzyme assay. (Note that these factors apply to the preparation of specimens for the investigation of enzyme characteristics as well as for enzyme assay.)

Since changes due to autolysis and denaturation must increase with time of storage, it will be apparent that the assays should be performed with a minimum of delay. If some delay is unavoidable, it may be necessary to store the specimens at very low temperatures (e.g. $-60°C$) to prevent loss of enzyme activity. However, freezing of tissues might increase the disruption of cells, so there can be no general rule about storage: the stability of the enzyme in question at various temperatures and the nature of the specimen or preparation (see Chapter 17.3) all need to be taken into account.

15.2.2 Enzyme Assay by Kinetic Determination of Catalytic Activity

By far the most convenient way to estimate the concentration of a particular enzyme in a preparation or fluid is to determine its catalytic **activity**, i.e. to find out how much substrate it is capable of converting to product in a given time under specified conditions. As discussed in Chapter 7.1.2, the initial velocity of an enzyme-catalysed reaction taking place under conditions where the Briggs-Haldane steady-state assumptions are valid is given by

$$v_o = \frac{k_2 [E_o][S_o]}{[S_o] + K_m}$$

where k_2 is the rate constant relating to product formation (i.e. k_{cat}).

Therefore, at constant $[S_o]$, $v_o \propto [E_o]$.

In other words, for any system where these assumptions are valid (see Chapter 7.1.3), the initial velocity of a reaction catalysed by an enzyme is directly proportional to the concentration of that enzyme at fixed substrate concentration. Hence there is a linear relationship between enzyme concentration and catalytic activity.

Although in theory this relationship is true provided the substrate concentration is fixed, regardless of the actual value, in practice it is more reliably valid when the substrate concentration is high enough to be approximately saturating. There are two main reasons for this. Firstly, integration of the Michaelis–Menten equation (see Chapter 7.2.1) shows that the linear, steady-state phase of the reaction is more prolonged as the degree of enzyme saturation is increased (all other factors being equal): in simple terms, this is because the rate of utilisation of substrate becomes less significant in relation to the total concentration of substrate present as $[S_o]$ is increased. For this reason more accurate estimates of v_o can be obtained at high rather than at low $[S_o]$ values. Secondly, at high $[S_o]$, $([S_o] + K_m) \triangleq [S_o]$, so $v_o \triangleq k_2 [E_o]$ and v_o does not vary with small changes in $[S_o]$; this means that reproducible results can be obtained from an assay system without it being necessary to fix the substrate concentration at an extremely precise value, provided $[S_o]$ is sufficient to almost saturate the enzyme. Hence the setting-up of a reliable assay system is more convenient the higher the chosen value of $[S_o]$.

According to the Michaelis-Menten equation, an enzyme is only completely saturated by substrate at infinite substrate concentration, so it is necessary to talk in terms of near-saturation rather than complete saturation. This equation also enables the v_o/V_{max} ratio to be calculated for each initial substrate concentration, provided the K_m value is known; for a simple system, this ratio will also be the fractional saturation of the enzyme (see Chapter 12.4).

$$\begin{aligned}
\text{If} \quad [S_o] &= \ 0.5 \ K_m, \quad v_o = 0.33 \ V_{max}, \\
\text{if} \quad [S_o] &= \qquad K_m, \quad v_o = 0.50 \ V_{max}, \\
\text{if} \quad [S_o] &= \ 5 \ K_m, \quad v_o = 0.83 \ V_{max}, \\
\text{if} \quad [S_o] &= \ 10 \ K_m, \quad v_o = 0.91 \ V_{max}, \\
\text{if} \quad [S_o] &= \ 50 \ K_m, \quad v_o = 0.98 \ V_{max}, \\
\text{and} \quad \text{if} \quad [S_o] &= 100 \ K_m, \quad v_o = 0.99 \ V_{max}.
\end{aligned}$$

It should be clearly understood that the v_o/V_{max} ratios and fractional saturation values are independent of $[E_o]$, provided $[S_o] \gg [E_o]$, as discussed in Chapters 7.1.1 and 7.1.2. However the actual values of v_o and V_{max} do vary with $[E_o]$, which is the whole point of enzyme assay by kinetic methods.

It might seem from the above discussion that a procedure being used for enzyme assay could always be improved by increasing the substrate concentration, without limit, but of course other factors have to be taken into consideration: these include the solubility of the substrate, the cost of the substrate, and the possibility of substrate inhibition (see Chapter 8.2.6). Hence the actual substrate concentration chosen must be a compromise.

If a reaction involves more than one substrate, then the concentrations of each must be fixed, preferably at near-saturating levels.

The rate of an enzyme-catalysed reaction may also depend on the concentration of a cofactor (or cofactors). As with substrates, each cofactor must be present at a fixed concentration, and preferably in excess, if a reliable and reproducible system for enzyme assay is to be obtained.

Another pre-requisite for a successful assay system is that the reaction being catalysed should be capable of being accurately monitored, i.e. there should be a change in optical, electrical or other properties as substrate is converted to product (see Chapter 18.1). If this is not the case, it may be possible to follow the course of the reaction indirectly by coupling it to one where some such change does take place (Chapter 15.2.3).

Regardless of this, the reaction should be carried out under fixed and suitable conditions of pH, ionic strength and temperature. Enzymes are often assayed at their optimal pH (Chapter 3.2.2), but this is not essential: an enzyme might not operate at its exact optimal pH *in vivo*, so there is no fundamental reason for investigating its activity at this pH *in vitro*; also, the pH optimum may vary with temperature and other factors. Nevertheless, the pH chosen for an assay system must be sufficiently near the optimum for an appreciable rate of reaction to take place; it must also be one where the enzyme is relatively stable, for enzyme stability can vary with pH. With some reversible reactions, the pH may influence which direction is favoured: for example, assays involving **lactate dehydrogenase** are best performed in the direction of lactate production at pH 7, but in the direction of pyruvate formation at pH 10.

The presence of salts may affect enzyme-catalysed reactions by shifting the equilibrium of any of the steps involved; they may also effectively reduce the concentration of a substrate by complexing with it. Hence enzyme assays must be performed under carefully controlled conditions of ionic strength and composition.

The choice of temperature is governed by two conflicting factors: reaction rate and enzyme stability. Specimens are usually stored at low temperatures prior to assay in order to prevent loss of enzyme activity (Chapter 15.2.1); however, if they were assayed at these same temperatures the rate of reaction would be extremely low (or even zero), so the sensitivity of the assay would be very poor. At higher temperatures the reaction would proceed at a faster rate, giving a more sensitive assay, but the stability of the enzyme would decrease (see Chapter 3.2.3). An important consideration is the time-scale of the assay procedure: it is essential that there is no significant change in the activity of the enzyme over the period in which v_o is calculated, but immaterial what happens after that. Therefore, in general, it might be possible to use a particular temperature for a short assay procedure but not for a longer procedure involving the same enzyme (Fig. 15.1).

The most common temperature at which enzyme assays are performed are 25°C (as recommended by IUB in 1961), 30°C (the 1964 IUB recommendation) or 37°C (as preferred by clinical chemists). Many enzymes are somewhat unstable at 37°C, but this temperature provides conditions which are nearest approximation to those found in human beings and other mammals, and reaction rates are faster than at 25°C or 30°C. The rates of many enzyme-catalysed reactions are approximately doubled if the temperature is increased by 10°C in this range. Hence, regardless of which temperature is chosen for an assay system,

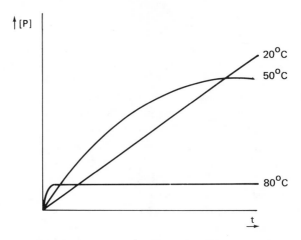

Fig. 15.1 Plots of product concentration against time for an enzyme-catalysed reaction at 20°, 50° and 80°C, without pre-equilibration of the enzyme at the reaction temperature, all other conditions being identical.

it is clear that it must be maintained to within about ± 0.2°C if the results are to be reproducible.

Enzyme activity is measured as the amount of substrate lost (or product gained) per unit time, and it should also be related in some way to the amount of specimen used for assay. In 1961 the enzyme commission of the IUB defined an **Enzyme Unit (U)**, later to be known as an **International Unit (IU)**, as the amount of enzyme causing loss of 1 μmol substrate per minute under specified conditions. Later, in 1973, the Commission on Biochemical Nomenclature introduced the **katal (kat)** as the **Système International (S.I.)** unit of enzyme activity: this is defined as the amount of enzyme causing loss of 1 mmol substrate per second under specified conditions. Both units are in current usage.

If, for convenience, the rate of reaction is being determined by a single reading rather than by continuous monitoring, it is necessary to demonstrate by a preliminary experiment whether a graph of [P] against t is linear over the whole of the incubation period; if it is not, the results cannot be expressed either as International Units or katals, and must be given in terms of the total incubation time (e.g. μmol substrate converted per 20 minutes).

Obviously there must be a direct relationship between the number of units of activity found and the amount of specimen assayed. Therefore, in order for the result of an assay to have any meaning, they must be expressed in terms of the amount of sample assayed, and, where appropriate, extrapolated back to the original tissue. So, for example, results might be given as International Units per litre blood plasma, or as katals per gram plant seed, and so on.

However even these units may not be ideal. For example, liver is likely to contain irregular deposits of glycogen and other non-protein material, so any results expressed as, say, katals per gram liver could vary according to which part of the same liver was cut off, weighed, homogenised and assayed. To minimise problems of this type, units of enzyme activity may be related to the **total protein** content of the sample being assayed, rather than to its weight or volume: this is termed **specific activity**, and may be expressed as, for example, International Units per mg protein, or as katals per kg protein. In the example being considered, a liver homogenate would be prepared and an aliquot used for enzyme assay; another aliquot of the same homogenate would be analysed for its total protein content (see Chapter 3.2.1), and so the total amount of protein in the sample used for enzyme assay could be calculated and related to the observed activity (see Problem 15.1).

Specific activity is quite different from **molar catalytic activity** (or **turnover number**, k_{cat}), the latter, by either name, being used to relate the observed activity of an enzyme to its known molar concentration, regardless of any other proteins which may be present. Molar catalytic activity may be expressed as, for example, katals per mole enzyme.

As discussed above, the conditions for the assay of an enzyme are chosen so that the concentrations of substrates, cofactors etc. are fixed and non-limiting. If this is successfully achieved, there will be a linear relationship between the initial velocity of the reaction and the concentration of enzyme in the incubation mixture. One way of testing if a system is satisfactory is to perform an assay and then repeat it using a different sample volume of the same enzyme preparation (and adjusting the volume of water or buffer added so that the total incubation volume is unchanged). The rates of reaction for the two assays should be different, but the calculated specific activity should be the same.

Enzyme assays based on the determination of catalytic activity can distinguish between **active** and **inactive** forms of enzymes, for they do not detect the presence of the latter. On the other hand, they cannot distinguish between **isoenzymes**, so the activity measured will be the sum of the contributions of all the active forms of the enzyme being assayed. Since these will not necessarily have the same molar activity, the relationship between activity and enzyme concentration could be less straightforward than suggested above. Even if it is known that only one isoenzyme is present, it cannot be assumed that its molar activity in, say, a homogenate is the same as that obtained in a pure preparation.

Hence kinetic assays are usually used to give an indication of the *relative* concentration of an enzyme in a preparation, without any attempt being made to interpret the results in terms of actual molar concentrations: the results are usually expressed simply in terms of units of activity.

15.2.3 Coupled Kinetic Assays

If easily-measurable changes take place as a particular enzyme-catalysed reaction proceeds, this reaction can readily be used for the kinetic assay of the enzyme which catalyses it (Chapter 15.2.2). Even if no such easily-measurable change takes place, it may still be possible to develop a kinetic assay for the enzyme by coupling the reaction to one where suitable changes do occur. For example, in the reaction catalysed by **alanine aminotransferase**,

$$\text{L-alanine} + \text{2-oxoglutarate} \rightleftharpoons \text{L-glutamate} + \text{pyruvate}$$

no reactant or product is coloured, nor does any one of them significantly absorb ultra-violet light. However, the reaction may be monitored spectrophotometrically if it is coupled to a second reaction, in which the pyruvate formed in the first reaction acts as a substrate for **lactate dehydrogenase (LDH)** and is reduced to lactate by the action of the coenzyme (and co-substrate) NADH:

$$\text{pyruvate} + \text{NADH} + \text{H}^+ \rightleftharpoons \text{lactate} + \text{NAD}^+.$$

NADH absorbs light at 340 nm, but NAD^+ does not (see Fig. 6.6), so the course of the reaction may be followed by monitoring at this wavelength. If conditions are chosen carefully, the rate of the second reaction will be an indicator of the rate of the first reaction.

Let us consider the general situation

$$A \xrightarrow{E_1} B \xrightarrow{E_2} C$$

where E_1 is the enzyme catalysing the **primary reaction** and E_2 the enzyme catalysing the **indicator reaction**. At zero time, the concentrations of B and C will be zero, and the concentration of A should be fixed and non-limiting (as discussed in Chapter 15.2.2). If there are any second substrates or cofactors for E_1 or E_2 (e.g. NADH is a co-substrate in the example where E_2 is LDH), then these should also be present initially at fixed and non-limiting concentrations. As the reaction proceeds, the concentration of B (pyruvate in the example where E_2 is LDH) will start to rise form its initial value of zero. In general, the rate of change of [B] is given by

$$\frac{d[B]}{dt} = v_1 - v_2$$

where v_1 is the velocity of the primary reaction and v_2 the velocity of the indicator reaction. As with the velocity of any single-enzyme system, v_1 should reach a constant value almost instantaneously and remain at this value over the period of interest (the next few minutes). In contrast, v_2 will have an initial value of zero (since [B] is initially zero) and will rise as [B] rises, according to the Michaelis–Menten equation

$$v_2 = \frac{V_{max}^B[B]}{[B] + K_m^B}$$

where V_{max}^B and K_m^B relate to the reaction catalysed by the enzyme E_2.

Therefore, over the period of interest

$$\frac{d[B]}{dt} = v_1 - \frac{V^B_{max}[B]}{[B] + K^B_m}.$$

For a satisfactory coupled assay system, the crucially important point is that [B] should reach a constant value very quickly. In other words, a steady-state should be set up for the overall system as quickly as possible and this should then remain in operation for the rest of the period of interest. Integration of the expression for $\dfrac{d[B]}{dt}$ shows that the time taken for this steady-state to be established (or at least very nearly established) is directly proportional to $\dfrac{K^B_m}{V^B_{max}}$. Now K^B_m is a constant characteristic of the enzyme E_2, but V^B_{max} is directly proportional to the concentration of E_2. Hence the larger the concentration of E_2, the quicker is the establishment of a near steady-state for the overall system, with $v_1 = v_2$. If there is so little E_2 present that V^B_{max} is less than v_1, then a steady-state can never be set up, [B] will continue to rise and v_2 can never reach the value of v_1. It is therefore important that sufficient of the indicator enzyme is added to a coupled assay system to ensure that there is a minimum of delay before a steady-state is established, and that the rate of the indicator reaction at steady-state is a reasonable approximation to the rate of the primary reaction. As with a single-enzyme system (Chapter 15.2.2), the validity of a coupled assay procedure can be checked by assaying different amounts of the same enzyme preparation; the specific activity should be found to be the same in each case.

One further factor which needs to be taken into account in the setting-up of a coupled assay procedure is the pH profiles of the various enzymes involved. Clearly it is essential that all the component enzymes are active at the pH used for assay, and the narrow ranges of activity of some enzymes pose problems in this respect. Consider, for example, the sequence:

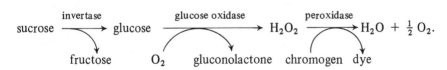

sucrose $\xrightarrow{\text{invertase}}$ glucose $\xrightarrow{\text{glucose oxidase}}$ H$_2$O$_2$ $\xrightarrow{\text{peroxidase}}$ H$_2$O + $\frac{1}{2}$ O$_2$.

A coupled assay procedure involving three enzymes, whilst more complicated than one involving two enzymes, may nevertheless be perfectly satisfactory provided the principles discussed above are adhered to. Hence, in theory, invertase could be assayed in presence of non-limiting amounts of sucrose, oxygen, glucose oxidase, peroxidase and a chromogen (e.g. guaiacum) and following the rate of appearance of the coloured dye. However, in practice, invertase is active around pH 5, glucose oxidase around pH 7-8 and peroxidase around pH 10.

Glucose oxidase and peroxidase are both sufficiently active at pH 8.5 to make practical a coupled system for the assay of glucose oxidase (or, more commonly, of the substrate glucose, according to the principles discussed in Chapter 17.2.2), but at no pH value are all three enzymes active enough to be coupled together.

15.2.4 Radioimmunoassay (RIA) of Enzymes

As a pre-requisite for RIA, the substance to be assayed must be an antigen (Ag) for which a specific antibody (Ab) can be obtained: such an antibody will have a specific site for the antigen. It must also be possible to obtain a pure specimen of antigen labelled with a radioisotope (Ag*).

The sample to be assayed is mixed with antibody and with a small amount of the labelled antigen. The following reactions take place, and are allowed to come to equilibrium:

$$Ag + Ab \rightleftharpoons Ag.Ab$$

$$Ag^* + Ab \rightleftharpoons Ag^*.Ab.$$

The antibody, with its bound antigen, is then separated from the free antigen and the distribution of radioactive isotope between the free and bound antigen fractions is investigated. It will be realised that the labelled antigen competes with the antigen in the sample for the available binding sites on the antibody, so the higher the concentration of antigen in the sample, the less radioactive antigen will be able to bind to the antibody and the greater will be the radioactive content of the free antigen fraction. In this way, the concentration of antigen in the sample can be estimated.

Antibodies for the RIA of enzymes may be prepared in the form of antisera by immunising rabbits with the required enzyme: for example, the blood of a rabbit immunised in the footpad with human pancreatic α-amylase contains sufficient antibodies within a few weeks to be usable as an antiserum. However, even so, there remains the problem of obtaining pure specimens of radioactively-labelled enzyme.

A further problem is that since the antigen is a protein like the antibody, it is more difficult to separate the free antigen from the antigen-body complex than would be the case if the antigen were small (when, for example, ion exchange resins might be employed). One way to overcome this problem is to use an antibody attached to an insoluble matrix. Another is to introduce further antigen–antibody interactions: for example, if an antiserum from rabbit is employed, the antigen–antibody complex might be precipitated with goat anti-rabbit γ-globulin.

The part of the enzyme which binds to the antibody is likely to be quite distinct from the active site, so RIA and catalytic assay of enzymes can give different information. Catalytically inactive forms of enzymes (e.g. proenzymes) may be detected by RIA if they contain the structural part which is recognised by the antibody. On the other hand, a RIA prodecure is likely to be specific for one particular isoenzyme (that used to produce the antibody). Hence it can be seen that catalytic assay and RIA procedures are not necessarily alternatives but can be used to complement each other. RIA's also have the advantage of being particularly sensitive (up to a thousand times more sensitive than catalytic assays). However their use has so far been restricted because of the problems of obtaining the required materials.

Antibodies produced as outlined above are always heterogeneous, i.e. they form a mixed population recognising a number of sites (determinants) on the antigen. However, Milstein and Köhler have shown that determinant-specific **monoclonal antibodies** may be produced by the following procedure: a mouse is immunised with an antigen and, several weeks later, cells from its spleen are removed and fused with mouse myeloma cells (cancer cells which can be cultured *in vitro*); hybrid cells with a desired immunological specificity may then be selected and cultured to produce a large number of identical copies (clones). These monoclonal antibodies are analytical reagents of extremely high sensitivity and specificity, and may be use in RIA or EIA (see Chapter 17.2.3) procedures. Milstein and Köhler were awarded the Nobel Prize for Medicine in 1984.

Antibodies (monoclonal or otherwise) may also be used to precipitate some isoenzymes from solution to allow others to be assayed by a kinetic procedure (see Chapter 19.1.1).

15.3 INVESTIGATION OF SUB-CELLULAR COMPARTMENTATION OF ENZYMES

15.3.1 Enzyme Histochemistry

The sub-cellular location of many enzymes may be revealed by microscopy, provided suitable fixation and staining procedures are followed.

Tissues are frozen to below $-20°C$ and sliced using a **cryostat** (a refrigerated microtome): this procedure ensures that the tissue is rigid, essential for obtaining thin slices (about $10\,\mu m$ thick) of satisfactory quality, and also minimises loss of enzyme activity. The slices are then usually fixed in **formaldehyde** (formalin), HCHO, to prevent any subsequent diffusion of proteins taking place. Formaldehyde brings about cross-linking between side chain amino groups of proteins.

$$
\begin{array}{ccc}
\text{Protein} & & \text{Protein} \\
| & & | \\
\text{NH}_2 & & \text{NH} \\
& & | \\
+ \text{ HCHO} \rightarrow & & \text{CH}_2 \\
& & | \\
\text{NH}_2 & & \text{NH} \\
| & & | \\
\text{Protein} & & \text{Protein}
\end{array}
$$

Most enzymes retain at least some activity after treatment; however, some do not, and these have to be investigated without prior fixation. In either case, the tissue section is treated with a buffered solution of a staining mixture containing the substrate of the enzyme under investigation; the staining process requires the conversion of this substrate to the appropriate product, and so can only take place in the regions where the enzyme is present. If possible, the product is immobilised after formation to prevent it diffusing away from the location of the enzyme.

For example, the substrate solution for the widely-used **lead–salt method** for **acid phosphatases** includes sodium β-glycerophosphate and lead ions in buffer at pH 5.5. Any acid phosphatase present in the tissue hydrolyses the glycerophosphate:

$$\beta\text{-glycerophosphate} \rightarrow \text{glycerate} + P_i$$

and the phosphate liberated reacts immediately with the lead ions to form insoluble lead phosphate ($PbPO_4$). This may be converted to a more conspicuous product, also insoluble, by developing with a solution of ammonium sulphide:

$$PbPO_4 \xrightarrow{(NH_4)_2S} PbS.$$

In this way, a black precipitate of lead sulphide may be visualised wherever acid phosphatase is located.

Many stains which show up clearly on **optical microscopy** are not suitable for use in **electron microscopy**, since their constituent atoms are no more opaque to an electron beam than are the atoms normally present in the tissue. Only atoms of large atomic weight, e.g. those of heavy metals, scatter an electron beam sufficiently to cause their locality to be significantly more opaque than the rest of the tissue. Thus the lead–salt method, discussed above, is one which can be applied to tissues being visualised by electron microscopy, since electron-opaque lead phosphate is precipitated in the regions where acid phosphatase is present. (Note that for electron microscopy, unlike optical microscopy, there is no advantage in developing to convert lead phosphate to lead sulphide, since both are equally electron-opaque.) However, ultra-thin sections (about 80 nm thick) are required for electron microscopy, so the sections prepared and stained as above must be dehydrated (replacing water with alcohol) and embedded in epoxy resin before being further sectioned on an **ultramicrotome**.

Such investigations have revealed that acid phosphatase is characteristically associated with lysosomes.

15.3.2 The Use of Centrifugation

The sedimentation characteristics of the various sub-cellular organelles are different, so it is possible to separate them by centrifugation of a tissue homogenate, and then to investigate which enzymes are associated with each cell fraction.

Any particle suspended in a liquid medium and being spun in a centrifuge is acted upon by a centrifugal field (G) which is determined by the angular velocity (ω) of the rotor and the radial distance (r) of the particle from the axis of rotation, according to the relationship

$$G = \omega^2 r = \left(\frac{2\pi.\text{RPM}}{60}\right)^2 r$$

where *RPM* is the number of revolutions per minute (rev. min^{-1}) of the rotor, ω normally being measured in radians per second (Fig. 15.2).

It is convenient to express the centrifugal field as a multiple of the gravitational constant (g), this being termed the **relative centrifugal field (RCF)**.

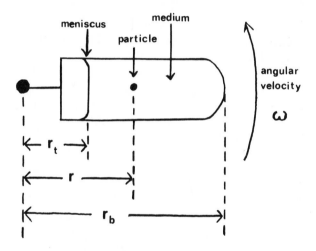

Fig. 15.2   Diagrammatic representation of a typical centrifugation procedure, as viewed from above. See text for discussion.

$$\text{RCF} = \frac{G}{g} = \left(\frac{2\pi.\text{RPM}}{60}\right)^2 \frac{r}{g}$$

But $g = 980 \text{ cm s}^{-1}$, so provided r is measured in centimetres,

$$\text{RCF} = 1.11 \times 10^{-5}.(\text{RPM})^2.r.$$

The sedimentation rate of the particle $= s.\omega^2 r$, where s is the sedimentation coefficient, i.e. the rate per unit centrifugal field. Its units, if the other terms are measured in the units stated above, are seconds, but a more practical unit is the Svedberg (S), of numerical value 10^{-13} seconds.

The sedimentation coefficient depends on the viscosity and density of the suspending medium and on the shape, size and density of the particle. For a spherical particle,

$$s = \frac{2(\rho_p - \rho)r_p^2}{9\eta}$$

where ρ_p = density of particle, ρ = density of medium, r_p = radius of particle and η = viscosity of medium.

The time (t) taken for the particle to sediment from the meniscus of the suspending medium to the bottom of the centrifuge tube is given by

$$t = \frac{9}{2(\rho_p - \rho)r_p^2} \cdot \frac{1}{\omega^2} \cdot \log_e \left(\frac{r_b}{r_t}\right) = \frac{1}{s} \cdot \frac{1}{\omega^2} \cdot \log_e \left(\frac{r_b}{r_t}\right)$$

where r_t = radial distance of liquid meniscus from the axis of rotation and r_b = radial distance of bottom of tube from axis of rotation (Fig. 15.2).

Thus it can be seen that all particles of a given size and density will have reached the bottom of the tube by a time determined by the applied centrifugal field. Note, however, that the density of the suspending medium is another relevant factor and that only those particles which are more dense than the medium will travel towards the bottom of the tube on centrifugation.

The simplest and most widely used method of separating the various sub-cellular organelles from each other is **differential centrifugation**, which developed from the work of Hogeboom and Schneider (1948). A tissue homogenate is prepared in a medium of low density (e.g. 0.25 M sucrose) and centrifuged in a series of stages, the centrifugal field of each step being higher than for the previous one. At the end of each stage the sedimented **pellet**, consisting of particles of similar sedimentation characteristics, is removed. A simplified scheme for the fractionation of a rat liver homogenate is shown in Fig. 15.3; similar schemes could be drawn up for the fractionation of homogenates of other animal and plant tissue.

Although heavy particles sediment faster than lighter particles, there is originally a homogeneous distribution of particles, so some relatively light particles must be present at the bottom of the tube when centrifugation commences. Hence each pellet must be contaminated by lighter particles which happen to be at the bottom of the tube throughout centrifugation. This problem can be overcome to some extent by resuspending each pellet in fresh homogenising medium and repeating the centrifugation. Another problem is that all sedimented particles tend to diffuse back up the tube towards less concentrated regions: this is of greatest significance when the applied centrifugal field is

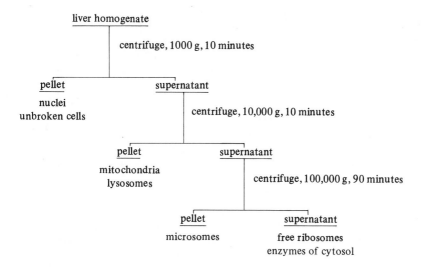

Fig. 15.3 Simplified scheme for the fractionation by differential centrifugation of a 10% rat liver homogenate in 0.25 M sucrose at 0°C. Note that microsomes are not organelles but particles produced during homogenisation, largely from endoplasmic reticulum, and constituting a specific sedimentation fraction.

low. A third problem with differential centrifugation is that some particles, e.g. mitochondria and lysosomes, show similar sedimentation characteristics ($s \simeq 2 \times 10^4 S$) because they are roughly the same size, even though they have quite different average densities.

Centrifugation techniques where the density of the suspending medium is not uniform throughout, i.e. **density-gradient centrifugation**, can help to overcome some or all of these problems. The simplest density gradients are step-wise (or discontinuous) ones, prepared by the careful layering of one solution on top of another in the centrifuge tube: for example, if a tissue homogenate or resuspended pellet in 0.25 M sucrose is layered on top of a pure solution of 0.5 M sucrose and centrifuged, the pellet formed will not be contaminated by lighter particles since no particles will be originally present at the bottom of the tube.

Even better separations of components can be achieved with continuous density gradients, prepared by the use of a **gradient-maker**. Sucrose or ficoll (a sucrose polymer) may be used as solute, the latter allowing the coverage of the same density range with smaller changes in osmolality.

In general, the aim of density-gradient centrifugation is to separate different groups of particles into zones along the density gradient, rather than to produce a pellet in the bottom of the tube. The **zonal rotor** (or **Anderson rotor**) is usually used in place of a simple centrifuge tube for these procedures. It is essentially a bowl with an assembly of vanes and ducts, designed so that a medium with a preformed density gradient may be introduced into it, after which a sample is applied to the top (least dense part) of the gradient; centrifugation is then performed, and finally the medium is run out and collected in fractions, still largely maintaining the original density gradient and the separation of the zones achieved during centrifugation.

Two main types of density-gradient zonal centrifugation may be performed: **rate-zonal** and **isopycnic-zonal centrifugation**. With **rate-zonal centrifugation**, the density gradient is shallow and covers a range below the density of the lightest of the particles being separated. The sample is applied to the top of the gradient and centrifugation is carried out for long enough to allow separation of the different zones, without allowing sufficient time for the heaviest particles to sediment completely. The purpose of the gradient is to minimise convection currents and thus prevent mixing of the zones.

Isopycnic (equal density) -zonal centrifugation employs a much wider density gradient, and centrifugation is continued until each particle sediments to the level at which its density is equal to that of the medium: it is then buoyant and will sediment no more.

Regardless of the method of separation, each fraction may then be investigated further. Enzymes still trapped in membrane-bound organelles (e.g. those inside mitochondria) may be liberated by techniques harsher than those used for tissue homogenisation (see Chapter 16.1.2). Further centrifugation can be used to seperate soluble enzymes liberated in this way from enzymes bound to the membrane of the organelle. Each fraction (or sub-fraction) may then be subjected to enzyme assay to find which enzymes are associated with it.

⋅ Two possible sources of error in the interpretation of such results should be pointed out: the presence of an enzyme in a certain fraction could be an artifact, due to the enzyme breaking away from a component of another fraction during homogenisation or fractionation; also, an enzyme might be present in a fraction but have no catalytic activity because of loss of cofactors or structural organisation during preparation of the fraction. Nevertheless, much useful information has been obtained from such investigations about the function of different sub-cellular compartments.

15.3.3 Some Results of the Investigation of Enzyme Compartmentation

The main enzymes of the cell **nucleus** are those involved in the replication and transcription of nucleic acids, including those which play a role in providing energy for these processes (e.g. the enzymes of glycolysis and tricarboxylic acid cycle).

The **matrix of mitochondria** contains the enzymes of the tricarboxylic acid cycle, together with enzymes of fatty acid oxidation, protein synthesis and other processes. Reducing equivalents produced as the tricarboxylic acid cycle operates are passed down the respiratory pathway and used to synthesise ATP by oxidative phosphorylation; the components of the respiratory pathway and oxidative phosphorylation are found associated with the **inner mitochondrial membrane**.

The **thylakoid membranes** of **plant chloroplasts** contain assemblies which utilise light energy to generate ATP and NADPH: the chloroplast matrix (or stroma) contains enzymes which use this ATP and NADPH to convert CO_2 to hexose (the **dark reaction** of **photosynthesis**), together with enzymes of protein and nucleic acid synthesis.

The **matrix of lysosomes** contains mainly hydrolytic enzymes which have optimal activity at acidic pH values; many of these enzymes are glycoproteins. The single **lysosomal membrane** contains ATPase and an NADH dehydrogenase.

Endoplasmic reticulum has been identified as the site of synthesis of phospholipids and of elongation of fatty acids. It is also important in many detoxification mechanisms.

The **cytosol** of cells contains the enzymes for the pathways of glycolysis and fatty acid biosynthesis. Also present are the enzymes of amino-acid activation (for protein biosynthesis).

The **plasma membrane** contains a Na^+, K^+-dependent ATPase, whose function is to transport sodium ions out of the cell in exchange for potassium ions, and so maintain a high intracellular concentration of K^+ and a high extracellular concentration of Na^+. The red blood cell Na^+-K^+ pump has an $\alpha_2 \beta_2$ structure, each α sub-unit traversing the membrane and possessing ATPase activity on the cytosol side. Also associated with the plasma membrane is adenylate cyclase, which is responsible for the production of cAMP from ATP: this enzyme is activated by the hormones adrenaline and glucagon, and plays an important role in glycogen breakdown (see Chapter 14.2.5) and other processes.

Enzyme-catalysed processes which are associated with intracellular membranes in eukaryotic cells are often linked with the plasma membrane in prokaryotic cells, where there are no intracellular membranes: examples include the components of the respiratory pathway and the enzymes of phospholipid synthesis, discussed above.

SUMMARY OF CHAPTER 15

Whole-organism studies give useful information about metabolic pathways, but can reveal little about individual enzymes. In order to investigate an enzyme in detail, it is necessary to isolate a tissue and obtain a suitable preparation. This inevitably results in a loss of cellular organisation, which should be borne in mind when the results are being interpreted. Tissue preparations are kept cold (below $4°C$) whenever possible to minimise autolysis and loss of enzyme activity.

Enzyme assays may be performed on tissue homogenates or other preparations: the aim is to determine the concentration (actual or relative) of the enzyme, whose characteristics must already be known. In a catalytic assay procedure, all other factors which influence the rate of reaction are made fixed and non-limiting so that the initial velocity is proportional to the concentration of enzyme present. From the rate of reaction observed and the amount of sample assayed, the activity of the enzyme preparation can be calculated. If the procedure involves the coupling of the primary reaction to a second, indicator, reaction, the concentration of the indicator enzyme must be made high enough to be non-limiting. In a radioimmunoassay procedure, the basis of the assay is the binding of the enzyme to a specific antibody, catalytic activity not being involved at all. The two types of assay procedure are complementary.

The sub-cellular compartmentation of enzymes may be investigated by enzyme histochemistry or by assay of the fractions obtained by differential or density-gradient centrifugation: in differential centrifugation, particles are separated mainly according to size, the largest and heaviest particles sedimenting fastest; in density-gradient centrifugation, the average density of particles is important, particularly in isopycnic-zonal centrifugation, where it forms the basis of separation.

FURTHER READING

Colowick, S. P. and Kaplan, N. O. (eds.), *Methods in Enzymology,* 1 (1955): Preparation and Assay of Enzymes; 70 (1980), 92, 93 (1983): Immuno-chemical Techniques, Academic Press.

Denton, R. M. and Pogson, C. I. (1976), *Metabolic Regulation* (Chapter 2: Practical aspects), Chapman and Hall.

Hopkins, C. R. (1978), *Structure and Function of Cells* (Chapters 2 and 3), Saunders.

Landon, J., Carney, J. and Langley, D. (1977), The measurement of enzymes by radioimmunoassay, *Annals of Clinical Biochemistry,* 14 (pages 90-99).

Lojda, Z., Gossrau, R. and Schiebler, T. (1979), *Enzyme Histochemistry,* Springer-Verlag.

Stryer, L. (1981), *Biochemistry,* 2nd edition (Chapters 23, 33, 35 and 36), Freeman.

Yelton, D. E. and Scharff, M. D. (1981), Monoclonal antibodies, *Annual Review of Biochemistry,* 50 (pages 657-680).

Williams, B. L. and Wilson, K. (1981), *Principles and Techniques in Practical Biochemistry,* 2nd edition (Chapters 1, 2 and 9), Arnold.

PROBLEM

15.1 Aliquots of an enzyme preparation were added to suitable incubation mixtures to give, in each case, a total volume of $5.0 \, \text{cm}^3$, and assayed at $25°C$. The following results were obtained:

Time	Decrease in substrate concentration ($mmol.l^{-1}$)	
(s)	$0.2\ cm^3$ enzyme $prep^n$.	$0.3\ cm^3$ enzyme $prep^n$.
30	0.17	0.25
60	0.33	0.49
90	0.49	0.74
120	0.64	0.96
150	0.80	1.20
180	0.93	1.38

The protein concentration of the enzyme preparation was found to be $720\ mg.l^{-1}$. Calculate the specific activity of the enzyme preparation and comment on the validity of the results.

CHAPTER 16

Extraction and Purification of Enzymes

16.1 EXTRACTION OF ENZYMES

16.1.1 Introduction

Before attempting to extract an enzyme from an organism, some preliminary considerations should be made.

For a commercial venture (see Chapters 17-20) the most important requirement would be to use a source which enabled large amounts of a suitable enzyme to be extracted by some convenient procedure: any source which fulfilled this requirement could be chosen. On the other hand, a programme of scientific investigation might require the extraction of a given isoenzyme from a specified source, no alternatives being permissible. In the present chapter, therefore, we will discuss principles of extraction which are of general application.

First of all, it is essential to know the location of the enzyme within the cell, and whether the enzyme is present in free solution or bound to a membrane: this information may be obtained by the use of the techniques discussed in Chapter 15.3. Only then is it possible to devise a suitable extraction procedure.

16.1.2 The Extraction of Soluble Enzymes

Soluble cytoplasmic enzymes are the simplest to extract, for any disruption of the plasma membrane will enable the enzyme to pass into the surrounding medium. Soluble enzymes present in the organelles of eukaryotic cells are also easy to liberate, but the disruption of intracellular membranes often requires harsher conditions than disruption of the plasma membrane. To minimise the extraction of unwanted enzymes, cell fractionation (Chapter 15.3.2) may be carried out prior to the disruption of organelles.

The extraction procedures used depend on the type of organism acting as source.

Animal tissue is removed as soon after death as possible and kept cold to minimise autolysis. Obvious non-enzyme matter (e.g. fat deposits and connective tissue) is removed and the remaining tissue treated to disrupt membranes using a technique appropriate to the location of the enzyme being extracted: homogenisation in an isotonic medium disrupts the plasma membrane but leaves most organelles intact (Chapter 15.1); the techniques of freezing and thawing, sonication and drying *in vacuo* disrupt all plasma and intracellular membranes.

The extraction of soluble enzymes from **higher plants** varies considerably according to the type of source material. Enzymes may be extracted from cereal flours simply by placing in an aqueous medium and stirring, and from soft storage organs (e.g. the potato) by mincing and pressing out the juice through muslin or cheesecloth. Extracts from leaves, dried seeds etc. can be obtained by grinding in a mortar and then treating with a suitable medium in a pestle homogeniser. Leaves and other plant tissue may also be homogenised directly in medium by the use of a Waring blender. Alternatively, leaves may be disrupted in a Waring blender in the presence of several volumes of cold acetone (to displace water) followed by careful drying to produce an **acetone powder**: such powders may subsequently be treated with suitable medium, and the extracts will contain less chlorophyll and resinous material than would otherwise be the case. Tougher tissues, such as fibrous roots, are disrupted by grinding in a mortar with sand or ground glass. This procedure, like that of homogenisation in a Waring blender, tends to disrupt organelles.

Micro-organisms, in common with plants, have a cell wall and so tend to be more resistant to disruption than are animal cells. In general, methods of extraction fall into two main categories: drying and mechanical disruption. Drying may cause the cell to become permeable, enabling soluble enzymes to be extracted. **Air-drying** (particularly of yeast) where the cells are allowed to dry in air for several days at around 30°C, and drying *in vacuo*, where a cell suspension is placed in a vacuum desiccator for one or more days, are both accompanied by some degree of autolysis. **Lyophilisation (freeze-drying)** procedures, where a frozen cell suspension is dried by sublimation *in vacuo*, is not accompanied by autolysis, and soluble enzymes are less easily extracted from lyophilised cells than from air-dried or vacuum-dried preparations. Micro-organisms may also be dehydrated with acetone to produce powders containing easily-extractable enzymes. Mechanical disruption of micro-organisms may involve grinding with carborundum beads or other abrasives, applying pressure (e.g. with Hughes or French presses), sonication, or a combination of these methods. Another procedure, which is becoming increasingly important, is to weaken the bacterial cell wall with enzymes such as lysozyme and glucuronidase, particularly where these are available in an insolubilised form and can be recovered: this treatment makes the cell more susceptible to disruption by relatively mild techniques, e.g. osmotic shock.

In general, the method of choice depends not only on the type of cell being disrupted and the sub-cellular location of the enzyme being extracted but also on the characteristics of this enzyme (e.g. its stability under the conditions of the extraction procedure): it must be admitted that the final choice is often made empirically, but a consideration of the principles discussed above (and in Chapter 16.1.3) can help to limit the number of possibilities.

16.1.3   The Extraction of Membrane-Bound Enzymes

Enzymes which are bound to plasma or intracellular membranes cannot normally be extracted into the surrounding medium simply by the disruption procedures discussed in Chapter 16.1.2. However, some enzymes are loosely-bound and some tightly-bound (many actually forming an integral part of the membrane), so there can be no general rule about their extraction.

Membranes, besides containing enzymes and other proteins, are made up largely of **amphipathic lipids**, i.e. lipids which contain hydrophilic and hydrophobic regions. The main classes of membrane lipids are phospholipids (fatty acid/ phosphoric acid esters of glycerol and sphingosine) and glycolipids (fatty acid esters, mainly of sphingosine, attached to a sugar). Some membranes in eukaryotic cells also contain sterols (Fig. 16.1). In each case the hydrophilic portion of the molecule is small compared with the hydrophobic part (note that in Fig. 16.1 the R-groups of the fatty acid residues are long hydrocarbon chains). An amphipathic

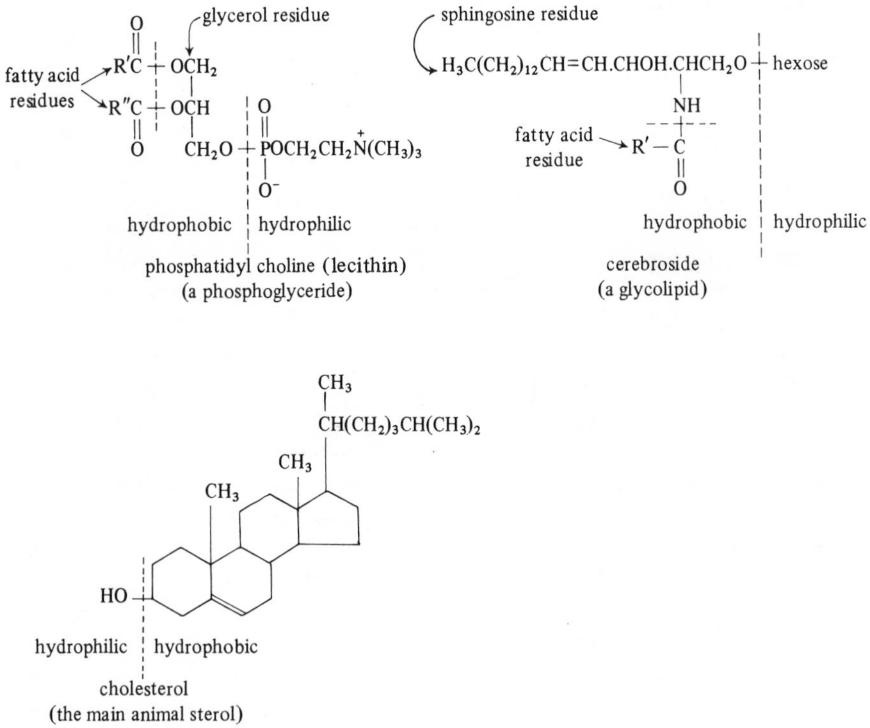

Fig. 16.1 Some important examples of lipids found in membranes.

lipid may therefore be represented as a molecule with a hydrophilic head and one or more hydrophobic tails, for example:

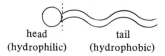

head
(hydrophilic)

tail
(hydrophobic)

Although the detailed structures of membranes have yet to be established beyond doubt, the model which best fits the available evidence (including that obtained by electron microscopy and nmr spectroscopy) is that of Singer and Nicolson (1972): this is termed the **fluid-mosaic model** (Fig. 16.2).

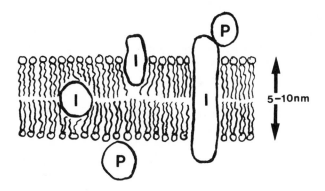

Fig. 16.2   Representation of a typical membrane, according to the fluid-mosaic model of Singer and Nicolson (P = peripheral protein and I = integral protein).

Proteins may be **peripheral** (or **extrinsic**), when they are bound to the surface of the membrane, or they may be **integral** (or **intrinsic**) when they are wholly or partly embedded in the lipid layers. Some evidence has been produced to show that the parts of polypeptide chains embedded in membranes are rich in non-polar amino acid side chains, which may form hydrophobic bonds with the lipids, whereas those parts which stick out, thus coming into contact with water, contain more polar side chains and, in some instances, are attached to hydrophilic carbo-hydrate units. It seems that at least some of the pores through the membrane may be lined with protein.

According to this model, lipids and proteins can move laterally about the membrane, to some extent, but find much more difficulty in moving in a transverse direction, from one membrane surface to the other.

Peripheral proteins are apparently linked by electrostatic or hydrogen bonds to the polar heads of lipids or to integral proteins and can easily be dissociated from the membrane (e.g. by treating with a solution of high ionic strength, such as 1 M NaCl, by freezing and thawing, or by sonication). Integral proteins, on the other hand, can only be extracted by breaking the hydrophobic interactions between lipid and protein (i.e. by dissociating a lipoprotein complex). Furthermore, for the extraction of an integral protein which is an enzyme, this must be done in such a way that enzyme activity is not lost during the process. A major problem is that an integral protein in its natural environment has some parts of its surface in contact with the hydrophobic membrane and other parts in contact with water, so structural changes are likely to take place no matter whether it is extracted into an organic or an aqueous medium. Again, the best method for each enzyme must be found largely by trial-and-error.

In some cases, **controlled autolysis** has been used to successfully free enzymes from membranes; this relies on the membrane structure being broken down and the enzyme released before the activity of the enzyme is lost, so it cannot be held to be a reliable procedure.

Some lipid–protein interactions are broken during the formation of **acetone powders**, so it may be possible to extract the required enzyme from these. Alternatively, the lipid–protein interactions may be broken by the action of **specific enzymes**: for example, phospholipase and triacylglycerol lipase have been used to extract alkaline phosphatase.

Of more general significance, **detergents** have been used to disrupt membranes and extract lipoprotein fragments into aqueous media as components of micelles (aggregates of amphipathic molecules arranged with their hydrophobic parts in the interior of the micelle and their hydrophilic parts on the surface). These detergents include natural bile salts such as cholate; non-ionic synthetic detergents like Triton X-100; and ionic synthetic detergents which may be zwitterionic, as in the case of CHAPS, acidic like sodium dodecyl sulphate (SDS) or basic, e.g. CTAB. In addition to the protein-detergent micelles, molecules of a detergent will themselves form micelles if present at a concentration aboe the critical micelle concentration (CMC). In general, the bile salts form the smallest micelles (molecular weight below 2,000 daltons) and have the highest CMC (1–10mM) while the non-ionic detergents form the largest micelles (molecular weight up to 100,000 daltons) and have the lowest CMC (in the order of 0.1mM). A knowledge of these characteristics may be helpful in the further purification of the extracted enzyme (see Chapter 16.2.2). The choice of detergent will be influenced by whether or not the enzyme is required in an active form, and what further purification procedures are intended. A detergent of particular importance in the investigation of membranes is SDS, which will sever most protein–lipid (and protein–protein) interactions and enable the molecular weights of the liberated proteins to be determined (see Chapter 16.3). However, for extraction of an active enzyme, non-ionic detergents are usually preferred.

Lipoprotein fragments may also be extracted by the use of **organic solvents**, and some free proteins may be liberated on subsequent dispersion in aqueous media: for example, xanthine oxidase has been obtained from lyophilised milk protein by a technique involving ether extraction. Such procedures might require evaporation of the organic medium and dissolving the enzyme in a detergent solution.

The solvent most widely used for the extraction of enzymes from membranes has been **n-butanol** (as introduced by Morton, 1950). This has both lipophilic and hydrophilic character, so can act like a detergent in removing lipoproteins from membranes and, further, tends to dissociate the protein and lipid components. Most membrane proteins are found in the aqueous phase of a butanol–water two-phase system, with the lipids in the butanol phase. Different groups of enzymes may be separated from membrane lipids by treatment with n-butanol under different conditions: for example, alkaline phosphatase has been extracted at pH 5-6, γ-glutamyltransferase at pH 7 and urate oxidase at pH 10. In general, however, organic solvents tend to cause some degree of protein denaturation, so are not ideal for the extraction of active enzymes.

It will be realised that enzymes can exist in cells complexed to nucleic acids or carbohydrates, but, unless they are also linked to lipids in membranes, their extraction will be relatively straightforward.

16.1.4 The Nature of the Extraction Medium

Once cells have been disrupted, by whatever method, the contents are extracted with several volumes of aqueous medium. In the case of enzymes which are an integral part of membranes the original extraction may be in organic solvent (Chapter 16.1.4) but the liberated enzyme must then be further extracted into aqueous medium.

In general, a considerable amount of debris (mainly membrane fragments) is likely to be present, so it is essential that the enzyme of interest can be easily separated from this by being soluble in the extraction medium. As discussed in Chapter 3.2.3, four main factors govern the solubility of proteins: **salt concentration, pH**, the **organic solvent content** and **temperature**. The temperature of the medium is usually kept below 4°C, despite the reductions in solubility that this entails, in order to minimise loss of activity of the enzyme.

Proteins are least soluble at their **isoelectric point**, so the extraction of an enzyme must be carried out at a pH value far from its isoelectric point, but neverthelesss at a value where the enzyme is stable. The salt and organic contents of the medium are chosen, largely on an empirical basis, to ensure that the enzyme is soluble but many other proteins are not, thus achieving a preliminary purification. For example, Cohn and co-workers (1951) showed that glutamate dehydrogenase, arginase, alkaline phosphatase and acid phosphatase from a bovine liver homogenate were soluble in 19% ethanol at pH 5.8 and ionic strength 0.02, but catalase, peroxidase and D–amino-acid oxidase were not; however, these three enzymes were soluble at pH 5.8 in the absence of ethanol. and at ionic strength 0.15; also, various proteinases which were insoluble in both of these media were soluble if the pH of the second medium was increased to 7.4.

The solubility of an enzyme can sometimes be increased by the addition of its **substrate**: for example, alkaline phosphatase is more soluble in the presence of β-glycerophosphate than otherwise. Hence the extraction of an enzyme can be facilitated by adding its substrate to the medium.

Stabilisers, such as **dithiothreitol (Cleland's reagent)** or **mercaptoethanol** are sometimes added to extraction media to prevent oxidation of sulphydryl groups and thus loss of enzyme activity. **Inhibitors** of proteolytic enzymes (e.g. DFP) may also be added to prevent these attacking the enzyme of interest. Polyols (e.g. glycerol) or chelating agents (e.g. EDTA) also have a use in stabilising solubilised enzymes.

16.2 PURIFICATION OF ENZYMES

16.2.1 Preliminary Purification Procedures

As mentioned in Chapter 16.1 a preliminary purification step is usually included in the procedure used to extract an enzyme. For example, some degree of cell fractionation is often carried out before extraction, so that only enzymes in the same sub-cellular compartment as the relevant enzyme are extracted. Also, the composition of the extraction medium is chosen so that the enzyme of interest is soluble but many others are not.

Extracted nucleic acids may be precipitated by treatment with basic substances such as streptomycin or protamine, or with $MnCl_2$ or $MgCl_2$.

All precipitates and cell debris are then removed by centrifugation and discarded. Polysaccharides may also be removed by high speed centrifugation, although the results are not always satisfactory. An alternative way to dispose of polysaccharides is to break them down into small units with enzymes such as α-amylase; nucleic acids may similarly be broken down by the use of nuclease enzymes.

The next stage of purification is usually to precipitate the enzyme of interest from solution, thus separating it from mono- and oligo-saccharides, nucleotides, free amino-acids etc. and from many other proteins which remain in solution. This may be achieved by altering the pH and the organic or salt concentrations of the medium. For example, the pH may be adjusted to the isoelectric point of the enzyme. Alternatively a high salt concentration can be introduced: ammonium sulphate is often used for this purpose, because it is extremely soluble in water. The salt concentration is usually increased in stages, the aim being to precipitate other proteins (which are then discarded) before precipitating the fraction containing the relevant enzyme. This is then redissolved in water, and residual ammonium sulphate removed by dialysis.

Although a considerable degree of purification can be achieved by these procedures, many other proteins will still be present because of overlap of solubility ranges. One possible way to remove some of these is to raise the temperature of the medium for a few minutes to a value where the enzyme being purified is known to be stable, but others might be denatured and precipitate from solution. Enzymes are often particularly stable in the presence of their substrates, so the relevant substrate can be added to increase the effectiveness of this procedure.

16.2.2 Further Purification Procedures

The crude extract, partially purified as described in Chapter 16.2.1 may be treated in a variety of ways to increase the purification of the relevant enzyme.

Adsorbent gels (e.g. zinc hydroxide) have been used to remove pigments from enzyme preparations. A mixture of enzymes may also be adsorbed by a suitable gel (e.g. aluminium hydroxide) and then fractionated by elution with buffers of increasing ionic strength. **Partition chromatography** has similarly been used to separate mixtures of enzymes: for example, ribonuclease was purified by Martin and Porter (1951) on columns packed with kieselguhr, using a two-phase system consisting of ammonium sulphate, ethyl cellosolve and water; the organic phase was mixed with the kieselguhr to provide the stationary phase, the sample was then applied to the top of the column and the aqueous phase run through. The different components of the sample were separated during their elution from the column according to their relative solubilities in the two liquid phases.

Such techniques have played an important part in the development of enzymology, but have now been superseded by others.

Electrophoresis is mainly an analytical procedure, since it is ideally suited to the separation of small amounts of material, but it has also been used for the purification of proteins. The rate and direction of migration of a protein in an electric field depends on its net charge at the pH used and also on the size of the molecule, since this imposes some resistance to movement. In **zone electrophoresis**, a mixture of proteins is introduced at a common point (the **origin**) and allowed to move in an electric field, usually in a horizontal direction, each molecule travelling in the same zone as others of the same charge and size. To minimise diffusion the whole process is usually carried out in a solid but porous **support medium** (e.g. starch gel) through which the buffer and proteins permeate. When separation has been achieved, as indicated by cutting a narrow longitudinal section from the support medium and staining for proteins (e.g. with Amido Black) then the rest of the support medium may be cut into strips *across* the direction of travel and the contents of each extracted by elution. Alternatively, zone electrophoresis may be carried out in a vertical direction in a closed column with, for example, powdered cellulose as the support medium; after electrophoresis has been completed, a tap at the bottom of the column is opened to enable the liquid contents to be run out and collected in fractions.

Some forms of electrophoresis capable of giving extremely high resolution, e.g. **isotachophoresis** and **isoelectric focusing** (see Chapter 16.2.3), which are employed at present mainly for analytical purposes, are likely to find increasing use in preparative work.

Cation exchange chromatography (see Chapter 2.4.2) on columns packed with carboxylated polystyrene has been used in the successful purification of stable, low molecular weight basic proteins such as lysozyme and ribonuclease. However, **cellulosic ion exchange chromatography** has proved of more general application to the separation of proteins. Such resins, in contrast to most other ion exchange resins, are relatively hydrophilic and have an open structure readily penetrated by large molecules such as proteins. The most frequently used are diethylaminoethyl (DEAE)-cellulose which is an anion exchanger, and carboxymethyl (CM) cellulose, a cation exchanger. The DEAE group, $-OC_2H_5\overset{+}{N}H(C_2H_5)_2$, is highly positively charged at pH 6-8, so DEAE-cellulose is most useful for the chromatography of proteins which are negatively charged in this range; similarly, CM-cellulose, cellulose-$OCH_2CO_2^-$, is most applicable for the separation of proteins which are positively charged at around pH 4.5.

Elution of proteins from the columns may be brought about by changes in either salt concentration or pH: as the concentration of salt (e.g. NaCl) increases, protein is displaced from DEAE-cellulose by the anion (Cl^-) and from CM-cellulose by the cation (Na^+); if the pH is altered over the relatively narrow working range, proteins are eluted as their isoelectric point is reached, since they then have no net charge with which to bind to the resin.

Ion exchange chromatography has been used in the purification of membrane proteins in non-ionic or zwitterionic detergents, as well as for the purification of soluble enzymes.

Gel filtration (molecular-sieve chromatography), another important technique in enzyme purification, is carried out with columns packed with swollen gels which separate components of a sample on the basis of molecular size: molecules too large to enter the pores of the gels will pass quickly through the columns; smaller molecules will pass through the column more slowly, the actual speed for each component being dependent on the ease with which its molecules can pass into the gels and be thus retarded. Gels commonly used include cross-linked dextrans (Sephadex), cross-linked agarose (Sepharose) and cross-linked polyacrylamide (Biogel). These are graded according to pore size, and thus to the size (and hence molecular weight) of protein molecule which can enter. For a successful separation, the enzyme of interest must not be too large to penetrate the gel, because it would then pass straight through the column together with all other proteins of similar and greater molecular size.

Eluting buffers should be of high ionic strength (e.g. 20 mM) to counteract the few charges which may be present on the gel: apart from that, the only criterion for the buffer is that the proteins are stable in it: no pH or salt gradient is employed, since the separation is by size alone. However, it should be noted that some of these gels are available with attached ion exchange groups (an example is DEAE-Sephadex) in which case the principles of elution are as for cellulosic ion exchange chromatography.

In the case of membrane proteins solubilised in detergents, gel filtration is widely used to separate protein-detergent micelles from excess detergent: this is particularly easy to achieve if detergent molecules are kept in the monomeric form, e.g. by ensuring that the detergent concentration is less than the CMC, or by adding a small amount of a bile salt.

Affinity chromatography is a bio-specific process and thus ideally suited for the separation of one protein from all others, including those which resemble it so closely in physical characteristics that separation by any other procedure is extremely difficult. The column is packed with an inert matrix (e.g. agarose) to which ligands for the required enzyme have been attached. These immobilised ligands may be, for example, substrate analogues, and must be covalently linked to the matrix in such a way that they can still bind to their enzyme, if this is present: this usually means that a **spacer arm** (e.g. a hydrocarbon chain) must be included between matrix and ligand, joining to the ligand at a point which does not interact with the enzyme, if steric hindrance to the formation of an enzyme-ligand complex is to be avoided. The attachment of the spacer arms to $-OH$ groups of the matrix is often achieved by the use of CNBr reagent (see Chapter 20.2.1).

The protein mixture is applied to the column, and the relevant enzyme is trapped by the immobilised ligands while all other proteins pass through and are discarded. The enzyme is then liberated from the column either by eluting with a **deforming buffer** at a pH which changes the characteristics of the enzyme and no longer allows it to bind to the immobilised ligand, or by the use of a **competitive counter-ligand**, which displaces the immobilised ligand on the enzyme. In both cases, the enzyme passes through the column and can be collected, now free of other proteins (Fig. 16.3). An example of an enzyme extracted by the use of such a technique is β-galactosidase with the substrate-analogue p-aminophenyl β-D-galactoside as ligand, the deforming buffer being 0.1 M borate at pH 10.

A closely related procedure is **immunoaffinity chromatography**, where the immobilised ligand is an antibody specific for the enzyme which is being purified. The use of an immobilised monoclonal antibody (see Chapter 15.2.4) as ligand gives even greater specificity than use of a more conventional antibody.

The bio-specific nature of these techniques means that a high degree of purification can be achieved; on the other hand, it also means that a polymer-ligand complex prepared for the purification of one enzyme is limited to that particular application.

Another rapidly developing form of affinity chromatography involves the use of reactive **triazine-based dyes**, e.g. Cibacron Blue F3G-A and dyes of the Procion series, as ligand. Under alkaline conditions the chlorotriazine dye is linked, usually directly, to a matrix such as agarose by a triazine bond; the

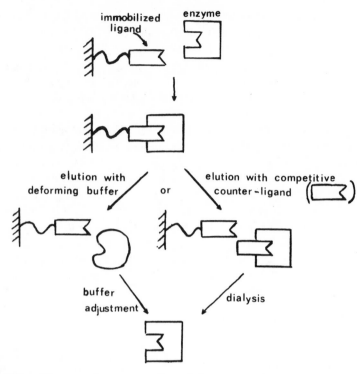

Fig. 16.3   Diagrammatic representation of the processes involved in affinity
chromatography; the treatment with deforming buffer and with competitive
counter-ligand are alternatives.

reaction involves a hydroxyl group on the matrix and a chloride on the dye. Such
immobilised dyes bind a wide range of NAD- and NADP-dependent dehydrogenases and other enzymes. They are, of course, less specific than biological
group-specific ligands, and their specificity is difficult to predict, but they have
many advantages: they have a greater protein-binding capacity, they are extremely
resistant to chemical and enzymic degradation, they can be used and re-used in a
variety of applications, and the triazine bond is more stable than the isouronium
linkage introduced by the CNBr activation of agarose. Elution of enzyme from
the ligand is usually achieved by changing pH or increasing salt concentration.

A major advance over the past few years has been the application of **HPLC
technology** to protein purification. The use of stainless steel columns and
high-quality robust packing material of particle size 10 μm or less has allowed
separations to be obtained similar to those for conventional low-pressure
chromatography but in minutes rather than hours. HPLC columns for preparative
work are wider than those for analytical use (perhaps 20 mm internal diameter),
which allows milligram or even gram amounts of protein to be applied; a fraction

collector is usually included in the system after the detector. **High-performance size-exclusion chromatography (HPSEC)** is based on the same principle as conventional molecular seive chromatography, but Sephadex, Sepharose and other non-rigid or semi-rigid gels can only withstand low pressures. Instead, HPSEC often utilises rigid spherical beads of porous silica with bonded hydrophilic polar groups. **High-performance ion exchange chromatography (HPIEC)** utilises amines as anion exchangers and sulphonic or carboxylic acids as cation exchangers, each bonded to some rigid support such as silica. **High-performance liquid affinity chromatography** utilises ligands bound to supports such as epoxy-silica:

$$-OH + (CH_3O)_3Si-(CH_2)_3OCH_2CH-CH_2 \longrightarrow R-O-\underset{|}{Si}-(CH_2)_3OCH_2CH-CH_2$$

$$\text{(ica)} \qquad \text{epoxy-silane} \qquad\qquad \text{epoxy-silica}$$

$$\downarrow \text{Ligand}-NH_2$$

$$R-O-\underset{|}{Si}-(CH_2)_3OCH_2\overset{\displaystyle OH}{\overset{|}{CH}}.CH_2NH-\text{Ligand}$$

Proteins may also be separated by **reversed-phase HPLC** on alkylsilica columns, the eluting solvents being buffered aqueous and organic mixtures.

When after the use of one or more of these procedures it is considered that the relevant enzyme has been separated completely from other proteins (see Chapter 16.2.3) mineral salts and any small molecules present may be removed by **dialysis**, if this is desired.

The purified enzyme preparation is likely to be quite dilute, so it might require concentrating before being used for the purpose for which it was prepared. One convenient method for doing this is **lyophilisation**, followed by the redissolving of the enzyme in a small volume of liquid. Another suitable procedure is **ultrafiltration**: this involves forcing the solvent molecules through a membrane of chosen porosity (e.g. by the application of nitrogen gas pressure) and thus separating them from the protein molecules, these being too large to pass through the pores.

The ideal way to complete a purification procedure is to **crystallise** the enzyme, if this is possible. Sometimes it may be achieved by adding not quite enough ammonium sulphate to cause precipitation of the enzyme and leaving this solution in a cold-room for several days. In general, successful conditions have to be determined empirically by attempting crystallisation at a variety of salt and organic solvent concentrations and pH values, and by showing much patience.

Membrane enzymes may be inactive after purification unless re-introduced into a phospholipid environment. For example, phospholipid may be added to purified protein-detergent micelles and concentrations adjusted so that the phospholipid replaces detergent in the micelles; released detergent, in the monomeric form, may then be removed by gel filtration, dialysis or ultrafiltration.

16.2.3 Criteria of Purity

At each stage of a purification procedure, an assay for the enzyme being purified should be performed on all fractions and its **specific activity** in each fraction determined (see Chapter 15.2.2). This will show, irrespective of preconceived ideas, precisely which fractions contain the important enzyme and will enable the degree of purification to be calculated. The *total* activity of this enzyme will be unchanged as a result of the purification step, unless some has been lost during purification; under no circumstances can it be increased. However, some of the contaminating proteins which were originally present in the same preparation as the enzyme should now be in different fractions. Hence, if the fractions containing the relevant enzyme are combined, the total activity of the enzyme in the combined fraction should be the same as that started with, whereas the total amount of protein in this fraction will be less than that originally present. The *specific* activity of the enzyme in the combined fraction should therefore be greater than that in the preparation before purification, and the increase in specific activity will be a measure of the purification achieved.

With each successive purification step, the specific activity of the fraction (or fractions) containing the enzyme should be greater than before until complete purification is achieved and the specific activity reaches a limiting value.

However, the finding of the same specific activity value before and after a purification step does not necessarily mean that the enzyme preparation is completely pure: it could simply mean that contaminating proteins have passed through the procedure in the same fraction as the enzyme. Similarly, crystallisation cannot be taken as proof that only one protein is present, for many mixed protein crystals have been found. Hence other criteria of purity have to be considered.

If a particular enzyme is known to be a possible contaminant, then a logical step is to carry out an assay for that enzyme and demonstrate its absence in that way. However, that would not exclude the possibility that other contaminants were present.

Analytical untracentrifugation is widely used to investigate the purity of enzyme preparations. With high centrifugal fields (e.g. 500,000 g) diffusion effects are largely overcome (see Chapter 15.3.2); hence, as a randomly distributed group of identical protein molecules sediment in an **analytical cell** under the effect of such a field, a sharp boundary is formed between the portion of the medium which has been cleared of protein molecules and that which still contains them. This boundary refracts light, and its position can be determined through a quartz window in the cell: for example, it shows up as a peak if the refractive index gradient at each point in the cell is measured using the **Schlieren optical system**. The movement of this boundary as sedimentation proceeds can be constantly monitored, and automatically recorded, and the data used to calculate the characteristics of the molecule in question (see Chapter 16.3). If the protein is pure, only one boundary should be observed: detection of extra boundaries suggests the presence of contaminants.

Electrophoresis in various forms is very effective in separating mixtures of proteins (see Chapter 16.2.2) so it is used extensively to investigate the purity of enzyme preparations. Of particular importance in this context is **discontinuous (disc) electrophoresis,** where the electrodes are placed in separate buffer reservoirs linked vertically by cylindrical glass tubes containing the support medium (usually polyacrylamide gels). The discontinuity is in the gels and buffers used, pH differences being used to concentrate the sample into a very narrow band as it passes through a large-pore **spacer gel** on its way to the **running gel** where the actual separation of the sample components takes place (Fig. 16.4). This running gel has smaller pores than the spacer gel, and separation is achieved by a combination of electrophoresis and gel filtration. On completion of the electrophoresis and after staining for proteins, it is easy to see if more than one band is present, for each is extremely narrow and well-defined.

Another relevant technique is **isoelectric focusing,** which is an example of **moving boundary electrophoresis** rather than zone electrophoresis. A pH gradient is set up between the electrodes by allowing an acid (e.g. phosphoric acid) to diffuse in from the anode end and a base (e.g. ethanolamine) from the cathode end. This can be stabilised by introducing a mixture of **ampholytes** (e.g. synthetic and natural amino-acids) selected so that their individual isoelectric points cover the required pH range. Each ampholyte will move in the electric field to the position in the pH gradient corresponding to its isoelectric point and then

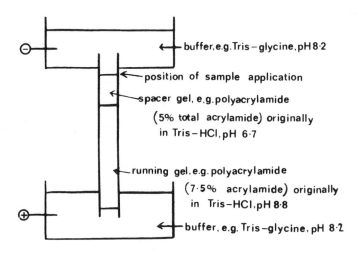

Fig. 16.4 Diagrammatic representation of an apparatus set up for discontinuous (disc) electrophoresis of proteins.

remain there, since it will no longer have a net charge. The protein sample is then introduced and electrophoresis continued, possibly for several days, until each component reaches a stationary position at its isoelectric pH. The zones containing each protein are very sharp as a result of this focusing and proteins whose isoelectric points differ by as little as 0.02 pH units can be distinguished by this method.

Isotachophoresis is another example of moving boundary electrophoresis, and it also has some similarities to discontinuous electrophoresis. For the separation of proteins at mildly alkaline pH, when most would be anions, a leading anion (e.g. phosphate) is added which has a faster mobility towards the anode than any of the sample anions. A trailing or terminating anion (e.g. glycine), with a slower mobility than any in the sample, is also added. The system is buffered by a common counterion, e.g. Tris. The leading and terminating anions are applied at different sides of the sample, the leading anion nearest the anode. A high voltage is applied and, when steady-state is achieved, all components will migrate towards the anode in discrete zones at the same velocity ('iso tacho' being Greek for 'same speed'), the sample anions arranged in order of their mobilities. Spacers similar to the ampholytes used in isoelectric focusing may be added to separate the zones of sample anions. Isotachophoresis may be carried out in a purely aqueous environment in a capillary tube or, particularly for preparative work, in columns packed with gel (e.g. polyacrylamide) of pore size large enough to minimise molecular seiving.

Analytical HPLC procedures may also be used to investigate the purity of a protein preparation.

No single technique, however sensitive, can establish enzyme purity because of the possibility that a contaminating protein might behave in an identical fashion to the important protein under the conditions used. However, if several characteristics are investigated, and the conditions of investigation varied, the chances of the contaminant avoiding detection are reduced. Therefore, evidence of purity must be obtained by the use of several of the above procedures before it can be concluded that the sample is indeed pure.

16.3 DETERMINATION OF MOLECULAR WEIGHTS OF ENZYMES

When an enzyme has been purified, its molecular weight may be determined.

Gel filtration, which separates molecules on the basis of size (see Chapter 16.2.2), provides one way of doing this. A packed column is calibrated by applying proteins of known molecular weight to the top of the column and determining the volume of buffer required for the elution of each protein: as each protein leaves the column there should be an absorbance peak at 280 nm in the column eluate, thus providing a simple way of monitoring the elution. From the data obtained, a graph may be drawn of elution volume against molecular weight. The protein of unknown molecular weight is then passed through the column and its elution volume determined exactly as for the marker proteins. Hence, by reference to the calibration graph, the molecular weight of the protein may be estimated. One source of error in this procedure is that it takes no account of molecular shape: in fact, a protein molecule which is somewhat elongated will pass down the column more slowly than a spherical one of the same molecular weight.

Ultracentrifugation (see Chapter 16.2.3) is also widely used for the determination of molecular weights of proteins. For a spherical protein of molecular weight M sedimenting through a medium in a centrifugal field, the centrifugal force per gram-mole is $M(1 - \bar{v}\rho)\omega^2 r$ where $\bar{v}$ is the partial specific volume if the molecule (i.e. the increase in volume in cm^3 when 1g solute is added to a large volume of solvent; the value for a protein is usually about 0.74). The other terms are defined as in Chapter 15.3.2. The centrifugal force is opposed by an equal frictional force per gram-mole of $\dfrac{RT}{D} \times$ sedimentation rate where R is the gas constant (in ergs per mole per degree), T is the absolute temperature and D is the diffusion constant of the molecule.

$$\therefore \ M(1 - \bar{v}\rho)\omega^2 r = \frac{RT}{D} \times \text{ sedimentation rate.}$$

But sedimentation rate $= s.\omega^2 r$ (see Chapter 15.3.2).

$$\therefore \ M(1 - \bar{v}\rho)\omega^2 r = \frac{RT}{D}.s.\omega^2 r.$$

$$\therefore \ M = \frac{RTs}{D(1 - \bar{v}\rho)}.$$

This is known as the **Svedberg equation**, after the pioneer of ultracentrifugation.

As discussed in Chapter 16.2.3 the rate of movement of the boundary separating the protein-free zone from the rest of the medium can be determined, and this is a measure of the sedimentation rate. Hence the sedimentation coefficient (s) can be calculated, since ω^2 and r are easily determined: for most proteins, s lies in the range 1–200 S. The diffusion constant (D) can similarly be calculated from moving boundary determinations, this time in the absence of a centrifugal field.

Once these and the other relevant factors have been determined, the molecular weight of the protein can be calculated by the use of the Svedberg equation. However, as with gel filtration, there is a possible source of error in the assumption that all protein molecules are spherical.

A technique which is commonly used to determine molecular weights of proteins, and which does not assume that the molecules are spherical is **SDS–gel electrophoresis**. SDS (sodium dodecyl sulphate, $CH_3(CH_2)_{11} SO_3^-Na^+$) is an anionic detergent whose hydrocarbon chain can become linked to hydrophobic regions in the interior of globular proteins, leaving the ionised sulphonate ($-SO_3^-$) group jutting out into the surrounding medium. Many SDS molecules can bind in this way to a single protein molecule, disrupting its natural shape and giving it a large negative charge, regardless of the net charge originally possessed by the protein. Large protein molecules will complex with more SDS than small ones, so all will have roughly the same charge/mass ratio, and all will have much the same shape. Therefore, if molecules of different proteins are complexed with SDS and subjected to zone electrophoresis in an unrestrictive medium, all should travel together, regardless of their original characteristics. However, if electrophorsis is performed in a support medium with a restrictive pore size (e.g. in polyacrylamide gel), the complexes from each of the different proteins will be separated, entirely on the basis of size. Since the size of each protein–SDS complex must be dependent on the original size of the protein molecule, a system can be calibrated by the use of marker proteins of known molecular weight; hence the molecular weight of a test protein may be determined. One point that should be borne in mind, however, is that SDS tends to break all weak polypeptide–polypeptide interactions, so, for an oligomeric protein, this technique may well give the molecular weight of the sub-unit rather than of the complete protein.

SUMMARY OF CHAPTER 16

Soluble enzymes may be extracted from cells simply by the disruption of membranes, the actual technique employed depending on the nature of the cell, the intracellular location of the enzyme and the stability of the enzyme. In each case, the pH, salt and organic solvent concentration of the medium must be carefully chosen to ensure that the enzyme being extracted remains in solution while the cell debris is removed by centrifugation.

Special procedures have to be employed to extract enzymes which form an integral part of membranes (e.g. extraction with a detergent): otherwise they would be discarded with the cell debris.

Techniques for the purification of the extracted enzyme include electrophoresis, ion exchange chromatography, gel filtration and affinity chromatography. Enzyme assay should be performed after each purification step, and the specific activity of each fraction determined. As purification proceeds, the specific activity of the enzyme preparation should rise to a limiting value. When this has been achieved, the purity of the preparation should be checked by a variety of techniques, including ultracentrifugation, electrophoresis, isoelectric focusing and by high performance liquid chromatography.

After purification, the molecular weight of the enzyme may be determined by gel filtration, ultracentrifugation or SDS–gel electrophoresis.

FURTHER READING

Chapman, D. (ed.) (1983), *Biomembrane Structure and Function,* Macmillan.
Colowick, S. P. and Kaplan, N. O. (eds.), *Methods in Enzymology,* **1** (1955): Preparation and Assay of Enzymes; **22** (1971): Enzyme Purification; **34** (1974): Affinity Chromatography; **104** (1984): Enzyme Purification, Academic Press.
Kerese, I. (ed.) (1984), *Methods of Protein Analysis,* Ellis Horwood.
Melling, J. and Phillips, B. W. (1975), Large-scale extraction and purification of enzymes, in Wiseman, A. (ed.), *Handbook of Enzyme Biotechnology,* Ellis Horwood.
Price, N. C. and Stevens, L. (1982), *Fundamentals of Enzymology* (Chapters 2 and 8), Oxford University Press.
Smith, R. L. and Oldfield, E. (1984), Dynamic structure of membranes by deuterium nmr, *Science,* **225** (pages 280–288).
Williams, B. L. and Wilson, K. (eds.) (1981), *Principles and Techniques in Practical Biochemistry,* 2nd edition (Chapters 1–4), Arnold.

PROBLEM

16.1 An attempt was made to purify further a malate dehydrogenase preparation by precipitating protein with ammonium sulphate at 55% saturation. The precipitate was redissolved in water, and this solution was found to have a protein content of 1.48 g.l^{-1}. A 1 in 500 dilution was made and aliquots of this were used for malate dehydrogenase assay, with excess malate and NAD^+, in a total incubation volume of 3.1 cm^3. When 10 μl diluted solution was used for assay, the initial velocity (measured as rate of increase of absorbance at 340 nm) was 0.11 units.minute^{-1} and when 15 μl were used the initial velocity was 0.165 units.minute^{-1}. The supernatant from the ammonium sulphate precipitation was found to have a protein content of 2.05 g.l^{-1}. A 1 in 1000 dilution was made, and assayed for malate dehydrogenase as above. When 10 μl diluted solution was used for assay, the initial velocity was 0.08 units.minute^{-1} and when 15 μl was used, the initial velocity was 0.12 units.minute^{-1}. Calculate the specific activities of the two fractions, and comment on the usefulness of the purification step. (Molar extinction coefficient of NADH $= 6.2 \times 10^3$ at 340 nm. Cells with a light path of 1 cm were used throughout.)

CHAPTER 17

Enzymes as Analytical Reagents

17.1 THE VALUE OF ENZYMES AS ANALYTICAL REAGENTS

Every enzyme catalyses a reaction which is both substrate-specific and product-specific (Chapter 4.1); because of this, enzymes are extremely valuable as analytical reagents. In particular, they can be used for the estimation of specific substances, possibly present at very low concentrations, in the presence of other, chemically similar, substances. Ordinary chemical reagents might not be able to distinguish between several of the components of a sample, and costly separation procedures might be required before a satisfactory analysis can be carried out; even then, several different products might be formed because of side-reactions, thus reducing the sensitivity of the method. In contrast, enzyme-based methods of analysis are both specific and sensitive, which means that little or no sample preparation is required in many cases.

Another characteristic of enzyme-catalysed reactions which makes them suitable for analytical applications is that they proceed under relatively mild conditions (e.g. at near neutral pH and at around room temperature): hence they are usually simple to set up, and can be used for the analysis of substances which would be unstable under more extreme conditions.

On the other hand, enzymes themselves are quite unstable. For this reason, supposedly identical enzyme preparations may be found to have different activities, as may the same preparation on different occasions, particularly if correct storage and handling procedures (Chapter 17.3) are not strictly adhered to. However, such variations in activity should not affect the results obtained by enzyme-based methods of analysis, provided the activity present in each case is a reasonable approximation to that specified, and provided appropriate calibration experiments are always carried out.

Another disadvantage in the use of enzymes is that they are often expensive and sometimes difficult to obtain. Hence it is important that they are not used in a wasteful fashion. One way of minimising wastage is by the use of immobilised enzymes (Chapter 20.2), which are now becoming increasingly available: these can be recovered intact at the end of a procedure and used again. Enzymes in this form are sometimes more stable than in free solution.

Enzyme-based analytical procedures may be designed to determine the concentrations of substrates, coenzymes, activators and inhibitors. All of these possible applications, together with enzyme assay, are included in the term **enzymatic analysis** and are discussed in the next section (Chapter 17.2).

17.2 PRINCIPLES OF ENZYMATIC ANALYSIS

17.2.1 End-Point Methods

End-point methods of enzymatic analysis, also called **total change** or **equilibrium methods**, allow the reaction to go to equilibrium, whereupon the amount of product formed is determined. Ideally, the nature of the reaction and the conditions should be such that the position of equilibrium is very much in favour of product formation, i.e. such that the reaction goes very close to **completion**, almost all of the substrate being converted to product. Since the concentrations of activators and inhibitors do not affect the situation at equilibrium (only its rate of attainment) they cannot be estimated by end-point methods, which are used exclusively for the analysis of **substrates**.

Analysis should be performed under conditions where the concentration of the added enzyme is high, to ensure rapid progress towards equilibrium, while the concentration of the substrate under investigation should be low enough for the reaction to be first-order with respect to this substrate (see Fig. 6.5). The amount of sample used for analysis is therefore chosen accordingly, from a knowledge (or estimate) of the likely upper limit of concentration of the appropriate substrate in the sample. The concentrations of any extra substrates (including co-substrates such as NAD^+) should be in excess, so they do not limit the reaction nor change its first-order characteristics. No products are initially present, so the concentrations of all products (including co-products such as NADH) at equilibrium will be dependent solely on the initial concentration of the substrate being analysed. Therefore, if one of these product concentrations can be measured (e.g. NADH by absorbance at 340 nm) and suitable calibration experiments performed, then the initial concentration of substrate may be calculated. If it is known that the reaction has gone to completion, then the final product concentration will be a measure of the initial substrate concentration, and calibration may not be required.

For any first-order (or pseudo first-order) reaction $S \rightarrow P$, the rate of reaction at time t, $-\dfrac{d[S]}{dt}$, is given by

$$-\frac{d[S]}{dt} = k[S]$$

where [S] is the concentration of S at time t, and k is a constant (see Chapter 6.3.1).

By integration,

$$t = \frac{1}{k} \log_e \left(\frac{[S_o]}{[S]} \right) = \frac{2.303}{k} \log_{10} \left(\frac{[S_o]}{[S]} \right)$$

where [S_o] is the concentration of S at zero time.

For an enzyme catalysed reaction $S \rightarrow P$ which goes to completion in the absence of any appreciable back reaction, a steady-state may be assumed to exist throughout. Therefore the rate of reaction at time t, $-\dfrac{d[S]}{dt}$, is given by

$$-\frac{d[S]}{dt} = \frac{V_{max}[S]}{[S] + K_m} \quad \text{(see Chapter 7.1.2).}$$

Note that, under these circumstances, this applies to the rate at any time, not just to the initial velocity, provided [S_o] is not substituted for [S]. At very low substrate concentrations $[S] \ll K_m$, and so $([S] + K_m) \doteq K_m$. Therefore, under these conditions

$$-\frac{d[S]}{dt} = \frac{V_{max}}{K_m} [S].$$

Hence by comparison to the general rate equation for a first-order reaction, it can be seen that $k = \dfrac{V_{max}}{K_m}$.

Therefore, for an enzyme-catalysed reaction progressing under first-order (or pseudo first-order) conditions in the absence of any appreciable back reaction,

$$t = 2.303 \frac{K_m}{V_{max}} . \log_{10} \left(\frac{[S_o]}{[S]} \right).$$

Now, K_m is a characteristic of the enzyme and should be known. Also V_{max} is proportional to the enzyme concentration, and may be calculated from the activity of the enzyme preparation and the amount of this used in the analytical procedure. Hence it is possible to determine the time for such an enzyme-catalysed reaction to reach 99% (i.e. by substituting $\dfrac{[S_o]}{[S]} = 100$, together with the calculated values for K_m and V_{max}, in the expression for t derived above). If, for a given procedure, the time for the reaction to reach 99% completion seems inconveniently long, it can be reduced by increasing the value of V_{max}, i.e. by adding more enzyme.

However, it will be apparent that this calculation applies only to those enzyme-catalysed reactions which are essentially irreversible. Most, in fact, will reach a position of equilibrium some way short of completion, and although end-point methods may still be applied to such reactions, their sensitivity and reliability will be considerably improved if the reactions can be driven towards completion in some way. One way of doing this is by **trapping the product** and thus continually disturbing the equilibrium: for example, aldehydes and ketones may be trapped by the use of hydrazine. Another possible way of displacing the equilibrium in the direction of completion is by the use of **analogues** of second substrates or coenzymes: for example, acetyl pyridine adenine dinucleotide (APAD$^+$) has this effect if used in place of NAD$^+$ to oxidise substrates. A third possibility is the use of **regenerating systems** to convert a co-product back to co-substrate, thus driving the main reaction towards completion: for example, pyruvate and LDH might be added to reform NAD$^+$ from NADH, as in the glutamate dehydrogenase system for the estimation of glutamate:

$$\text{glutamate} + \text{H}_2\text{O} + \text{NAD}^+ \xrightarrow{\substack{\text{glutamate} \\ \text{dehydrogenase}}} \text{NADH} + \text{H}^+ + 2\text{-oxoglutarate} + \text{NH}_3$$

$$\text{lactate} \xleftarrow{\quad\text{LDH}\quad} \text{pyruvate}$$

(Note, however, that in this system, the final NADH concentration does not give a measure of the original glutamate concentration.)

In general, an enzyme-catalysed reaction being used for substrate analysis by an end-point method should have a product (or co-product) whose concentration can easily be determined. If this is not the case the reaction may be **coupled** to another to enable the analysis to be performed, a product of the **primary** reaction acting as substrate for the **indicator** reaction. Consider, for example:

$$A \xrightarrow{E_1} B \xrightarrow{E_2} C$$

where A is the substrate whose concentration is being determined, E_1 is the enzyme of the primary reaction and E_2 the enzyme of the indicator reaction. The initial concentrations of B and C should be zero, and any additional substrates of E_1 or E_2 present in excess. For a satisfactory analytical procedure, both reactions should quickly reach equilibrium, preferably at or near completion. Since, under these conditions, the rate of the primary reaction is $\frac{V_{max}^A}{K_m^A}.[A]$ and that of the indicator reaction $\frac{V_{max}^B}{K_m^B}.[B]$, where V_{max}^A and K_m^A relate to E_1, and V_{max}^B and K_m^B to E_2 (see above), it is important that the enzyme concentrations are chosen such that $\frac{V_{max}^B}{K_m^B} > \frac{V_{max}^A}{K_m^A}$. This ensures that the B being formed from

A is quickly converted to *C*. If the coupled procedure is satisfactory, the final concentration of *C* (or of a co-product of the indicator reaction) is a measure of the initial concentration of *A*.

The main advantage of end-point methods of analysis is that they do not require constant attention: a procedure may be set up and left to come to equilibrium, after which the final product concentration can be determined at any convenient moment. However, now that automatic monitoring and recording devices are widely available, this advantage over other methods of analysis is considerably less important than it once was.

17.2.2 Kinetic Methods

At the present time, most procedures for enzymatic analysis involve steady-state kinetic methods: the initial velocity of the reaction is determined, at fixed temperature and pH, and used to calculate the concentration of the substance being investigated. Such procedures may be used for the analysis of enzymes (i.e. enzyme assay), substrates, co-enzymes, activators and inhibitors.

As discussed in Chapter 15.2.2, the initial velocity is directly proportional to **enzyme concentration** if all other relevant factors are made fixed and non-limiting. This is the general principle of enzymatic analysis by kinetic methods: it is arranged that all substances which influence the rate of reaction are present in fixed and non-limiting concentrations, with the exception of the one being analysed, whose concentration may thus be determined from the observed reaction rate.

The relationship between **substrate concentration** and initial velocity is apparent from the Michaelis–Menten equation (Chapter 7.1):

$$v_o = \frac{V_{max}[S_o]}{[S_o] + K_m} = \frac{k_2[E_o][S_o]}{[S_o] + K_m}.$$

Therefore, at constant $[E_o]$ in the absence of inhibitors and at fixed concentrations of any activators and additional substrates, there is a hyperbolic relationship between v_o and $[S_o]$ (Fig. 7.1). This could be used to determine substrate concentration, provided suitable calibration experiments were performed.

However, the best analytical procedures are those where there is a linear relationship between the concentration of the substance being analysed (in this case $[S_o]$) and the experimentally-determined property which is dependent on it (in this case v_o): this is because the procedure is then equally sensitive over the whole of the calibrated concentration range; also it enables errors in individual calibration values to be easily detected.

At low substrate concentrations there is a linear relationship between v_o and $[S_o]$, since if $[S_o] \ll K_m$, then $([S_o] + K_m) \simeq K_m$ and $v_o = \dfrac{k_2[E_o][S_o]}{K_m}$.

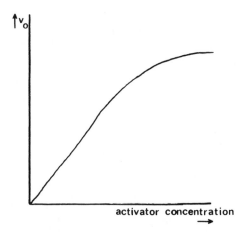

Fig. 17.1 Typical plot of initial velocity against activator concentration at fixed and non-limiting concentrations of enzyme and all substrates and coenzymes, for a system which cannot function in the absence of activator.

Hence, the calibrated values for substrate analysis should be restricted to the approximate range 0 to $0.1\,K_m$. The test sample should be diluted if necessary and the amount used for analysis chosen so that the concentration of the appropriate substrate *in the reaction mixture* lies in the required range: this is straightforward if the likely range in the sample is known; otherwise a series of different dilutions may have to be made to ensure that one gives readings in the calibrated range.

Another advantage of restricting the investigation of substrate concentrations to this range is that it avoids possible confusion resulting from substrate inhibition at high concentrations (Chapter 8.2.6).

Co-enzymes which act as co-substrates for enzyme-catalysed reactions may be analysed in exactly the same way as any other substrate.

Activators may also be determined in much the same fashion. These substances increase the catalytic performance of enzymes: some enzymes can function to a limited extent in the absence of their activators, whereas others cannot. In either case there is usually a linear relationship between initial velocity and activator concentration over the lower part of the concentration range, at fixed and non-limiting concentrations of enzyme and all substrates and co-enzymes; at higher activator concentrations, the enzyme becomes fully activated and initial velocity reaches a maximum value (Fig. 17.1).

A system may be calibrated, as in Fig. 17.1, and used to determine the activator concentration in a test solution. However, as before, it is best to work only in the lower linear part of the concentration range and to dilute the sample as necessary to bring the test results into this range.

An example of the use of such a system is the determination of Mn^{2+} by means of its activation of **isocitrate dehydrogenase**: this provides an extremely sensitive method of analysis, enabling extremely low concentrations of Mn^{2+} ions to be accurately determined.

Enzyme **inhibitors** may also be determined by sensitive procedures of this kind. As discussed in Chapter 8, many different types of enzyme inhibitors exist. However, as far as enzymatic analysis of inhibitors is concerned, the type of inhibition is immaterial. A system is calibrated at fixed and non-limiting concentrations of enzyme and all substrates, co-enzymes and activators, and a calibration curve plotted. This may take either of two forms: initial velocity may be plotted against inhibitor concentration (Fig. 17.2a): alternatively, percentage inhibition may be used instead of initial velocity (Fig. 17.2b).

$$\text{Percentage inhibition} = \frac{v_o \text{ (uninhibited)} - v_o \text{ (inhibited)}}{v_o \text{ (uninhibited)}} \times 100$$

where all initial velocities are determined under identical conditions, apart from inhibitor concentration.

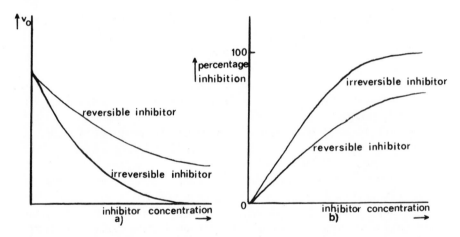

Fig. 17.2   Plots of (a) initial velocity against inhibitor concentration, and (b) percentage inhibition against inhibitor concentration, for typical reversible and irreversible inhibitors, at fixed and non-limiting concentrations of enzyme and all substrates, coenzymes and activators.

Irreversible inhibitors by definition, give 100% inhibition if they are present in excess, but reversible inhibitors do not give complete inhibition, no matter how high their concentrations. The inhibitor concentration giving 50% inhibition, under specified conditions, is sometimes termed the I_{50} **value**.

Once a system has been calibrated, it may be used to determine the concentration of an inhibitor in a test solution. Again, only the lower part of the concentration range, where there is a more or less linear relationship between initial velocity and inhibitor concentration, should be used. One such procedure, involving liver **carboxyl esterase**, enables small amounts of fluoride to be accurately determined in the presence of large amounts of phosphate, a factor which complicates many other procedures for the analysis of fluoride.

However, although enzymes are highly specific towards substrates, co-enzymes etc., they may be affected by a large number of inhibitors. This should be borne in mind in all forms of enzymatic analysis, particularly where the test sample contains many different components. Many metal ions, for example, are enzyme inhibitors, so it is not possible to use enzymatic analysis to determine the concentration of one of these if others which inhibit the same enzyme are also present. In enzymatic analysis in general, it is important to know what substances may be present in significant amounts in the sample, in addition to that being analysed.

The **coupling** of reactions for the purpose of enzyme assay was discussed in Chapter 15.2.3; similar coupling procedures may be required in any type of enzymatic analysis where the rate of the main reaction under investigation cannot be determined directly by any convenient method. Consider the general situation:

$$A \xrightarrow{E_1} B \xrightarrow{E_2} C$$

where E_1 is the enzyme catalysing the **primary reaction** (i.e. the main reaction under investigation) and E_2 the enzyme catalysing the **indicator reaction**. The initial concentrations of B and C should always be zero, and the initial concentrations of any additional substrates of E_1 and E_2 should be fixed and non-limiting.

In the case of enzyme assay, the primary reaction is usually made to proceed under zero-order conditions which makes the design of the coupling procedure relatively straightforward (see Chapter 15.2.3). However, substrate (or co-substrate) analysis is normally performed under conditions where the primary reaction is of first-order with respect to the substrate (or co-substrate) being analysed, which makes the relationship between the primary and indicator reactions less easy to establish. The initial velocity of the primary reaction at steady-state is given by the Michaelis–Menten equation:

$$v_o = \frac{V_{max}^A \cdot [A_o]}{[A_o] + K_m^A}$$

which under first-order conditions (where $[A_o] \ll K_m^A$) simplifies to:

$$v_o = \frac{V_{max}^A}{K_m^A} \cdot [A_o].$$

To ensure that B does not accummulate and that the rate of the indicator reaction is a measure of the rate of the primary reaction over the period of interest, the concentration of E_2 (and hence V_{max}^B) is made sufficiently large so that

$$\frac{V_{max}^B}{K_m^B} \gg \frac{V_{max}^A}{K_m^A}$$

where V_{max}^A and K_m^A relate to E_1, and V_{max}^B and K_m^B to E_2. To give a reasonable margin of safety, the ratio $\dfrac{V_{max}^B}{K_m^B} : \dfrac{V_{max}^A}{K_m^A}$ is usually made in the order of 100:1 for substrate determinations and for most other forms of enzymatic analysis.

Another important requirement for coupled reactions is that the enzymes involved are active in the same pH range (see Chapter 15.2.3).

The simplest, though most time-consuming way to perform a kinetic method of enzymatic analysis is to remove samples at timed intervals from the reaction mixture. The reaction in each sample tube is terminated immediately on removal, e.g. by rapid change of temperature or pH, or by the use of a protein precipitant such as perchloric acid, and then the product concentration in each is determined. From this data, the initial velocity can be calculated.

However, the use of such laborious **manual** procedures has now largely been discontinued. The methods used at the present time are mainly **instrumental** ones and, in particular, ones where the course of the reaction can be monitored and recorded automatically (see Chapter 18). Since the initial velocity alone is required, some of the more complex (and expensive) instruments can determine this and display results within a minute or so of the start of the reaction.

17.2.3 Immunoassay Methods

The determination of **enzyme concentration** by **radioimmunoassay (RIA)** was discussed in Chapter 15.2.4.

Enzymes may also be involved in other immunoassay procedures, although it must be stressed that the role of the enzymes in these is a secondary one: they are used to replace radioisotopes as markers, since they are not such a hazard to health and can be detected by techniques which are more generally available. Any enzyme which has a sensitive and convenient assay procedure can be employed. There are currently two main types of these **enzyme-immunoassay (EIA)** procedures: **ELISA** and **EMIT**.

ELISA (**enzyme-linked immunosorbant assay**) resembles RIA in that it involves a heterogenous system which must undergo separation as part of the analysis procedure. For example, a pure specimen of antigen may be labelled by attaching it to an enzyme (just as it is labelled with a radioisotope in RIA) in such a way that the activity of the enzyme is not affected. Then, specified amounts of antibody and enzyme-labelled antigen are mixed with the sample which contains an unknown amount of the antigen. The free antigen (Ag) and enzyme-labelled antigen (E.Ag) then compete for binding to the available antibody (Ab), according to the following reactions:

$$Ag + Ab \rightleftharpoons Ag.Ab$$

$$E.Ag + Ab \rightleftharpoons E.Ag.Ab.$$

After equilibrium has been reached, the antibody-bound fractions are separated from the rest (the use of immobilised antibodies, i.e. antibodies attached to an insoluble matrix, enables this to be done conveniently) and the enzyme activity in one or both of the fractions is determined. From the distribution of the enzyme between the free and antibody-bound fractions, the concentration of antigen in the sample may be calculated: the more antigen in the sample, the greater will be the enzyme activity of the free antigen fraction after equilibration. Steroid hormones such as progesterone and cortisol have been determined by procedures of this type, and so have proteins such as insulin. Enzymes which have been used as labels include peroxidase, alkaline phosphatase and β-galactosidase.

An alternative form of ELISA involves attaching the enzyme to the *antibody* and the use of an immobilised antigen. Specified amounts of enzyme-labelled antibody and immobilised antigen are mixed with the sample: the antigen in the sample (Ag) then competes with the immobilised antigen (X.Ag) for the available enzyme-linked antibody (Ab.E) as follows:

$$Ag + Ab.E \rightleftharpoons Ag.Ab.E$$

$$X.Ag + Ab.E \rightleftharpoons X.Ag.Ab.E.$$

After equilibration, the immobilised fraction is separated from the rest and the distribution of the enzyme is determined. This enables the antigen content of the sample to be calculated, there being an inverse relationship between the enzyme activity of the insoluble fraction and the amount of antigen in the sample. An assay for α-fetoprotein has been based on this principle.

EMIT (enzyme multiplied immunoassay technique) is a homogenous procedure, since no separation of components is required. As with some types of ELISA, an enzyme is attached to a specimen of antigen to act as a label. However, in contrast to ELISA, the subsequent binding of the enzyme-labelled antigen to the antibody results in a significant change in activity of the enzyme: in most types of EMIT procedure, enzyme activity is lost completely on binding to the antibody, either as a result of steric hindrance or conformational changes. In such an assay, specified amounts of antibody and enzyme-labelled antigen are mixed with the sample, The antigen in the sample (Ag) then competes with the enzyme-labelled antigen (E.Ag) for the available antibody (Ab) according to the following reactions:

$$Ag + Ab \rightleftharpoons Ag.Ab$$

$$E.Ag + Ab \rightleftharpoons E'.Ag.Ab.$$

The enzyme-antigen-antibody complex formed (E'.Ag.Ab) has no catalytic activity, so the total activity present is contributed by E.Ag. Hence the antigen content of the sample may be calculated from the total enzyme activity at equilibrium: the more antigen there is in the sample, the less E.Ag is able to bind to the antibody, so the greater will be the total enzyme activity.

EMIT procedures are more convenient than ELISA ones, since they do away with the separation step. However, the technical problems of obtaining enzyme-labelled antigen suitable for EMIT are considerable, particularly since the reasons for the loss of activity on binding to antibody are not fully understood and may vary from case to case. Because of this, only a limited number of EMIT procedures are currently available, but this number is increasing rapidly.

EMIT has so far been used principally for the determination of relatively small (i.e. non-protein) molecules, e.g. barbiturates. The enzymes which have been involved include lysozyme and malate dehydrogenase.

17.3 HANDLING ENZYMES AND COENZYMES

Enzymes easily become denatured and lose catalytic activity, so careful storage and handling is essential if this is to be prevented. In general, temperatures higher than 40°C and extremes of pH (i.e. below pH 5 and above pH 9) should be avoided at all times. If the pH of an enzyme solution is being changed, continuous mixing helps to prevent denaturation taking place around the added drops of acid or alkali. However, too vigorous mixing causing the formation of froth is counter-productive, since many enzymes are denatured at surfaces.

At room temperature, most enzymes are denatured by organic solvents such as ethanol or acetone, except where these are present at very low concentrations. As with changes of pH, changes in organic solvent concentration should be performed with extreme care to avoid localised denaturation of enzymes.

Some dry enzymes are quite stable at 0-4°C, whereas others need storing at lower temperatures. Enzyme preparations should also be stored at low temperatures, but not necessarily frozen, since freezing and thawing may cause loss of activity: this depends on the nature of the preparation. Crystalline enzymes suspended in ammonium sulphate solutions are best stored at 0-4°C, as are lyophilised enzyme preparations; however, the latter may be stored frozen if it is certain that they contain no moisture. In contrast, solutions of enzymes should always be stored frozen at temperatures as low as possible, but repeated freezing and thawing should be avoided: for this reason, such solutions are best divided into small portions for storage. Enzyme preparations are sometimes stored in aqueous solutions of glycerol, since this enables them to be kept at temperatures below $0°C$ without freezing taking place.

Most coenzymes must be stored as solids at $0-4°C$, preferably in a desiccator, since many are subject to hydrolysis. Some coenzymes must also be protected from light or oxygen. NADH, for example, is destroyed by light to quite an appreciable extent, while also reacting with any atmospheric water to form inhibitors of dehydrogenases. In solution, NADH is particularly unstable in acid conditions, whereas NAD^+ is alkali-labile. The sulphydryl group of Coenzyme A can be oxidised by atmospheric oxygen, and its pyrophosphate bond is easily hydrolysed if moisture is present. Some of the more stable coenzymes may be stored in frozen solution, but, as with enzyme solutions, repeated freezing and thawing should be avoided.

All enzyme solutions, and other solutions used in connection with them, should be free from contamination by micro-organisms and metal ions. They should be made up in fresh glass-distilled water and checked regularly for the presence of micro-organisms. All glassware should be clean and the final wash before use should be with distilled water rather than with an organic solvent, to avoid any risk of causing enzyme denaturation.

SUMMARY OF CHAPTER 17

Enzymes have great value as analytical reagents because of their specificity of action. Enzyme-based methods of analysis are very sensitive and are usually carried out under mild conditions. Such procedures may be used to determine the concentration of a substrate, co-enzyme, activator or inhibitor of the enzyme in question. The general name for investigations of this type, including enzyme assay, is **enzymatic analysis**.

The principle of enzymatic analysis by kinetic methods is that the initial velocity of reaction is dependent solely on the concentration of the substance being analysed, regardless of whether it is an enzyme, substrate, coenzyme, activator or inhibitor, provided all other relevant factors are made fixed and non-limiting. End-point methods of analysis, on the other hand, involve measurement of product (or co-product) concentration when a reaction has gone to (or near to) completion and can only be used for the analysis of substrates. Both types of analysis may involve the coupling of the primary reaction to an indicator reaction.

Enzymes may also be used as an alternative to radioisotopes as markers in immunoassays. Such procedures, termed enzyme–immunoassays, have been used for the determination of a variety of proteins and hormones.

Enzymes easily lose their catalytic activity, so require careful handling:

exposure to high temperatures, extremes of pH and high concentrations of organic solvents should be avoided at all times. In common with coenzymes, they require storage at low temperatures.

All solutions used in enzymatic analysis procedures must be free from contamination by micro-organisms and metal ions.

FURTHER READING

Bergmeyer, H. U. (ed.) (1978), *Principles of Enzymatic Analysis* (Sections I–III), Verlag Chemie.

Colowick, S. P. and Kaplan, N. O. (eds.), *Methods in Enzymology*, **92** (1983): Immunochemical Techniques, Academic Press.

Guilbault, G. C. (1970), *Enzymatic Methods of Analysis* (Chapter 1), Pergamon Press.

Maggio, E. T. (ed.) (1980), *Enzyme Immunoassay*, CRC Press.

O'Sullivan, M. J., Bridges, J. W. and Marks, V. (1979), Enzyme immunoassay: a review, *Annals of Clinical Biochemistry*, **16** (pages 221–240).

PROBLEMS

17.1   An enzyme used in an end-point method of analysis has a K_m of 9×10^{-4} mol.l^{-1} with respect to the appropriate substrate and its activity in the reaction mixture is 2.0 IU.cm^{-3}. If the reaction is largely irreversible and is proceeding under first-order conditions, calculate the time required for it to reach 99% completion.

17.2   Sketch, from memory or deduction, the following plots for enzyme-catalysed reactions which obey the Michaelis–Menten equation. In each case, state clearly what restrictions must be imposed on other relevant factors to enable the plot you draw to be obtained.

(a) v_o against $\dfrac{v_o}{[S_o]}$;

(b) v_o against $[E_o]$;

(c) v_o against coenzyme (e.g. NADH) concentration;

(d) $[P]$ against t (under first-order and under zero-order conditions);

(e) v_o against $[S_o]$ in the absence and presence of a competitive inhibitor;

(f) $\dfrac{1}{v_o}$ against $\dfrac{1}{[S_o]}$ showing the effects of substrate inhibition;

(g) v_o against activator concentration, for a system which does not function in the absence of activator;

(h) v_o against $[I_o]$ for an irreversible inhibitor.

CHAPTER 18

Instrumental Techniques available for use in Enzymatic Analysis

18.1 PRINCIPLES OF THE AVAILABLE DETECTION TECHNIQUES

18.1.1 Introduction

A variety of detection procedures are used, or have been used in the past, for enzymatic analysis by either end-point or kinetic methods; they may also, of course, be used for the investigation of enzyme characteristics. The most important ones are discussed below.

In most cases, the instrument detecting the changes which take place as the reaction proceeds can be coupled to a recorder, enabling the course of the reaction to be followed in detail without requiring constant human attention. Also, the enzyme involved may be in an immobilised form (see Chapter 20.2) to allow them to be recovered at the end of the procedure. However these refinements, whose use largely depends on the finances available, do not affect the principles of the detection techniques.

18.1.2 Manometry

If any of the substrates or products of an enzyme-catalysed reaction is a gas, the course of the reaction may be followed by manometric techniques, as developed by Warburg (1924). The basic system must be air-tight, and consists of a reaction vessel attached to a manometer (Fig. 18.1). The sample and reagents are put into different compartments in the reaction vessel and mixed together at zero time; then, as the reaction proceeds, the uptake or evolution of gas is indicated by the movement of the fluid in the manometer tube.

However, it will be realised that as the manometer fluid moves towards or away from the reaction vessel, the level of liquid in the two arms of the manometer will no longer be the same; this will introduce pressure changes which will affect the measurement of the progress of the reaction. Various features are therefore added to the basic system to ensure that all readings are made under identical conditions.

In **constant volume manometry**, the manometer is usually connected to a reservoir, the access being controlled by a tap, so that fluid may be added to or

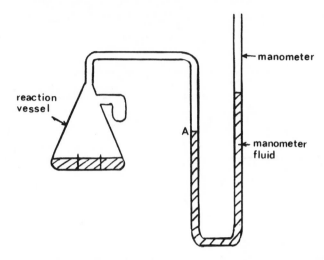

Fig. 18.1 Diagrammatic representation of the basic set-up for manometry. Note that the reaction vessel should be immersed in a constant-temperature water bath.

removed from the manometer as required. The amount of fluid in the manometer is always adjusted prior to taking a reading so that the level in the limb nearest the reaction vessel is at a fixed point (e.g. point A in Fig. 18.1). The difference between the fluid levels in the two limbs of the manometer then gives an indication of the pressure in the reaction vessel; this changes as gas is taken up or evolved, and so may be used to follow the course of the reaction.

In **constant pressure manometry**, the volume of the reaction chamber is changed by measurable amounts (e.g. by connecting it to a micrometer, as in the **Gilson differential manometer**) sufficient to keep the manometer fluid level at a fixed point (e.g. point A in Fig. 18.1) as the reaction proceeds. The amount of fluid in the manometer is not changed, so this procedure ensures that the levels in *both* arms remain fixed, and no pressure changes take place. The course of the reaction may thus be followed in terms of the volume changes required to maintain constant pressure.

Manometric methods may be used to investigate reactions where a gas is consumed, e.g. that catalysed by glucose oxidase, where molecular oxygen is a substrate. They may also be used when a gas is produced, e.g. to monitor CO_2 production in reactions catalysed by decarboxylases. Reactions where an acid is produced (including all reactions involving NAD^+ reduction to $NADH + H^+$) may be investigated by manometry by adding bicarbonate to the reaction mixture and measuring the evolution of CO_2.

However, such techniques usually involve a margin of error greater than 5%, so they are rarely used today for enzymatic analysis, since more accurate and reliable procedures are now available.

18.1.3   Spectrophotometry

If a substrate or a product absorbs light at a characteristic wave length, then the reaction can be monitored by following the changes in absorbance at this wavelength. The absorbance (A) is given by

$$A = \log_{10} \left(\frac{I}{I_o} \right)$$

where I_o is the intensity of the incident light and I is the intensity of the light transmitted through the sample. The absorbance is related to the concentration of the light-absorbing substance in the sample by the **Beer–Lambert law**:

$$A = \epsilon.c.l$$

where ϵ is the molar extinction coefficient, c is the molar concentration and l is the length of the light-path through the sample (usually measured in centimetres).

This is generally true for any wavelength of visible light (wavelength range 400–700 nm) or ultra-violet light (200–400 nm). The wavelength required can be selected from the spectrum obtained when light is passed through a prism or a diffraction grating. In simpler and cheaper instruments (**colorimeters**), a waveband, rather than a particular wavelength, is selected by passing the light through a coloured-glass filter: such instruments are less reliable than spectrophotometers and cannot be used in the ultra-violet region.

For enzymatic analysis, the temperature of the reaction mixture should be maintained to within ±0.2°C, which necessitates the use of a jacketed cell-holder connected to a circulating water bath.

Procedures which involve the calculation of initial velocity should preferably be ones where an increase (rather than a decrease) of absorbance with time is observed: they will then be operating in the most sensitive part of the absorbance scale (the lower part) during the crucial early stages of the reaction. Such procedures are best followed by instruments which give direct absorbance readings rather than by **null-point instruments** which require separate adjustments for every reading. **Direct-reading instruments** may be coupled to recorders so that the whole course of the reaction may be traced out. **Double-beam instruments**, which automatically compare the absorbance of the reaction mixture to that of a reagent blank at each point, are preferable in these circumstances, since changes of absorbance with time of the blank may be taken into account without having to re-set the instrument during the analysis.

Spectrophotometric methods are particularly useful for monitoring reactions which involve NAD^+ or $NADP^+$ as coenzyme (see Chapter 6.5.1), or ones which can be coupled to such reactions.

There are two main problems associated with spectrophotometric methods. Firstly they can only be used satisfactorily over the lower part of the absorbance range: at high absorbance values there is, in addition to reduced sensitivity, often some departure from the Beer–Lambert law because of interactions between solute molecules in concentrated solution. Secondly, errors may be introduced when there is a high background absorbance or when the sample contains material in suspension (e.g. blood cells or micro-organisms), thus scattering the incident light. For this reason, crude samples may often have to be partially purified (e.g. by dialysis, selective precipitation and/or centrifugation) before one of its components may be investigated by spectrophotometric procedures.

18.1.4 Spectrofluorimetry

Compounds are said to be **fluorescent** when they absorb light of one wavelength and then emit light of a longer wavelength. At low concentrations, the intensity of fluorescence (I_f) is related to the intensity of the incident light (I_o) of appropriate wavelength by the relationship:

$$I_f = I_o . 2.3 . \epsilon . c . l . q$$

where ϵ is the molar extinction coefficient, c the molar concentration, l the length of the light-path and q the quantum efficiency (i.e. the number of quanta fluoresced divided by the number of quanta absorbed). The fluorescence is usually measured at right angles to the incident beam, to avoid detection of any transmitted light. The wavelength of the fluorescence required for measurement, together with the appropriate wavelength for the incident beam, are usually selected by the use of diffraction gratings.

In general, spectrofluorimetry is several times (theoretically up to about a hundred times) more sensitive than spectrophotometry. However, fluoresence effects vary with temperature, so it is particularly important to maintain a fixed temperature throughout.

NADH and NADPH exhibit fluorescence, absorbing light at 340 nm and re-emitting it at about 460 nm. Therefore, any reactions utilising these as coenzymes may be followed by the use of sensitive spectrofluorimetric procedures. Spectrofluorimetry is also widely used for the investigation of hydrolytic reactions, using synthetic non-fluorescent substrates which are esters of highly-fluorescent alcohols or amines: for example, Guilbault and Kramer (1963) devised an assay for **triacylglycerol lipase** based on the following reaction:

$$\text{dibutyryl fluorescein} \xrightarrow{\text{lipase}} \text{fluorescein}.$$
$$\text{(non-fluorescent)} \qquad\qquad \text{(fluorescent)}$$

In a similar way, fluorescein diacetate may act as a substrate for esterases, and derivatives of highly-fluorescent 4-methylumbelliferone are often used in the assay of esterases, glycosidases, phosphatases and sulphatases.

Tyrosine and tryptophan fluoresce at 330–350 nm, and since residues of these are usually present in enzymes, they would give a high background emission in the ultra-violet region. Hence, it is advisable to restrict the use of spectrofluorimetry in enzymatic analysis to the detection of substances which emit light in the visible region. Apart from this, the main problem associated with spectrofluorimetric techniques is **quenching**, where the extra energy possessed by the excited molecules after absorption of a photon of light is transferred to another molecule or group (e.g. to iodide or dichromate ions, if these are present) instead of being released as another photon. **Concentration quenching** may also occur, due to some net absorption of either the incident or the emitted light taking place.

18.1.5 Electrochemical Methods

Many different types of electrochemical procedures have been used in enzymatic analysis.

Often these involve **potentiometric techniques**, where an electrical potential is generated which is dependent on the concentration and properties of substances in the test solution; the change in potential as the reaction proceeds can be taken as a measure of the rate of the reaction (see Chapter 18.2.3). It should be noted that a potential, which is caused by a transfer of electrons between an electrode and a solution, cannot be measured in isolation: the half-cell consisting of the electrode dipped in the test solution must be connected to a reference half-cell, of fixed potential, to enable the potentials of the two half-cells to be compared.

Of particular importance are techniques involving **ion-selective electrodes**, where the potential generated can be related to the concentration of a specific substance. The best known ion-selective electrode is the **glass electrode** used in the **pH meter**: in this, an internal electrode (e.g. silver wire coated with silver chloride) is dipped in a solution containing HCl at known pH. This is separated from the test solution by a thin glass wall and, in a way not fully understood, a potential is generated which is dependent on the pH difference across the glass. The potential of the entire half-cell, which includes the contribution of the internal electrode as well as that resulting from the pH difference, may be measured by linking the internal electrode to an external reference electrode (e.g. a calomel electrode) and completing the circuit by means of a salt bridge (Fig. 18.2). The complete cell consists of the following interacting compartments:

| Ag | AgCl | HCl | glass | test sample | | saturated KCl acting as a salt bridge | external reference electrode. |

The potentials produced at each of these junctions will be constant for a particular instrument, the one exception being the important one between glass

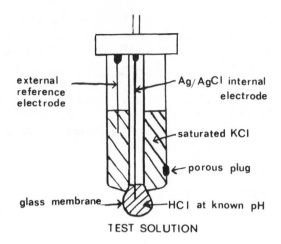

Fig. 18.2 Diagrammatic representation of a combined glass electrode and external reference electrode, as used in the pH meter.

and the test sample. Thus, a suitably calibrated instrument may be used to determine the pH of the test sample.

Many enzyme-catalysed reactions involve the production of an acid or base from a neutral compound (or vice versa), so the course of such reactions could be followed by measuring the change in pH as the reaction proceeds. However, this is rarely done because a change in pH would cause the activity of the enzyme to change. The best way of monitoring the course of such reaction is by the use of a **pH-stat**. This usually consists of a pH meter coupled to a motor-driven burette syringe and a circuit-braker designed to switch off the burette motor when the pH of the reaction mixture is at the original pre-set value. As the reaction proceeds and the pH of the reaction mixture starts to change, the motor switches on and forces alkali (or acid) from the burette into the reaction vessel, with continuous stirring, to restore the pH to its original value, while a recorder automatically makes a trace of the amount added from the burette in relation to time. Thus the pH is maintained at an approximately constant value throughout, and the rate of addition of alkali (or acid) required to achieve this is a measure of the rate of reaction. Such techniques may be used, for example, to monitor the hydrolysis of esters.

Other types of electrochemical procedure include those based on **polarography** or **voltammetry**, where an increasing voltage is applied between two electrodes immersed in the test solution, whose composition determines the current which flows at each instant.

If a *fixed* voltage is applied between the electrodes, the current depends on the electrical conductance of the solution, which is the sum of the contribution of all the ions present (since these must be the carriers of the current through the solution); it is also the reciprocal of the electrical resistance between the electrodes. For those enzyme-catalysed reactions where there is a change in the number of charged species as substrates are converted to products, the course of the reaction may be followed by determining the change of electrical conductance with time; this is termed **conductometry**. Such procedures usually make use of alternating rather than direct current, to avoid electrolysis taking place, which would tend to reduce the applied voltage. Since the reaction mixture must be buffered, which will contribute a background conductance, a low concentration of a buffer of low intrinsic conductivity (e.g. Tris) is usually employed.

An example of a reaction which is ideally suited to enzymatic analysis by conductometric procedures is that catalysed by urease:

$$O=C(NH_2)_2 + H_2O \rightleftharpoons 2NH_4^+ + CO_3^{2-}.$$
urea

An uncharged molecule is converted into ions as the reaction proceeds, so the conductance must increase. This has been used for the analysis of the substrate, urea, and gives a linear response up to a concentration of about $22\,\mu\text{mol.l}^{-1}$, above which there is interaction with buffer ions; samples containing urea at higher concentrations than this should be diluted prior to analysis.

A specific electrode which is based on similar principles is the **oxygen electrode**. In the type developed by Clark (1956) there is a central cathode of gold or platinum surrounded by an annular silver anode and separated from it by an epoxy resin casing (Fig. 18.3). Oxygen dissolved in the reaction mixture (which

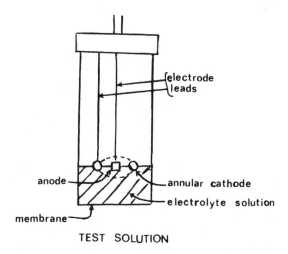

TEST SOLUTION

Fig. 18.3 Diagrammatic representation of a typical oxygen electrode.

should be in a thermostatted, air-tight container and stirred continuously) can freely diffuse to the electrodes through a membrane (e.g. Teflon or polypropylene) and an electrolyte solution (usually buffered KCl). A constant voltage of between 0.5 and 0.8 V is applied across the electrodes, providing a supply of electrons at the cathode which can reduce oxygen atoms:

$$4e^- + O_2 + 2H^+ \rightarrow 2H_2O$$

whereas at the anode there is a corresponding transfer of electrons *to* the electrode:

$$4Ag + 4Cl^- \rightarrow 4AgCl + 4e^-.$$

Thus a current flows between the electrodes which is proportional to the pO_2 of the sample. By monitoring how this current changes with time, the course of a reaction involving oxygen production or consumption (e.g. that catalysed by glucose oxidase) can be followed. This is a far more sensitive procedure than manometry (Chapter 18.1.2), a possible alternative.

As a refinement, the oxygen electrode (or other specific electrode) may be coupled with an immobilised enzyme to form an **enzyme electrode** for the analysis of the appropriate substrate. For example, Hicks and Updike (1967) trapped glucose oxidase in polyacrylamide gel and layered it over the membrane of an oxygen electrode, thus constructing an instrument which could be used to analyse D-glucose by a sensitive, specific and convenient procedure.

In general, electrochemical methods have an advantage over certain other procedures (e.g. spectrophotometry) in that contaminating substances can be present in suspension without affecting the analysis.

18.1.6 Microcalorimetry
Enzymatic analysis by microcalorimetry is becoming increasingly important because of the sensitivity, freedom from interference and almost universal possibilities of application of these techniques. In most reactions, heat (enthalpy) is gained or lost as the reaction proceeds (see Chapter 6.1). In microcalorimetry, the reaction is carried out under controlled conditions in a calorimeter and the temperature change is monitored. The main reason why these procedures have been little used in the past is that the temperature change is small (usually 10^{-2} to 10^{-4} °C) and well within the range of possible fluctuations in the ambient temperature. Hence extremely accurate thermostatting or excellent insulation is required, together with very sensitive temperature sensors. Suitably accurate temperature measurements may be obtained by the use of **thermistors** (temperature-sensitive resistors consisting of sintered mixtures of metal oxides in glass or plastic casings) or of **thermopiles** (thermocouples connected in series).

Secondary reactions with the buffer can be used to increase the signal and thus improve the sensitivity of the procedure: for example, a proton produced by the primary reaction can be trapped by buffer with evolution of heat; Tris is very useful in this respect, for its heat of protonation is large. McGlothlin and Jordan (1975) used microcalorimetry to estimate blood glucose by means of a hexokinase-catalysed reaction in Tris buffer at pH 8:

$$\text{glucose} + \text{ATP} \rightarrow \text{glucose 6-phosphate} + \text{ADP} + \text{H}^+ \quad \Delta H = -28 \text{ kJ.mol}^{-1}$$

$$\text{Tris} + \text{H}^+ \rightarrow \text{Tris} - \text{H}^+ \qquad\qquad \Delta H = -47 \text{ kJ.mol}^{-1}.$$

It will be noted that the secondary reaction liberates more heat than the primary reaction.

18.1.7 Radiochemical Methods

The use of a radioactively-labelled substrate can be valuable in enzymatic analysis: the isotopes most commonly used for labelling purposes are hydrogen-3 (tritium), carbon-14, phosphorus-32, sulphur-35 and iodine-131. All of these isotopes emit β-radiation (electrons) as they decay.

After the enzyme-catalysed reaction has progressed for a specified period, it is terminated. The substrate is then separated from the product, usually by chromatography or electrophoresis, and the product concentration is determined indirectly by measuring the radioactivity of the product fraction.

High-energy β-emission, e.g. from ^{32}P or ^{131}I, may be detected and counted directly using a **Geiger–Müller counter**. Low-energy β-emission, e.g. from ^3H, ^{14}C or ^{32}S, is usually monitored by **liquid scintillation counting**: some organic solvents fluoresce when bombarded with β-radiation, giving out light of very short wavelength; this is not suitable for efficient detection, but it may be converted into light of longer wavelength by the fluorescence of one or more **fluors**, or scintillants (e.g. 2,5-diphenyloxazole, PPO), and detected by a photomultiplier.

A typical example of enzymatic analysis by a radiochemical procedure is the method of Reed and colleagues (1965) involving the cholinesterase-catalysed hydrolysis of $[^{14}\text{C}]$-acetylcholine:

$$^{14}\text{CH}_3.\text{CO}_2\text{CH}_2\text{CH}_2\overset{+}{\text{N}}(\text{CH}_3)_3 \rightarrow {}^{14}\text{CH}_3.\text{CO}_2\text{H} + \text{HOCH}_2\text{CH}_2\overset{+}{\text{N}}(\text{CH}_3)_3.$$

$[^{14}\text{C}]$-acetylcholine $[^{14}\text{C}]$-acetic acid   choline

After incubation, the unreacted substrate and the choline may be removed by an ion exchange resin, and the radioactivity of the acetic acid fraction determined.

In general, radiochemical procedures are extremely sensitive (far more sensitive than photometric ones). They can be used in a straightforward fashion for enzyme assay, and, by means of isotope dilution techniques, for other forms of enzymatic analysis also. However, they have certain disadvantages when compared to other possible techniques. These include the health hazard inevitable in the handling of radioisotopes, and the separation steps which have to be performed. Also, it is impossible to use them to monitor continuously the course of a reaction, and any estimate of the initial velocity which is obtained can only be based on a limited number of points (possibly only *one*). A further possibility of error can result from a **quenching** of the radiation between source and detector: this is particularly likely in the case of tritium, whose emission is so weak, it can be absorbed by paper.

18.1.8 Polarimetry

Because of the stereospecificity of enzymes (Chapter 4.1), they will often attack only one isomeric form of a substrate, and the product formed from it may or may not be optically active; even if it is, the optical activity might be quite different from that of the substrate. Therefore, many enzyme-catalysed reactions (e.g. those catalysed by amino-acid racemases) are likely to involve a change of optical rotation as the reaction proceeds.

The angle of rotation of the plane of plane-polarised monochromatic light passing through a solution of an optically active substance is proportional to the length of the light-path, the concentration of the substance and its specific rotation. Hence the change in the angle of rotation as a reaction proceeds may be used to monitor the course of the reaction: this is termed **polarimetry**. However, the sensitivity of such methods is usually poor compared to other available procedures. In some cases, it may be possible to magnify the change in rotation taking place by the formation of a complex: for example, lactate reacts with molybdate to form a complex which has a high specific rotation, thus enabling the LDH-catalysed conversion of lactate to pyruvate to be monitored by a polarimetric method. However, because of the generally low sensitivity involved, such procedures are rarely used in enzymatic analysis.

18.2 AUTOMATION IN ENZYMATIC ANALYSIS

18.2.1 Introduction

As discussed in Chapter 18.1, most of the detectors used to monitor the course of enzyme-catalysed reactions can be coupled to automatic recorders. However, if each sample still has to be mixed with reagents and introduced to the detector by hand, the overall process can only be said to be **semi-automatic**. Here in Chapter 18.2 we are concerned with **fully-automatic** procedures, where simultaneous or sequential analysis can be carried out on a batch of samples without any manual steps being involved other than in sample preparation and in loading the batch on to the analyser, and possibly in the calculation of the results from recorder traces. Many analysers are now linked to computers, to enable the results to be calculated and assessed automatically. As always, the choice of instrument and of its range of facilities is limited by financial considerations.

Most fully-automatic techniques used for enzymatic analysis give a result based on an estimate of the initial velocity of the reaction: in the case of relatively simple instruments, this may be based on just one or two readings; more complex (and expensive) instruments monitor the reaction throughout and give the best statistically-derived estimate of initial velocity. Regardless of this, the analysis must always be carried out at constant pH and temperature.

When an instrument allows only a limited number of readings to be made for each analysis, it is advisable, before using it for a particular application, to perform preliminary experiments to ensure that the progress of the reaction is linear with respect to time over the period of interest. Under suitable conditions, most enzyme-catalysed reactions quickly reach a linear, steady-state phase which is then maintained for a few minutes before the rate of reaction starts to drop (Chapter 7.2.1). However, some reactions (e.g. that catalysed by creatine kinase) have quite an appreciable lag-phase (i.e. one or more minutes) before the steady-state phase is reached; even the others may, in practice, take seconds to reach the linear phase, because complete mixing of the sample and reagents cannot be achieved instantaneously. The duration of the linear-phase depends mainly on the relative concentrations of enzyme and substrate (see Chapter 15.2.2). Hence, by preliminary investigations, it should be possible to choose analytical conditions which ensure that a *two-point* procedure gives a reasonable estimate of initial velocity provided the test substance falls within a specified concentration range (Fig. 18.4). However, ideal conditions for a *one-point* procedure, which assumes linearity from zero time to the time of the reading, can never be achieved if there is any appreciable lag-phase.

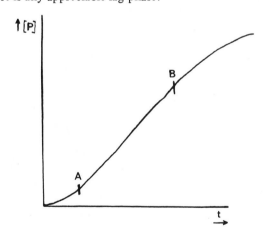

Fig. 18.4   Typical plot of product concentration against time for an enzyme-catalysed reaction. Conditions for a two-point procedure should be chosen by preliminary experiments so that the readings fall between *A* and *B*.

Although linearity over the period of interest should be aimed at in both one- and two-point procedures, reasonable results may nevertheless be obtained even if there are slight departures from linearity, provided suitable standards of known composition are included among each batch of test samples. These standards may be solutions made up from solids, or they may be samples which have been analysed previously by a reliable method.

18.2.2 Fixed-Time Methods

In fixed-time methods of analysis, the change in experimental reading (e.g. absorbance) over a fixed time interval is recorded. If the reaction has linear characteristics over this time interval, then the observed change is directly proportional to the initial velocity, and hence can be related to the concentration of the substance being analysed (Chapter 17.2.2).

Such analysers may be termed **discrete**, when each sample is mixed with reagents by means of automatic pipettes and passed into a detector at an appropriate time, according to programmed instructions. Discrete analysers (e.g. the Boehringer-Mannheim Hitachi 705, 712 and 737, The Technicon RA 1000 and the Roche Cobas Mira systems) can work efficiently on large batches for single tests (e.g. analysing glucose in perhaps 200 samples per hour) or can be programmed to carry out 12 or more different tests on each sample with a throughput of up to 1000 tests per hour. The stress on automatic pipetting systems is considerable, and some early discrete analysers were susceptible to mechanical breakdown, but such problems have now largely been overcome. Detection is usually by spectrophotometry, but ion-selective electrode modules may be fitted to some types. Most modern discrete analysers are **selective**, i.e. more tests can be done on some samples than on others, as considered appropriate. Another modern development is to have **random access** facility, allowing the operator, through the linking computer, to be in control throughout. For instance, samples added while the instrument is already in operation can be given priority, if required. The computer may also be able to arrange the quickest way of carrying out all the required tests on the samples.

An alternative approach to fixed-time analysis is in the use of **continuous-flow** methods (as in the Technicon AutoAnalyser or SMA systems). If a solution is pumped through a tube at a fixed flow-rate, each molecule will take the same time to pass between two given marks on the tube. In continuous-flow analysis, a mixture of reagents is pumped continuously through a tube, usually by means of a peristaltic pump, and, at regular time intervals, one of a series of samples is automatically introduced into the reagent line. Suitable mixing coils and constant-temperature heating-baths ensure that each sample is mixed with the reagent and that, as they are pumped along the tube, a reaction takes place under controlled conditions. Eventually the reaction mixture reaches a detector and the progress of the reaction is determined. The set-up ensures that a fixed time elapses between the mixing of each sample with reagent and the arrival of the resulting mixture at the detector. Usually, air (or other gas) is pumped into the reaction tube along with the sample and reagents, producing large bubbles which segment the

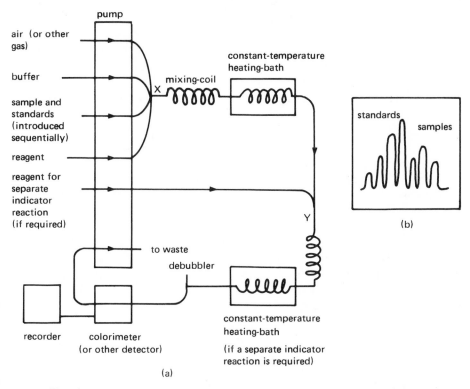

Fig. 18.5 (a) Simplified representation of a set-up for continuous-flow analysis;
(b) typical recorder tracing from a continuous-flow analysis procedure.

reaction mixture into lengths of no more than a few centimetres each. These pass along the tube separately, with air in between, and so diffusion between reaction mixtures from different samples is prevented. Alternatively, a very narrow tubing may be used to minimise diffusion.

A typical set-up is shown in Fig. 18.5a: this is simplified slightly in that it shows only one reagent line, whereas it is common for several different reagents to be pumped separately until they are mixed together with the sample; also, the mixing is usually staggered slightly so that no more than three lines ever come together at the same point.

In some applications, the concentration of a product of the **primary**, enzyme-catalysed reaction is determined by a separate, **indicator** reaction. For example, in an assay of **alkaline phosphatase** involving the use of disodium phenyl phosphate as substrate:

$$\text{disodium phenyl phosphate} \xrightarrow[\text{pH 10}]{\substack{\text{alkaline} \\ \text{phosphatase}}} \text{phenol} + \text{sodium phosphate}$$

the phenol liberated must be analysed after the primary reaction has been stopped, because of the nature of the indicator reagents. This can be done satisfactorily if a one-point procedure is being employed, whereas such a coupled system could not be used for monitoring *continuously* the course of the primary reaction.

In the King and Armstrong (1934) method, the phenol is made to react with Folin–Ciocalteu reagent and the absorbance determined at 680 nm (see Chapter 3.2.1). However, this reagent also reacts with the phenolic amino-acid, tyrosine, which is present in a bound form in alkaline phosphatase, so the enzyme must be removed prior to treatment with the reagent. In a continuous-flow analysis procedure, this may be achieved by placing a **dialyser** at point Y (of Fig. 18.5 a), allowing the phenol, but not the protein, to pass into the stream of indicator reagent.

In the King and Kind (1954) method, on the other hand, the phenol is made to react with a reagent (4-amino antipyrine) which is insensitive to phenolic amino-acids. Therefore, without any need for a dialyser, the antipyrine reagent may be introduced at Y, stopping the primary reaction and initiating the indicator reaction, whose absorbance is subsequently read at 520 nm.

In either case, the procedure may be calibrated by placing phenol standards of known concentration amongst the batch of samples: typical results would be as indicated in Fig. 18.5 b. The incubation time of the enzyme-catalysed reaction is effectively the flow-time from X to Y (of Fig. 18.5 a). Provided no dialyser is present at Y, this may easily be determined by introducing a coloured dye into the sample line and observing its movement. However, it is usually considered unnecessary to know the precise incubation time, particularly since it is often in the order of 10 minutes, meaning that the reaction is likely to have proceeded beyond the linear phase. Hence it is best to calibrate the procedure for each analysis by the use of standards of known alkaline phosphatase activity, this being a relatively stable enzyme. Such standards (e.g. standard sera) may be obtained commercially, but of course have to be stored carefully to avoid loss of enzyme activity (Chapter 17.3).

Many continuous-flow procedures have an advantage over the ones discussed above in that there is no need to perform a separate indicator reaction. The reaction catalysed by **glucose oxidase**, for example, can be monitored by coupling with a peroxidase-mediated dye formation, both reactions taking place simultaneously (see Chapter 15.2.3). Hence the flow circuit is simpler than the general one shown in Fig. 18.5 a, having no reagent line joining at point Y and no second heating-bath. This is also true for enzyme-catalysed reactions which do not need coupling to others: for example, the assay of **alkaline phosphatase** by the use of colourless p-nitrophenyl phosphate as substrate involves the formation of a product (p-nitrophenol) which is yellow in alkaline conditions, the absorbance being read at 410 nm. The system may be calibrated with p-nitrophenol standards and the incubation time determined as the time taken for a coloured dye to travel from the point of mixing (point X) to the colorimeter; however, as before, the system may be calibrated by the use of alkaline phosphatase standards, eliminating errors due to non-linearity.

In procedures where enzymes are being used as **reagents** (e.g. the determination of blood glucose by the use of the glucose oxidase/peroxidase coupled method mentioned above), a dialyser is often included at a point in the sample line before it merges with the enzyme line (i.e. before the point X in Fig. 18.5a). This ensures that the only protein in the reaction mixture is the enzyme being used as a reagent, since no protein from the sample would be able to pass through the dialysis membrane. Calibration of such procedures is easier than for enzyme assay, since there is not usually any problem in obtaining standards of reliable, precisely-known composition.

Single-point continuous-flow procedures of this type can analyse in the region of 60 samples per hour, which is low compared to the performance of a discrete analyser. However, the sample stream may be split and identical fractions pumped simultaneously into a variety of reagent streams. Thus each sample may be analysed for several different factors at the same time, with little extra labour involved. However, such a system is wasteful of reagents and each test will be carried out on each sample, whether necessary or not.

Two-point continuous-flow systems may give slightly more accuracy than one-point systems for enzymatic analysis, since it is easier to ensure that the reaction is being investigated only in the linear-phase. The first detector is placed in the flow-line at a point immediately after the first mixing coil, while the second detector is situated after the heating-bath. The two detectors would preferably be components of a single instrument, e.g. two electrodes of an electrochemical set-up, or two sensors of a differential-spectrophotometer. Ideally, the flow-time between the two detectors should be sufficiently short (e.g. 30 seconds) to keep within the linear phase, and the sample should be pumped for a sufficient length of time to enable it to occupy (together with the reagent) the whole of the system. When this has been achieved and the system stabilised, the difference between the two detector readings will be a measure of the rate of reaction.

For example, Blaedel and Hicks (1962) used a two-point continuous-flow system with a differential-spectrophotometer as detector to determine glucose concentrations by a method involving glucose oxidase/peroxidase; 15 samples per hour could be analysed.

Since the incubation time is short, the difference in readings is very small, so very accurate detection techniques are required for such procedures. Since they have neither the convenience of the one-point methods nor the comprehensive qualities which come from continuous monitoring (see Chapter 18.2.4), two-point continuous-flow methods are not widely used at the present time.

18.2.3 Fixed-Concentration Methods

In fixed-concentration methods of analysis, the time for an experimental reading (relating to the concentration of some substance participating in the reaction) to change by a specified amount is measured automatically. As with fixed-time methods, the readings should ideally be restricted to the linear-phase of the reaction, as indicated by preliminary experiments. Under these conditions, the time taken is inversely proportional to initial velocity.

Procedures based on this principle have usually been two-point methods, and must involve discrete rather than continuous-flow analysis. For example, Pardue and Malmstadt (1961) developed an automatic potentiometric procedure for the glucose oxidase system, using the product, hydrogen peroxide, to oxidise iodide, an almost instantaneous process in the presence of molybdate as catalyst:

$$D\text{-glucose} + O_2 \longrightarrow H_2O_2 + \text{gluconolactone}$$

$$H_2O_2 + 2I^- + 2H^+ \xrightarrow{\text{MoO}_4} I_2 + 2H_2O.$$

A cell is set up by placing one platinum electrode in the reaction mixture and another in a reference solution of fixed iodine concentration; the two solutions are in electrical contact through a small aperture, but unable to mix together. The voltage produced between the electrodes changes as iodine is formed from iodide, and the time taken for it to change by a specified amount is taken to be inversely proportional to the rate of the primary reaction.

As with two-point fixed-time methods, these procedures are used to a very limited extent.

18.2.4 Methods Involving Continuous Monitoring

Several automated rate-monitoring systems, specifically designed for enzymatic analysis, have been available for some years: these include the LKB Kinetic Analysis System and the Beckman Enzyme Activity Analyser. Although they differ in design, all are discrete analysers which can handle in excess of 40 samples per hour, working sequentially. The main difference from other types of discrete analyser (see Chapter 19.2.2) is that when a sample is mixed with reagents, the ensuing reaction is monitored continuously (or almost continuously, readings being taken at no more than 3 second intervals) by means of a spectrophotometer, the results being fed directly into a computer. Constant temperature is maintained throughout.

Any lag-phase reaction (see Chapter 18.2.1) can be eliminated from the calculations either by imposing a delay before the first reading is taken or by programming the computer to recognise the various phases of the reaction. When the linear-phase has been observed for 15–20 seconds, enough readings will have been taken for the computer to calculate its slope and give a statistical assessment of the results, after which the reaction mixture is ejected from the detector and a new one introduced. Warnings are often printed if there is any significant departure from linearity over the period of interest, or if the optical limits of the instrument are being exceeded.

With the development of microprocessor technology, several **spectrophoto-meters** now have kinetic attachments. For example, the Kinetics Analysis Accessory for the Beckman DU-7 spectrophotometer can make up to 1200 readings per minute on a single sample or can monitor simultaneously the reaction in a series of samples, making readings every few seconds; the absorbance against time curves can be visualised on a screen, and linear regression analysis carried out on selected portions of the curves. Some routine discrete analysers now also have a kinetics capability. The Coulter Dacos system, for instance, has a rotating optical system which can scan up to 120 reactions continuously, with a through-put of up to 450 tests per hour and random access facility.

Fast (Centrifugal) Analysers can also be used for enzymatic analysis, and some have been available for several years (e.g. the Roche Cobas Bio and the Baker CentrifiChem System). In these, 15–30 samples are analysed simultaneously in cuvettes on a spinning, horizontal centrifuge plate. The cuvettes are on the outer portions of the plate, and each is aligned with specific depression on the inner portion which originally contain sample and reagents (Fig. 18.6). The inner plate can be removed for ease of loading, and to allow one batch of sample to be loaded while another is being analysed.

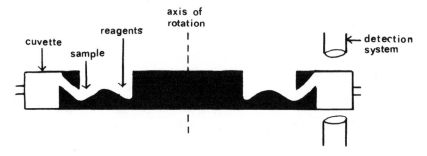

Fig. 18.6   Diagrammatic representation of a fast (centrifugal) analyser set-up (side view).

When centrifugation commences, each sample is mixed with its reagent as they are carried together by centrifugal force into a cuvette, thus starting a reaction. As the plate revolves, each cuvette passes through a detecting system (usually a spectrophotometer) linked to a computer, so readings can be taken and stored for each reaction mixture on each circuit (i.e. at about 100 msec intervals). Also on each circuit, a dark current reading is taken, when the light beam is interrupted by a solid part of the rotor. An oscilloscope may be linked to the detector, to enable the progress of the reaction in each cuvette to be visualised. Regardless of this, a suitable computer can process the readings to give the best estimate of the initial velocity of each reaction, and print out the results. A trap may then be opened at the bottom of each cuvette, and the contents emptied by centrifugal force, preparing the instrument for the next batch of samples. It is possible to analyse in the order of 200 samples per hour by such a procedure.

Recently-developed systems, such as the Roche Cobas Fara, can perform fluorimetric, nephelometric and turbidimetric measurements in addition to spectrophotometry, and may also be fitted with ion-selective electrode modules; some, for example the Baker Encore, have research modes particularly useful for studying enzyme kinetics.

SUMMARY OF CHAPTER 18

The course of enzyme-catalysed reactions may be monitored by a variety of techniques, of which the most common at the present time are spectrophotometric, spectrofluorimetric and electrochemical procedures. Another relevant technique, which is of increasing importance, is microcalorimetry. Radiochemical procedures may also be used, but although they are extremely sensitive, they do not permit the rate of the reaction to be monitored continuously.

Fully-automatic procedures of enzymatic analysis may involve fixed-time methods (either discrete or continuous-flow analysis), fixed-concentration methods or continuous-monitoring methods. The last-mentioned type includes the use of discrete, rate-monitoring analysers and fast (centrifugal) analysers, both of which may be coupled to computers to allow automatic processing of results.

FURTHER READING

Barker, S. A. and Somers, P. J. (1978), Enzyme electrodes and enzyme-based sensors, in Wiseman, A. (ed.), *Topics in Enzyme and Fermentation Biochemistry,* 2, Ellis Horwood.

Bergmeyer, H. U. (ed.) (1978), *Principles of Enzymatic Analysis,* Verlag Chemie.

Fersht, A. (1985), *Enzyme Structure and Mechanism,* 2nd edition (Chapter 6: Practical kinetics), Freeman.

Guilbault, G. G. (1970), *Enzymatic Methods of Analysis,* Pergamon Press.

Northam, B. E. (1981), Whither automation?, *Annals of Clinical Biochemistry,* 18 (pages 189–199).

Rosalki, S. B. (1980), Quality control of enzyme determinations, *Annals of Clinical Biochemistry,* 17 (pages 74–77).

Saunders, R. A. and Burns, R. F. (1977), A survey of enzyme reaction rate analysers readily available in the United Kingdom, *Annals of Clinical Biochemistry,* 14 (pages 185–200).

Williams, B. L. and Wilson, K. (eds.) (1981), *Principles and Techniques of Practical Biochemistry,* 2nd edition, Arnold.

CHAPTER 19

Some Applications of Enzymatic Analysis in Medicine and Industry

---

19.1 APPLICATIONS IN MEDICINE

19.1.1 Assay of Plasma Enzymes

Assays of some of the enzymes present in blood plasma or serum are carried out routinely in most clinical chemistry laboratories and these play an important role in diagnosis. (For those not familiar with medical terminology, plasma is the fluid from unclotted blood, serum the fluid from clotted blood: in most respects, the two are identical.) Although a few plasma enzymes have a clearly-defined role in that particular location (e.g. in coagulation), the majority do not: they are simply enzymes which have leaked out of blood or tissue cells. For each such enzyme present in plasma, a balance is normally set up between its rate of arrival by leakage from cells and its rate of removal by catabolism or excretion. Therefore, for each plasma enzyme, there is a normal concentration range and hence a **normal range of activity**, which can be determined.

If the cells of a particular tissue are affected by disease in such a way that many of them no longer have intact membranes, then their contents will leak out into the blood stream at an increased rate and the enzymes associated with those cells will be found in the plasma in elevated amounts. Since many enzymes or isoenzymes are characteristically associated with the cells of certain tissues, plasma enzyme assay can help to identify the location of damaged cells. The results should, of course, correlate with the symptoms and case history of the patient, and with other biochemical findings. Some important examples of the application of plasma enzyme assay to diagnosis will now be discussed.

Lactate dehydrogenase (LDH) is easily assayed, since it catalyses a reaction involving $NAD^+/NADH$ as coenzyme. It is found in most cells of the human body, and so an increased activity of this enzyme in plasma does not, in itself, give much clue as to what is wrong with the patient. However, the total activity is in fact the sum of the activities of five isoenzymes (see Chapter 5.2.2), and these are associated with different tissues: in particular, liver and skeletal muscle cells contain mainly the M_4 form, while those of heart muscle contain mainly H_4

and MH_3; all of the isoenzymes are found in kidney cells and erythrocytes (red blood cells), the H_4 form being the most prominent.

If plasma is subjected to electrophoresis (e.g. on cellulose acetate strips) at pH 8.6, the LDH isoenzymes may then be located by means of a specific stain, e.g. a mixture of lactate, NAD^+ and a chromogen, which will only form a coloured product where LDH is present to catalyse the first step of the reaction:

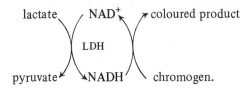

In this way, plasma is found to contain all five isoenzymes (Fig. 19.1), and an abnormal pattern can be of help in establishing a diagnosis. Various different properties of the LDH isoenzymes can enable their relative proportions in plasma

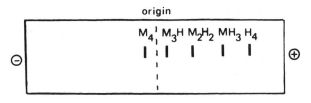

Fig. 19.1   Separation of the isoenzymes of LDH achieved by electrophoresis at pH 8.6.

to be assessed without separation having to be performed: H_4 is heat-stable and its activity is unaffected if it is kept at 60° for 30 minutes, whereas that of M_4 is largely destroyed; H_4 can act as a hydroxybutyrate dehydrogenase (HBD), i.e. can use hydroxybutyrate as substrate, whereas M_4 has little activity in that respect; also, H_4, but not M_4, is strongly inhibited by oxalate (0.2 mmol.l^{-1}), whereas M_4, but not H_4, is markedly inactivated by urea (2.0 mmol.l^{-1}). Immunological reagents may, for instance, be used to precipitate all isoenzymes containing M sub-units, leaving only H_4 in solution. Hence, if the total LDH activity is found to be higher than normal, there are several ways of establishing whether this is predominantly due to an excess of H_4 (as could be seen in the case of heart disease, haematological disorders or renal disease) or to an excess of M_4 (which would indicate skeletal muscle or liver disease).

Aspartate aminotransferase (AST), formerly known as glutamate oxalo-acetate transaminase (GOT), an enzyme widely distributed throughout the body, catalyses the reaction:

$$\text{aspartate} + \text{2-oxoglutarate} \rightleftharpoons \text{oxaloacetate} + \text{glutamate.}$$

It may be assayed by coupling, via the product, oxaloacetate, to an indicator reaction catalysed by malate dehydrogenase which involves NAD⁺/NADH as coenzyme. Markedly raised plasma activities (10–100 times normal) of AST usually indicate severe damage to the cells of heart (as in myocardial infarction) or liver (as in viral hepatitis or toxic liver necrosis): moderate increases of activity are found in many diseases.

Alanine aminotransferase (ALT), formerly known as glutamate pyruvate transaminase (GPT), catalyses the reaction:

$$alanine + 2\text{-}oxogluturate \rightleftharpoons pyruvate + glutamate.$$

This enzyme may be assayed by coupling, via the product, pyruvate, to an LDH-catalysed indicator reaction. ALT is found in high concentrations in liver cells, and in much smaller concentrations elsewhere. Hence, a markedly raised plasma activity indicates a severe liver disease, usually viral hepatitis or toxic liver necrosis.

Alkaline phosphatase, the name given to a group of isoenzymes which catalyse the hydrolysis of organic phosphates at alkaline pH, is found principally in bone, liver, kidney, intestinal wall, lactating mammary gland and placenta. Different isoenzymes are associated with each of these, and they may be partially separated by electrophoresis at pH 8.6 and visualised with a specific stain such as calcium α-naphthyl phosphate mixed with a diazonium salt (Fig. 19.2).

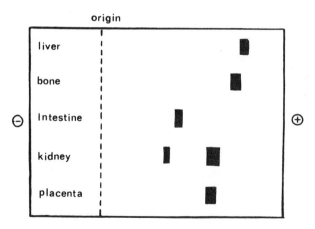

Fig. 19.2   Separation of the isoenzymes of alkaline phosphatase achieved by electrophoresis at pH 8.6.

The kidney isoenzymes are not normally present in plasma to any appreciable extent, even in patients with kidney disease, and the placental isoenzyme is only found in the plasma of pregnant women. The total activity of alkaline phosphatase is easy to determine, because the relatively low specificity of each of the isoenzymes means that an artificial substrate whose hydrolysis is easy to monitor be used in the assay procedure (see Chapter 18.2.2). As with LDH, if the total activity is found to be in excess of normal, the isoenzyme or isoenzymes responsible may be identified from the electrophoresis pattern, considered together with the results of other tests. Unlike LDH, no hybrid isoenzymes are found, and at least some of the alkaline phosphatase isoenzymes may be products of the same gene, the differences being due to post-genetic modification. Placental alkaline phosphatase is very heat-stable, as is the similar Regan isoenzyme found in some patients with cancer, whereas the bone isoenzyme is the least stable at elevated temperatures. Hence the effect of heat (e.g. 65°C for 30 minutes) on the total enzyme activity can help to indicate which are the predominant isoenzymes present. Inhibition studies may also be used, for the intestinal and placental forms are strongly inhibited by L-phenylalanine; also, urea (2 mmol.l^{-1}) inactivates the bone isoenzyme more strongly than the liver form, and this in turn more strongly than the ones from intestine and placenta.

Alkaline phosphatase in the liver is found associated with the cell membrane which adjoins the biliary canaliculus, and so high plasma concentrations of the liver isoenzyme indicate cholestasis (i.e. blockage of the bile duct, with a resulting back flow through the liver into the blood stream) rather than simply damage to the liver cells. In view of the generally similar characteristics of the bone and liver isoenzymes, confirmation of high plasma concentrations of the liver form may be achieved by the investigation of other enzymes whose plasma activities are increased in cholestasis: these include **5'-nucleotidase** and **γ-glutamyl transferase (GGT)**, also called γ-glutamyl transpeptidase; high plasma activities of the latter enzyme are also found in other forms of liver disease.

Creatine kinase (CK), also known, with a degree of tautology, as creatine phosphokinase (CPK), catalyses the reaction:

$$\text{creatine} + \text{ATP} \rightleftharpoons \text{creatine phosphate} + \text{ADP}.$$

It may be assayed by a variety of procedures, e.g. by coupling via the product, ADP, to a reaction catalysed by pyruvate kinase, and thence via the further product, pyruvate, to an LDH-catalysed indicator reaction.

The enzyme is found mainly in the heart and skeletal muscle, and in brain. It is a dimer and, like the tetrameric LDH, may be made up from two types of polypeptide chain (B or M) in any combination. Thus, three isoenzymes may be found: *BB*, the main form of brain, but also found to some extent in nerve tissue, thyroid, kidney and intestine; *MB*, found in heart muscle and the diaphragm, but not present in skeletal muscle; and *MM*, found in both heart and skeletal muscle. On electrophoresis, the *BB* isoenzyme travels nearest to the anode; this form is not normally present in plasma. In addition to electrophoresis, immunological techniques and triazine dye affinity chromatography are proving useful in the analysis of the individual CK isoenzymes.

The plasma activity of the *MB* isoenzyme is a very good indicator of a possible myocardial infarction, whereas the *MM* form is found in high concentrations in plasma when skeletal muscle cells are damaged (e.g. in the early stages of muscular dystrophy). Skeletal muscle disorders of this type are also characterised by increased plasma activities of the appropriate isoenzyme (isoenzyme *A*) of **fructose 1,6-bisphosphate aldolase**, often called simply aldolase (see Chapter 10.1.1).

Hence, to summarise our discussion so far, liver cell damage is indicated by increased plasma activities of AST, ALT, LDH (M_4) and GGT, the magnitude of the increases reflecting the degree of damage; alkaline phosphatase and 5'-nucleotidase are useful mainly as indicators of cholestasis, when the GGT activity will also be raised. Skeletal muscle damage may result in the presence of the plasma of excess CK (MM), fructose 1,6-bisphosphate aldolase (*A*), and, to a lesser extent, LDH (M_4), ALT and AST; levels are usually higher in the early stages of muscular dystrophy than they are when the muscle has become wasted. Severe damage to heart cells, as in myocardial infarction, is characterised by increased plasma activities of CK, AST and LDH: they start to rise in that order, and subsequently fall in the same order; within a week of an infarction, plasma activities of all three may be back to normal, but the effects may still be seen with H_4 abnormally prominent amongst the LDH isoenzymes, resulting in the HBD activity still being above normal limits.

Cardiac failure may cause secondary liver damage, so there could be some confusion as to the interpretation of the results of plasma enzyme assay. However, it would be apparent from the HBD and CK-MB activities whether or not a myocardial infarction had occurred, and the extent of the liver damage would be indicated by the ALT activity.

Diseases in parts of the body not previously mentioned may also be diagnosed with the help of plasma enzyme assay.

The digestive enzyme **trypsin, triacylglycerol lipase** and **α-amylase** are produced in the pancreas and these are found in increased amounts in plasma in certain diseases of the pancreas, particularly acute pancreatitis. Of the three, α-amylase is the simplest to assay, since it utilises starch as substrate and produces reducing sugars as products. An unusual feature of pancreatic α-amylase is that it is one of the few proteins small enough to pass through the glomerulus of the kidney and be excreted in the urine. In a rare condition called macroamylasaemia, the plasma α-amylase forms complexes with other proteins (possibly immunoglobulins) and can no longer be excreted in this way, which causes the plasma α-amylase activity to rise. However, macro-amylasaemia could not be confused with acute pancreatitis, because of the very different clinical symptoms.

An **acid phosphatase** isoenzyme is found in large amounts in the prostate gland, and its assay in plasma is used in the diagnosis of prostatic carcinoma. Acid phosphatase isoenzymes are also found in liver, red cells, platelets and bone. They catalyse the same reaction as alkaline phosphatase, but at lower pH values, and may be distinguished from each other by the action of various inhibitors: the prostate and red cell forms are inactivated by ethanol, the red cell (but not the prostate) form by formaldehyde, and the prostate (but not the red cell) form by L-tartrate. As will be apparent, the total plasma acid phosphatase activity is raised in a variety of conditions, but this is of little use in diagnosis as there are better indices for each of them.

Cholinesterase is another enzyme which is usually assayed in plasma only in specific circumstances, although it could have a more general application: plasma concentrations are low in liver disease, parallelling those of the routinely-measured albumin. The enzyme catalyses the hydrolysis of choline esters, such as acetylcholine:

$$\text{acetylcholine} + H_2O \rightleftharpoons \text{acetic acid} + \text{choline.}$$

It may conveniently be assayed by pH-stat or electrochemical procedures. The 'true' **acetylcholinesterase** is found in nerve tissues and red blood cells; however, it is not this isoenzyme but the **pseudocholinesterase** of liver, heart and intestine which is normally present in blood plasma. Acetylcholinesterase will act on acetyl-β-methylcholine as substrate, but not on benzoylcholine; pseudocholinesterase has the opposite specificity.

The main clinical significance of plasma cholinesterase is that its presence is assumed when the muscle relaxant **scoline** (suxamethonium; succinyl dicholine) is administered, for otherwise the effect would be permanent. Patients with low plasma activities of cholinesterase (e.g. as a result of liver disease or of poisoning by organophosphorus insecticides) experience severe breathing difficulties for several hours after treatment with scoline, so a preliminary assay may be carried out to see if it is safe to administer the drug. One further problem is that a rare condition exists where the subject has a modified pseudocholinesterase as a result of an inherited genetic mutation. The plasma activity in such individuals is usually low, and they react adversely to scoline administration. The condition may be investigated by determining the **dibucaine number** (i.e. the percentage inhibition by dibucaine) of the plasma cholinesterase: subjects with the normal enzyme have a high dibucaine number (75-85), while those with one type of abnormal enzyme have a low one (15-30). Alternatively, some patients may show reduced inhibition by fluoride. Such inhibition studies are useful in detecting homozygotes and heterozygotes (see Chapter 19.1.2) in the family of a patient known to have a cholinesterase deficiency, with a view to identifying those individuals at risk from scoline administration.

In general, most assays of plasma enzymes are carried out by kinetic methods, since these are the most convenient. However, radioimmunoassay procedures (Chapter 15.2.4) are of increasing importance, because they are specific for a particular isoenzyme.

19.1.2 Enzymes and Inborn Errors of Metabolism

An inborn error of metabolism is characterised by the loss of activity of a specific enzyme as a result of a genetic mutation. Inborn errors are rare but severe conditions, often causing mental retardation or even death in infancy.

The **preliminary diagnosis** is usually made by observing a build-up in plasma or urine of the metabolic intermediate which is the substrate for the defective enzyme: it cannot be metabolised in the usual way, and so leaks out of the cell in relatively large amounts. The diagnosis may then be **confirmed** by assay of the enzyme in a sample of the tissue where it is metabolically most important (often the liver). In some instances, it may be possible to assay the same enzyme in more accessible cells (e.g. in red or white blood cells, or in skin), but, because of the known existence of isoenzymes (see Chapter 19.1.1), it can never be assumed that the defect will be expressed wherever activity of that enzyme is usually found.

If, because of previous family history, it is considered possible that a particular pregnancy might result in the birth of a baby with an inborn error of metabolism, it may be possible to obtain a **pre-natal diagnosis** by assaying the enzyme in amniotic fluid cells, which are mainly derived from skin. Among the many disorders which have been diagnosed in this way before birth is **Tay-Sachs disease**, a deficiency of **hexosaminidase A** activity.

With most inborn errors, the disease is transmitted from generation to generation by an **autosomal recessive** mode of inheritance. The main features of autosomal inheritance are that every individual has a gene for the synthesis of the relevant enzyme on each of a pair of chromosomes, one coming from the father and one from the mother, and that the inheritance pattern does not depend on the sex of the individual. If the mode is recessive, the affected patient must have inherited the same abnormal gene from each parent, for otherwise the disease would not express itself as a clinical abnormality. Such a **homozygous** patient would usually have less than 10% of the normal activity of the enzyme, whereas both **heterozygous** parents, each having one normal and one abnormal gene, would be able to synthesise active enzyme from their normal gene and would thus have about 50% of the normal activity, an amount usually sufficient for a normal life. There is a 1 in 4 chance that a child of heterozygous parents will be homozygous for that particular defect, and a 1 in 2 chance that the child would be heterozygous like the parents. Probably the best-known example of an autosomal recessive inborn error is **phenylketonuria**, where the patient has a severely limited ability to convert phenylalanine to tyrosine, because of a reduced activity of **phenylalanine 4-monooxygenase**, better known as **phenylalanine hydroxylase**.

Autosomal dominant diseases, in contrast to autosomal dominant characteristics such as eye colour, are *not* found, presumably because they are self-limiting: with such a mode of inheritance, the heterozygous individual would be clinically affected, so would be unlikely to grow to sufficiently-normal maturity to have children and pass on the abnormal gene.

However, **X-linked dominant** inborn errors *are* found. In these, the mutated gene forms part of the chromosome pair which determines the sex of the individual. Females possess two X chromosomes, whereas males possess one X chromosome and one, much shorter, Y chromosome. Therefore, where a gene on the X chromosome has no counterpart on the Y chromosome, it would seem likely that females would normally produce twice as much as males of the protein coded for on that gene. This does not happen, because, according to the **Lyon hypothesis**, a random cut-off procedure operates in females: shortly after conception, when the girl-to-be consists of just a few cells, one of the X chromosomes is inactivated in each cell. Now, if she happens to be heterozygous for an X-linked disease, this means that some normal genes are inactivated and so are some abnormal ones. The process is quite random. So, if most of the inactivated genes are abnormal ones, the girl who develops from those few cells will produce her enzyme largely from normal genes, and will therefore be clinically normal (termed **favourable Lyonisation**). However, it is just as likely that most of the inactivated genes are normal ones, a situation which would ultimately lead to the birth of a severely affected patient (termed **unfavourable Lyonisation**).

So, X-linked dominant diseases are inherited through the mother alone, who is heterozygous but not necessarily clinically abnormal. However, a heterozygous daughter may not be so lucky: furthermore, a boy who inherits the abnormal gene from his mother is **hemizygous** because the gene has no counterpart on his Y chromosome, so he must be severely affected. Each daughter of a heterozygous mother has a 1 in 2 chance of being heterozygous too, and each son has a 1 in 2 chance of being hemizygous.

An example of an inborn error which appears to be transmitted by an X-linked dominant mode is **ornithine transcarbamoylase (OTC) deficiency**, also known as **hyperammonaemia**, a defect of urea cycle metabolism. Another X-linked, but recessive, disease is the **Lesch-Nyhan syndrome**, a disorder of purine metabolism resulting from a deficiency of **hypoxanthine phosphoribosyl transferase** activity.

In most inborn errors, it is likely that the enzyme is present in an altered form as a result of the mutation, rather than being absent altogether. In the case of the Lesch-Nyhan syndrome, Wyngaarden, in 1974, used radioimmunoassay to demonstrate that the enzyme–protein was present, even though no catalytic activity could be detected.

In contrast to the situation with the inherited disorders of haemoglobin (Chapter 2.4.5), it has not so far been possible to determine the structure of any of these abnormal enzymes: this is because the human body contains much smaller amounts of each enzyme than it does of haemoglobin, and in much less accessible places. However, it is reasonable to suppose that the same general features would be found, i.e. that the mutation often results in the substitution of only a single amino-acid residue by another, and that different substitutions in a particular protein may be found in different families, since not all mutations will be identical. The latter phenomenon is termed **genetic heterogeneity**.

Therefore, although a group of patients may be classified as having the same disease because they have a reduced activity of the same enzyme, the nature of the alteration may not always be the same. For example, it is known that the characteristics (e.g. K_m) of the residual enzyme activity in OTC deficiency varies from patient to patient: although it might be argued that different proportions of normal and abnormal enzyme in different heterozygous females might account for this, at least to some extent, only genetic heterogeneity could explain the fact that some affected males have relatively mild symptoms whereas most have died shortly after birth. In the case of autosomal recessive diseases, a 'homozygous' patient could have *two* forms of the abnormal enzyme, since the relevant gene from each parent might have different mutations.

Treatment of inborn errors of metabolism usually involves reducing the dietary intake of substances which cannot be metabolised in the normal way because of the defect: the aim is to prevent their concentrations building up to high levels within cells, this being one of the main features responsible (directly or indirectly) for the clinical symptoms. Such methods of treatment can be extremely effective in certain instances (e.g. with **phenylketonuria**), but of course the special diets have to be maintained throughout life. In many other inborn errors, dietary treatment has no effect: this may be due to endogenous synthesis of the metabolites in question, or it could be because the end-product of the blocked metabolic pathways are vitally important to normal body function and cannot be provided in any other way.

Ideally, complete treatment would be achieved if some normal enzyme could be introduced at the appropriate site. In the case of the enzymes of the gastrointestinal tract, this may easily be achieved by oral administration. For example, where there is a **lactase** (β-galactosidase) deficiency in the cells of the intestine, which may occur as an inborn error or as a secondary feature of other diseases, ingested lactose (in milk) cannot be broken down to galactose and glucose, and severe discomfort results. This condition may be treated by removing milk products from the diet, or, more conveniently if more expensively, by adding to it β-galactosidase extracted from micro-organisms to perform the task normally done by the body's lactase.

However, where the main site of action of the enzyme in question is not in such direct contact with the external environment (if, for example, it is in the liver or the central nervous system), the problems involved in its introduction are immense. Protein is not absorbed intact into the body from the gastro-intestinal tract, and although it could be infused directly into the blood stream, it would be fairly quickly destroyed by antibodies if it was of non-human origin. In any case, it would not usually be taken up by the cells of the appropriate tissue.

In some cases, the **attachment of sugar units** to enzymes (particularly lysosomal ones) prior to infusion increases their chances of reaching the appropriate destination: the sugars presumably trigger some specific transport system in membranes.

Another possible approach would be to trap the enzyme in artificial membrane-enclosed structures called **liposomes** (see Chapter 20.2.1). Cells of a particular tissue might then be able to take these up from the blood stream, provided their membranes were of suitable composition.

Yet another possible long-term treatment for inborn errors of metabolism involves **organ or tissue transplantation**: for example, fibroblasts from a normal subject of identical tissue-type to the patient may be transplanted by subcutaneous injection. However this and the other approaches to enzyme replacement therapy are still in the very early stages of development.

In passing, it is worth pointing out that enzymes may be used in the treatment of other diseases besides inborn errors of metabolism. For example, enzymes such as **urokinase**, extracted from human urine, can be infused into the blood stream of patients at risk from a **pulmonary embolism** (a fragment of a blood-clot lodging in the pulmonary artery): these enzymes stimulate a **cascade system** (see Chapter 14.2.5) responsible for the production of active **plasmin**, a proteolytic enzyme which digests **fibrin**, the main structural component of blood-clots. Some enzymes may also be used to restrict the growth of **cancer** cells by depriving them of essential nutrients: for example, **L-asparaginase** may be used in the treatment of several types of **leukaemia**, since the tumour cells, in contrast to normal cells, have a requirement for exogenous L-asparagine.

19.1.3 Enzymes as Reagents in Clinical Chemistry

D-glucose in blood and other physiological fluids is commonly analysed by means of procedures involving glucose oxidase (see Chapters 15.2.3 and 18.1.5). These methods are specific for the β-form of D-glucose, but, except when solutions are freshly prepared, β-D-glucose and α-D-glucose will be present as an equilibrium mixture in samples and standards alike, so this apparent complication can be ignored. The reaction catalysed by glucose oxidase is utilised in a test-strip (Clinistix) for the screening of urine specimens: glucose oxidase and the reagents for a colour-producing indicator reaction are impregnated into a paper strip, so that the intensity of the blue colour obtained when it is dipped into a specimen gives an indication of the D-glucose content. This can be of use in the diagnosis of **diabetes mellitus**. D-glucose may also be analysed by a hexokinase-catalysed reaction (see Chapter 18.1.6), but this enzyme can utilise any D-hexose as substrate.

Blood **lactate** and **pyruvate** are usually determined by means of LDH-catalysed methods (see Chapter 19.1.1), and blood **urea** is sometimes analysed by procedures involving urease (see Chapter 18.1.5).

The activation of various enzymes by Mg^{2+} ions has been used, though not extensively, as the basis of procedures for the determination of plasma Mg^{2+}. One such method involves the reaction catalysed by isocitrate dehydrogenase:

$$\text{isocitrate} + \text{NADP} \rightleftharpoons \text{2-oxoglutarate} + \text{NADPH} + H^+.$$

Another Mg^{2+}-activated enzyme is luciferase, which utilises luciferin (firefly extract) as substrate, according to the reaction:

$$\text{luciferin} + O_2 + \text{ATP} \rightleftharpoons \text{oxyluciferin} + \text{ADP} + P_i.$$

Oxyluciferin exhibits a green **chemiluminescence**, so the reaction may be followed on a spectrofluorimeter with the light source blocked off. Luciferase-catalysed procedures are used chiefly for the analysis of **ATP** (e.g. from blood platelets), but may also be used to determine the Mg^{2+} concentration or the pO_2.

19.2 APPLICATIONS IN INDUSTRY

We will now mention just a few of the many applications of enzymatic analysis in industry.

In the foodstuff industry, the activity of certain enzymes may be determined before and after **pasteurisation** and **sterilisation** procedures, to check whether these have been properly carried out. For example, the **alkaline phosphatase** (see Chapter 19.1.1) present in milk is inactivated within the same temperature range as is required for pasteurisation, so the alkaline phosphatase activity at the end of the process gives an indication of its effectiveness.

Similarly, the degree of **bacterial contamination** of foodstuffs can be estimated by the assay of microbial enzymes not normally present in food. For example, milk should contain only small amounts of **reductases**, but bacteria produce large amounts. Reductases may be easily assayed because they catalyse the reduction of methylene blue to colourless leuco-methylene blue under anaerobic conditions. A test-strip (Bactostrip) incorporating 2,3,4-triphenyltetrazolium salts provide a convenient way of testing for the presence of bacteria, a red formazan dye being produced as a result of the action of reductase.

Enzyme assay may also be used to determine whether **stored plant products** are suitable for use as foodstuffs. For example, α-**amylase** should be present in relatively low amounts in stored wheat seeds. However, if sprouting (i.e. germination of the stored crop) takes place, as may happen if the crop is left to stand under conditions which are too moist, or if there is prolonged rain at harvest time, then there is a greatly increased production of α-amylase and some proteolytic enzymes. This causes breakdown of the reserve starch and proteins, so flour produced from the sprouted wheat is not suitable for baking purposes. Hence the α-amylase activity of the seeds may be determined (see Chapter 19.1.1) to give an indication of the degree of sprouting.

Once a flour has been produced, its amylase content may again be assayed to give an indication of the amount of starch breakdown which can be expected to take place when a dough is prepared (see Chapter 20.3.1).

Similarly, the amylase activity of malt (sprouted barley), and of malt flour, is often determined.

Just as enzyme assay is used to diagnose diseases in human beings (Chapter 19.1), so it may be applied to the investigation of **diseases in plants**. For example, it has been found that an injury (either mechanical or pathogenic) results in a marked, localised increase in the activity of **glucose 6-phosphate dehydrogenase**, but not of **glucose phosphate isomerase**, indicating diversion of glucose breakdown from glycolysis to the pentose phosphate pathway.

Enzyme assay is also used for research into such processes as the **browning of plant products**, which presents problems during the conversion of fruits and vegetables into drinks and preserves; the browning process involves the cyanide-resistant uptake of oxygen, which oxidises phenols to quinones, thus initiating the sequence of reactions resulting in the formation of dark melanins.

The determination of D-glucose by enzymatic analysis (see Chapter 19.1.3) is widely performed in the foodstuff and other industries. The method involving **glucose oxidase** is specific for D-glucose, whereas most chemical methods would not be able to distinguish D-glucose from any other monosaccharides present. The concentrations of various other sugars, and individual amino-acids, may also be determined by specific methods of enzymatic analysis (e.g. the disaccharide sucrose by an **invertase**-catalysed procedure). In winemaking, the concentration of malic acid is sometimes determined by a method involving **malate dehydrogenase**.

Many industrial processes utilise micro-organisms, and enzymatic analysis may be employed to ensure that the concentrations of vital ingredients in the growth media are within required limits (see Chapter 20.1). For example, **glycerol kinase** is obtained commercially from *Candida mycoderma*, and it is essential that the glycerol content of the growth medium is high enough throughout to induce glycerol kinase synthesis by the organism. Hence the glycerol concentration may be monitored, and more glycerol added as required. A suitable procedure involves the reaction catalysed by **glycerol dehydrogenase**:

$$\text{glycerol} + \text{NAD}^+ \rightleftharpoons \text{dihydroxyacetone} + \text{NADH} + \text{H}^+.$$

Finally, an important example of the use of enzymatic analysis for the determination of inhibitors is the **cholinesterase**-catalysed procedure for the analysis of organophosphorus insecticides (see Chapter 19.1.1).

SUMMARY OF CHAPTER 19

In medicine, the assay of blood plasma enzymes can be extremely useful in helping to establish a diagnosis in a sick patient. The concentrations, and hence activities, of many enzymes in plasma increase when the cells of a tissue are damaged, because there is an increased rate of leakage of enzymes out of these cells. Since many enzymes or isoenzymes are associated with particular tissues, the pattern of plasma enzyme activities can indicate which tissues are damaged. The diagnosis of an inborn error of metabolism can be established by demonstrating the reduced activity of a particular enzyme in a tissue. In some instances, it may be possible to obtain a pre-natal diagnosis by assay of enzymes in amniotic fluid cells, Enzymes are becoming increasingly important in the treatment of inborn errors and other diseases.

Enzyme assay is important in the foodstuff industry in monitoring the processing and storage of food, and as a means of detecting contamination by micro-organisms.

Enzymes are widely used as reagents in clinical chemistry and in industry.

FURTHER READING

Bergmeyer, H. U. (ed.) (1974), *Methods of Enzymatic Analysis* (Section A: General Introduction), **1**, 2nd edition, Verlag Chemie/Academic Press.

Cockburn, F. and Gitzelmann, R. (eds.) (1982), *Inborn Errors of Metabolism in Humans,* MTP Press.

Foster, R. L. (1980), *The Nature of Enzymology* (Chapter 6). Croom Helm.

Henry, R. J., Cannon, D. C. and Winkelman, J. W. (eds.) (1974), *Clinical Chemistry: Principles and Technics,* 2nd edition (Chapter 21), Harper and Row.

Hunter, I., Attock, B. and Palmer, T. (1983), A novel combination of techniques for the assay of lactate dehydrogenase isoenzymes, *Clinica Chimica Acta,* **135** (pages 73–82).

Landon, J., Carney, J. and Langley, D. (1977), The measurement of enzymes by radioimmunoassay, *Annals of Clinical Biochemistry,* **14** (pages 90–99).

Moss, D. W. (1982), *Isoenzymes,* Chapman and Hall.

Palmer, T., Oberholzer, V. G., Burgess, E. A., Butler, L. J. and Levin, B. (1974), Hyperammonaemia in 20 families: biochemical and genetical survey, *Archives of Disease in Childhood,* **49** (pages 443–449).

Ryman, B. E. and Tyrrell, D. A. (1980), Liposomes, in Campbell, P. N. and Marshall, R. D. (eds.), *Essays in Biochemistry,* **16**, Academic Press.

Singer, S. (1978), *Human Genetics,* Freeman.

Vitale, L. J. and Simeon, V. (eds.) (1980), *Industrial and Clinical Enzymology,* Pergamon Press.

Whitby, L. G., Percy-Robb, I. W. and Smith, A. F. (1984), *Lecture Notes on Clinical Chemistry,* 3rd edition (Chapter 8), Blackwell.

CHAPTER 20

Biotechnological Applications of Enzymes

20.1 LARGE-SCALE PRODUCTION OF ENZYMES

The general principles of enzyme extraction and purification were discussed in Chapter 16. Here we will be concerned chiefly with factors affecting the choice of source material and with those involved in the design of large-scale procedures, where the enzyme is extracted from at least 1 kilogram of cells.

Although enzymes have been used in certain industrial processes for centuries, their precise role, or even their identity, was not known over most of this period: they were often utilised as components of intact cells (e.g. yeasts in the baking and brewing industries) or as extracts containing a mixture of many enzymes (e.g. malt, containing **amylase** and **proteolytic enzymes**, also used in baking and brewing). The first enzyme to be made commercially available in a partially-purified form (by Hansen, 1874) was the acid protease, **rennin (chymosin)**: this was as rennet, a crude preparation obtained from the fourth stomach of young calves, used to curdle milk in cheese production. Despite the importance of enzymes from animal and plant sources, their large-scale extraction poses technical, economical and possibly ethical problems. Hence the trend has been towards increased utilisation of micro-organisms as sources of enzymes.

Because of the large number, and wide range, of micro-organisms available, a species can usually be found which contains an enzyme bearing a reasonable resemblence to one from animal or plant sources, or even one more advantageous for a particular application (see Chapter 20.3.1). If there is not, it may be possible to use genetic manipulation to make a micro-organism produce the required enzyme, although this is largely a matter for future development. Of particular importance in this context are **plasmids**, which are small, circular, cytoplasmic molecules of DNA, acting as extrachromosomal genes in bacteria; sometimes, but not always, they may be able to attach themselves to chromosomes, when they are known as **episomes**.

Plasmids are not essential for the functioning of a particular organism, but, when they are present, they confer important additional properties upon it (e.g. resistance to certain antibiotics). It is possible to extract and purify plasmids, and to insert extra genes into the circle. These altered plasmids can then be taken up again from the medium by the micro-organism, when the proteins coded for by the extra genes will be synthesised. In addition to plasmids, λ **phage** may be used as vehicles (termed **vectors**) to introduce extra genes into bacteria.

Such **genetic engineering** to produce **recombinant DNA** makes use of a variety of enzymes, particularly **restriction endonucleases** and **DNA ligases**. For example, the plasmid pBR322 can be cleaved at a single specific site by the EcoRI restriction enzyme. This makes a staggered cut in the double helix, leaving complementary single-stranded ends. If the DNA fragment to be inserted was removed from its original location on a large DNA molecule by means of the same enzyme, then it would have single-stranded ends complementary to those of the cut plasmid, and the two can be spliced by means of DNA ligase. The new circular plasmid so formed can be taken up to a small degree by *E. coli* bacteria, and since the plasmid contains genes which convey resistance to the antibiotics ampicillin and tetracycline, the bacteria containing the plasmid can easily be selected. In this and similar ways, synthetic, prokaryotic and eukaryotic genes can be incorporated into *E. coli* and other bacteria. Recombinant DNA technology can be used to increase the yield of enzymes already produced by the micro-organism (e.g. *β*-**galactosidase** and **aspartate transcarbamoylase** by *E. coli*), or to produce completely different enzymes. Such techniques may even be used to produce enzymes of modified structure from suitably modified or synthesised genes: this is termed protein engineering.

However, it is not usually sufficient simply to find a micro-organism capable, naturally or otherwise, of synthesising the required enzyme: a mutant strain is often sought which will have a minimum of controls restricting the synthesis of the enzyme (see below), and will not synthethise certain other substances.

Once a suitable strain of a suitable micro-organism has been found (or produced), a culture may be grown to produce a large supply of the enzyme (loosely termed **fermentation**). In the same way, animal or plant cells could be cultivated (see Chapter 15.1) to enable large-scale extractions of their enzymes to be performed, but this has yet to be done on a wide-spread, commercial basis.

Some enzymes, e.g. *Aspergillus* **proteases**, may be obtained from a **semi-solid culture**, where the low water content and high degree of aeration (at the surface) apparently favours production of these enzymes. However, the vast majority are produced in aseptic **submerged culture**, where the conditions of pH, temperature, degree of aeration etc. can be finely controlled. The synthesis of most of the enzymes of commercial significance is related to cell growth, but can lag behind this, particularly in the case of hydrolytic enzymes and/or enzymes whose production is controlled by plasmids. In general, in a single-batch fermentation procedure, the number of cells present may increase for about a day (termed the **growth phase**) until a limiting value is reached, but enzyme production is likely to continue at significant levels for several days after this.

The culture medium must contain the essential nutrients needed by the micro-organism, together with any **inducer** necessary for the synthesis of the required enzyme (see Chapter 3.1.5); alternatively, it may be possible, by growing cells under conditions where the substrate of the enzyme is present in only limited amounts throughout, to isolate and culture a mutant strain which does not require induction. However, regardless of induction, the synthesis of many catabolic enzymes is **repressed** if the cells are grown rapidly on readily-utilisable carbohydrate or protein; this **catabolic repression** can be overcome if the rate of utilisation of the carbon source is controlled, either by feeding small amounts into the system at regular intervals, or by adding it in a form (e.g. an ester) which is slowly hydrolysed to yield the nutrient. Similarly, **feedback repression** of biosynthetic enzymes (and some others) may occur in the presence of metabolic end-products, so the build-up of these should be avoided if the synthesis of such an enzyme is required. The problems of repression may also be overcome by the use of mutants.

Oxygen is often a limiting nutrient in fermentation procedures, and the capacity for absorbing oxygen can be assessed by filling a fermenter with a solution of reducing agent (e.g. sodium sulphite) and operating it under the normal conditions. In general, it is important that the culture medium is homogenous with respect ot pO_2, pH, temperature and nutrient concentrations, so the mixing procedure (usually by propeller or flat-bladed turbine) must be adequately designed; this presents one of the main problems in converting a small-scale procedure to a large-scale one.

Most enzymes obtained commercially from microbial fermentation procedures are hydrolases. These are usually **extracellular enzymes**, i.e. they pass into the culture medium and do not have to be extracted from the cell. Increased yields of these enzymes may be obtained by introducing surfactants (e.g. Tween 80) into the culture medium, although the precise reason for this is not known.

Extracellular enzymes may be separated from the cells by filtration or centrifugation, once fermentation is complete. Continuous fermentation processes could be employed, but these are more difficult to design and control than are single-batch ones.

Intracellular enzymes (e.g. **glucose isomerase**) must be released by cellular disruption: many of the techniques discussed in Chapter 19.1.2 are unsuitable for large-scale procedures, and those most commonly employed in this context are **liquid shear**, obtained by passing a cell suspension through a needle valve at very high pressure, and **grinding with abrasives**. The cell debris can then be removed by centrifugation.

The enzyme solution is usually concentrated by evaporation or ultrafiltration. The enzyme could also be purified by the procedures discussed in Chapter 16.2.2, but it is often unnecessary and uneconomical to do this, provided no materials are present in the solution which affect the enzyme's usefulness. Stabilisers are often added, as are biocides, to deal with contaminating micro-organisms, and sometimes the solution is spray-dried to yield an impure solid.

Obviously, the precise treatment given to the enzyme solution will be determined largely by the nature of the application for which it was prepared: for example, enzymes which will be used for processing food must be free from toxic impurities.

20.2 IMMOBILISED ENZYMES

20.2.1 Preparation of Immobilised Enzymes

An immobilised enzyme is one which has been attached to or enclosed by an insoluble support medium (termed a **carrier**) or one where the enzyme molecules have been cross-linked to each other, without loss of catalytic activity (Fig. 20.1). More than one of these factors may be present.

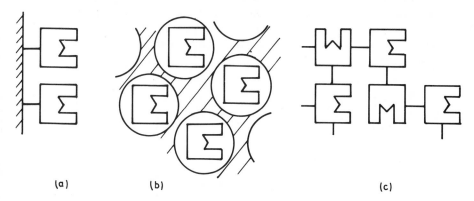

(a)   (b)   (c)

Fig. 20.1 Diagrammatic representation of enzyme immobilisation by (a) attachment to an insoluble support medium, (b) entrapping by an insoluble support medium, and (c) cross-linking of enzyme molecules.

Physical adsorption on to an inert carrier is a very simple procedure for immobilising an enzyme, for it requires just the mixing of the enzyme solution with the carrier. In 1916, Nelson and Griffin showed that **invertase** could be adsorbed on to activated charcoal without any change in enzymic activity, thus producing the first immobilised enzyme, although they made no subsequent use of it. Other inorganic materials which may be used as carriers include clay, alumina and silica. The weak linkages established between enzyme and carrier (mainly van der Waals and hydrogen bonds) have little effect on catalytic activity. However, because the bonds are so weak, the enzyme can easily be desorbed from the carrier: this can be brought about by changes in pH, ionic strength or substrate concentration. Also, the adsorption process is non-specific, so many other substances will become attached to the carrier as the immobilised enzyme is used.

Ionic binding provides a slightly more specific way of attaching an enzyme to a carrier: therefore, many ion exchange resins, e.g. DEAE–Sephadex and

CM-cellulose, have been used as support media. The enzyme will remain bound to the carrier provided pH and ionic strength are maintained at suitable values.

Covalent binding can provide even more permanent linkages between enzyme and carrier. However, it is essential that the conditions used for the formation of the covalent bonds are sufficiently mild that little, if any, catalytic activity is lost; also, the active site of the enzyme must remain free from covalent attachments, so it is sometimes protected by a substrate or substrate-analogue during the immobilisation procedure.

The enzyme functional groups most commonly linked by covalent bonds to a carrier are free α- or ϵ-amino groups, but sulphydryl, hydroxyl, imidazole or free carboxyl groups may also be involved.

Many procedures depend on the coupling of phenolic, imidazole or free amino groups on an enzyme to a **diazonium** derivative of a carrier. For example, the method developed by Campbell and colleagues in 1951 for linking albumin to diazotized p-aminobenzyl cellulose to form an immobilised antigen has been widely used to prepare immobilised enzymes:

$$\text{cellulose-OCH}_2\text{-}\underset{\text{p-aminobenzyl cellulose}}{\bigcirc}\text{-NH}_2 \xrightarrow{\text{NaNO}_2/\text{HCl}} \text{cellulose-OCH}_2\text{-}\underset{\text{(diazonium derivative)}}{\bigcirc}\text{-}\overset{+}{N}\equiv N$$

$$\downarrow \text{enzyme}$$

$$\text{cellulose-OCH}_2\text{-}\underset{\text{(immobilised enzyme)}}{\bigcirc}\text{-N=N-enzyme.}$$

In 1953, Grubhofer and Schleith, using the same principle, but starting with diazotized polyaminopolystyrene, were the first to immobilise enzymes with a view to their future application: the enzymes in question were **pepsin**, **ribonuclease**, **carboxypeptidase** and others.

A **peptide** bond may be formed between a free amino group in an enzyme and a carboxyl group in an insoluble carrier, if the latter group is present as an isocyanate, acid azide or other reactive derivative. For example, Brandenberger, in 1957, prepared an isocyanate derivative of polyaminopolystyrene by treating it with phosgene ($COCl_2$), and then used it to immobilise **catalase**:

$$\left[\begin{array}{c}\text{-CHCH}_2\text{-}\\ \bigcirc \\ \text{NH}_2\end{array}\right]_n \xrightarrow{\text{COCl}_2} \left[\begin{array}{c}\text{-CHCH}_2\text{-}\\ \bigcirc \\ \text{NCO}\end{array}\right]_n \xrightarrow[\text{(enzyme)}]{\text{H}_2\text{N.E}} \left[\begin{array}{c}\text{-CHCH}_2\text{-}\\ \bigcirc \\ \text{NHCONH-E}\end{array}\right]_n$$

polyamino-
polystyrene   (isocyanate derivative)   (immobilised enzyme)

Many enzymes have been immobilised by reacting with the azide derivative of a carrier, whose formation may involve treatment with hydrazine (NH_2NH_2). For example:

$$cellulose\text{-}OCH_2CO_2H \xrightarrow[HCl]{CH_3OH} cellulose\text{-}OCH_2CO_2CH_3$$

CM–cellulose   (methyl ester)

$$\Big\downarrow NH_2NH_2$$

$$cellulose\text{-}OCH_2CON_3 \xleftarrow{NaNO_2/HCl} cellulose\text{-}OCH_2CONHNH_2$$

(azide derivative)   (hydrazide derivative)

$$\Big\downarrow H_2N.E$$

$$cellulose\text{-}OCH_2CONH\text{—}E.$$

(immobilised enzyme)

Peptide bonds between enzyme and carrier may also be formed by the use of condensing reagents such as carbodiimides. For example:

$$cellulose\text{-}OCH_2CO_2H + \begin{array}{c} R' \\ | \\ N \\ \| \\ C \\ \| \\ N \\ | \\ R'' \end{array} \longrightarrow cellulose\text{-}OCH_2CO_2 . \begin{array}{c} R' \\ | \\ NH \\ | \\ C \\ \| \\ N \\ | \\ R'' \end{array}$$

CM–cellulose   a carbodiimide

$$\Big\downarrow H_2N.E$$

$$cellulose\text{-}OCH_2CONH\text{—}E + R'NHCONHR''.$$

(immobilised enzyme)

Another category of covalent bond formation is the **alkylation** of phenolic, sulphydryl or free amino groups by reactive groups in the carrier. For example, bromoacetyl cellulose has been used to immobilise a variety of enzymes:

$$cellulose\text{-}OCOCH_2Br \xrightarrow{enzyme} cellulose\text{-}OCOCH_2\text{-}enzyme.$$

bromoacetyl cellulose   (immobilised enzyme)

Yet another widely used method for enzyme immobilisation was developed by Axén and colleagues in 1967, and involves the reagent, cyanogen bromide (CNBr). This reagent, which is also used for the preparation of immobilised ligands for affinity chromatography (see Chapter 16.2.2), activates polysaccharide hydroxyl groups. The most significant of several reaction sequences which may take place is as follows:

$$
\begin{array}{ccccc}
\mathrm{-OH} & \xrightarrow{\text{CNBr}} & \begin{array}{c}\mathrm{-O}\\ \\ \mathrm{-O}\end{array}\!\!\!\Big\rangle\mathrm{C{=}NH} & \xrightarrow{\text{H}_2\text{N.E}} & \begin{array}{c}\overset{\text{NH}}{\underset{\parallel}{}}\\[-2pt]\mathrm{-OC.NHE}\\ \mathrm{-OH}\end{array}\\
\mathrm{-OH} & & & &
\end{array}
$$

polysaccharide　　(imidocarbonate derivative)　　(immobilised enzyme)

An enzyme which possesses an accessible sulphydryl group not essential for activity can easily be linked to a similar group on a carrier by the formation of a **disulphide bridge**. The commercial significance of this procedure is that it is reversible, so inactivated or unwanted enzyme may be removed from the carrier under reducing conditions, and then be replaced by fresh enzyme. The only other immobilisation technique where such easy replacement is possible is ionic binding.

In general, the conditions required for the covalent attachment of an enzyme to an insoluble carrier are such that some loss of activity is inevitable. However, very little activity change often takes place if a covalent attachment is brought about by means of **chelation** rather than by a chemical reaction: a wide range of carriers (e.g. cellulose, glass and nylon), if treated with salts of transition metals (e.g. titanium, vanadium or iron chlorides), then washed and dried, can chelate enzymes; strong metal bridges (see Chapter 11.4.1) are formed between hydroxyl oxygen atoms of the carrier and amino nitrogen atoms on the enzyme.

The **entrapping** (or **occlusion**) of an enzyme is commonly achieved within the lattice of a **polymerised** gel. The most widely used is polyacrylamide, which may be synthesised from an aqueous solution of acrylamide and N,N–methylene-bisacrylamide (BIS) in the presence of initiators and accelerators, as outlined (without stoichiometry) below:

$$
\begin{array}{lll}
\mathrm{CH_2{=}CH} & \mathrm{CH_2{=}CH} & \mathrm{-CH_2.CH.CH_2CH.CH_2.CH.CH_2-}\\
\quad\mid & \quad\mid & \qquad\mid\qquad\quad\mid\qquad\qquad\mid\\
\quad\mathrm{CONH_2}\; + & \quad\mathrm{CO} & \qquad\mathrm{CO}\quad\;\;\mathrm{CONH_2}\quad\mathrm{CO}\\
& \quad\mid & \qquad\mid\qquad\qquad\qquad\quad\mid\\
\text{acrylamide} & \quad\mathrm{NH} & \qquad\mathrm{NH}\qquad\qquad\qquad\mathrm{NH}\\
& \quad\mid\quad\xrightarrow{\text{initiators}} & \qquad\mid\qquad\qquad\qquad\quad\mid\\
& \quad\mathrm{CH_2}\quad\xrightarrow{} & \qquad\mathrm{CH_2}\qquad\qquad\qquad\mathrm{CH_2}\\
& \quad\mid\quad\;\text{accelerators} & \qquad\mid\qquad\qquad\qquad\quad\mid\\
& \quad\mathrm{NH} & \qquad\mathrm{NH}\qquad\qquad\qquad\mathrm{NH}\\
& \quad\mid & \qquad\mid\qquad\qquad\qquad\quad\mid\\
& \quad\mathrm{CO} & \qquad\mathrm{CO}\qquad\qquad\qquad\mathrm{CO}\\
& \quad\mid & \qquad\mid\qquad\qquad\qquad\quad\mid\\
& \;\mathrm{CH_2{=}CH} & \mathrm{-CH_2CH.CH_2CH.CH_2.CH.CH_2-}\\
& \quad\text{BIS} & \qquad\qquad\qquad\;\mathrm{CO}\\
& & \qquad\qquad\qquad\;\mid\\
& & \qquad\qquad\qquad\;\mathrm{NH}\\
& & \qquad\qquad\qquad\;\mid
\end{array}
$$

polyacrylamide

The degree of cross-linking, and hence the average pore size, is dependent on the proportions of acrylamide and BIS present initially, for only the latter can cross-link. If an enzyme is present in the solution, it will be entrapped as the gel is formed.

Alternatively, an enzyme may be entrapped within the semi-permeable membrane of a **microcapsule**. This is most commonly done by the **interfacial polymerisation** method developed by Chang and colleagues around 1964: an aqueous solution containing the enzyme and a hydrophilic monomer is added to a water-immiscible organic solvent and an emulsion is formed. More of the same organic solvent, but containing a suitable hydrophobic monomer is then added and the solutions are well mixed. Polymerisation of the monomers will take place at each interface between organic and aqueous solvents, and, in this way, enzyme molecules in aqueous solution are enclosed in polymer membranes. For example, nylon microcapsules may be formed if the hydrophilic monomer is 1,6-hexamethylene diamine, the hydrophobic monomer sebacoyl chloride, and the organic solvent a cyclohexane–chloroform mixture. Such microcapsules are typically 10–100 μm in diameter.

The entrapping of an enzyme, either by gel formation or microencapsulation, is of general application and should not affect the activity of the enzyme. However, in some instances, free radicals generated during the polymerisation procedure may cause some loss of enzymic activity. Also, since the entrapped enzymes cannot escape because of their size, it follows that a very large substrate will not be able to diffuse in to reach the enzymes; hence, this immobilisation procedure is not suitable for proteolytic enzymes or others whose substrates are macromolecules. Each polymerised gel will inevitably contain a range of pore sizes, for the cross-linking cannot be exactly regular: therefore, in contrast to the situation with macrocapsules, there is likely to be some degree of leakage of enzyme. Another advantage of microencapsulation is that each enzyme is in much closer contact with the surrounding solution than are those entrapped in the interior of gels.

Another encapsulation technique which can be used for enzyme immobilisation is the formation of **liposomes**: amphipathic lipids such as phosphatidyl choline and cholesterol (see Chapter 16.1.3) are dissolved in chloroform and spread as a film over the walls of a rotating flask; an aqueous solution of enzyme is added and rapidly dispersed, and liposomes are formed as lipid membranes enclose the water droplets. Some of the enzyme will remain in the aqueous solution within the liposomes, while the rest will be incorporated into the membrane.

Immobilisation by **cross-linking** molecules of enzyme is most commonly brought about by the action of glutaraldehyde, whose two aldehyde groups form Schiff's base linkages with free amino groups:

$$
\begin{array}{ccc}
 & & \text{E.N} \\
 & & \| \\
\text{E.NH}_2 & \text{CHO} & \text{CH} \\
 & | & | \\
 & + \ (\text{CH}_2)_3 & \rightarrow \ (\text{CH}_2)_3 \\
 & | & | \\
\text{E.NH}_2 & \text{CHO} & \text{CH} \\
 & & \| \\
 & & \text{E.N}
\end{array}
$$

Since several free amino groups are likely to be present on each enzyme molecule, a cross-linked network will be formed. This procedure was first used by Quiocho and Richards, in 1964, to immobilise **carboxypeptidase A**. Other reagents with two functional groups (i.e. **bifunctional reagents**) of relevance to enzyme immobilisation include derivatives of bis-diazobenzidine,

$$\overset{+}{N}_2 - \hexagon - \hexagon - \overset{+}{N}_2$$

which act by means of diazo coupling.

Although cross-linking between identical enzyme molecules can result in immobilisation at high enzyme concentrations, it is not an ideal method since many molecules simply act as supports for others. Hence, cross-linking is often performed in conjunction with other methods of immobilisation. It can, for example, be used to prevent the leakage of enzymes from a polymerised gel (see above), or to trap the enzyme around pre-formed polymer molecules (e.g. starch). Also, bifunctional reagents may be used, not only to link molecules of enzyme to each other, but also to link them to an inert carrier: thus, glutaraldehyde has often been used to attach enzymes to amino groups on carriers such as aminoethyl (AE)-cellulose or aminoalkylated porous glass.

As with any other method which involves the formation of covalent bonds by chemical reaction, cross-linking is usually performed under conditions which cause some loss of enzymic activity.

Immobilised enzymes are usually prepared in particle form, but enzymes may be attached to, or entrapped within, carriers in the form of membranes, tubes or fibres, according to the requirements of a given application. Also, in the case of intracellular enzymes, it may be more economical to **immobilise the intact cell** rather than to perform an extraction step. This is satisfactory provided none of the other enzymes present affect the proposed application, and provided the appropriate substrates and products can freely pass through the cell membranes. The most common method for immobilising microbial cells is to entrap them in polyacrylamide gel, as done by Mosbach and colleagues in 1966. The use of immobilised cells is particularly advantageous where is is known that the enzyme of importance would be unstable during or after extraction, a not uncommon situation.

20.2.2 Properties of Immobilised Enzymes

The properties of an immobilised enzyme may be different from those of the same enzyme in free solution, and depend both on the method of immobilisation and on the nature of the insoluble carrier. A reduction in specific activity may occur as an enzyme is immobilised, particularly if a chemical process is involved, since the conditions might cause some denaturation to take place (see Chapter 20.2.1). Furthermore, the carrier creates a new microenvironment for the enzyme, and could thus influence its activity in a variety of ways. For example, the characteristics of the enzyme might be altered if the active site undergoes some conformational change as a result of chemical or physical interactions between enzyme and carrier. Also, the carrier could affect the characteristics of the enzyme-catalysed reaction by imposing some steric hindrance, by preventing free diffusion of substrate to all molecules of the enzyme, or by forming electrostatic interactions with molecules of substrate or product (see below).

The **stability** of an enzyme on heating or storage can increase, decrease or stay the same when it is immobilised, depending on how the new microenvironment affects its tendency to denature: few clear-cut trends have been observed. Another relevant factor is the ease of attack by substances which might degrade the enzyme: for example, autodigestion of proteolytic enzymes is reduced if the molecules are protected from each other by immobilisation; similarly, steric hindrance protects many other immobilised enzymes from attack by proteolytic enzymes.

The **pH optimum** can change by as much as 2 pH units when the enzyme is immobilised, largely due to the effect of the new microenvironment. Goldstein and colleagues (1964-70), and others, have demonstrated that the pH optimum of many enzymes moves to a more alkaline value if the carrier is anionic, and to a more acid value if it is cationic, due to changes in the degree of ionisation of amino-acid residues at the active site. The effect is not observed if the ionic strength is high, presumably because the salt ions interfere with the **electrostatic field** produced by the carrier.

The **apparent K_m** may be similarly affected by the **electrostatic field** of the carrier, being significantly decreased when a carrier is used which is of opposite charge to that of a substrate. For example, when positively charged benzoyl-L-arginine ethyl ester is used as substrate, the apparent K_m of **ficin** immobilised by covalent attachment to CM-cellulose, a polyanionic carrier because of the presence of unreacted carboxyl groups, is ten times less than that of the free enzyme. Goldstein and colleagues have explained this on the basis of the electrostatic interaction causing the substrate to be present in the region of the carrier, and hence of the enzyme, at higher concentrations than elsewhere in the solution. The converse is also true for the situation where the carrier has the same charge as the substrate: the apparent K_m for **creatine kinase** immobilised on CM-cellulose is ten times greater than that for the free enzyme, the substrates ATP and creatine being negatively charged. Again, these effects are not seen at high ionic strength.

The **apparent K_m** may also be affected by **diffusion factors**. Lilly and colleagues, in 1968, postulated that around each particle or membrane is an unstirred layer (or layers) of solution, 10-100 μm thick. The substrate contained in such a layer will be quickly utilised by an enzyme immobilised within the particle or membrane, so the subsequent reaction velocity will depend on the rate at which substrate from the bulk solution can diffuse through the unstirred layer and reach the enzyme. In general, the steady-state substrate concentration in the vicinity of the enzyme will be less than that in the bulk solution, and so the reaction velocity will be less than that expected from the external substrate concentration. The thickness of the diffusion layer depends on the rate at which the bulk solution is stirred, so that it may be possible to increase the reaction velocity by more vigorous stirring.

However, restricted diffusion of the substrate through a membrane or gel lattice can also reduce the rate of reaction. Furthermore, the ease of diffusion of the product away from the enzyme can also be a significant factor, particularly if the reaction being catalysed is subject to product inhibition.

For one or more of these reasons, the apparent K_m of an immobilised enzyme is often significantly higher than for the same enzyme in free solution. For example, Goldman, Goldstein, Katchalski and colleagues (1964–71) have shown that this is true for **papain** and **alkaline phosphatase** immobilised in collodion membranes, and, furthermore, the apparent K_m increases with increasing membrane thickness.

The rate of diffusion of substrate to the enzyme will rise to a limiting value as the substrate concentration in the bulk solution is increased. If this limiting value is reached before an immobilised enzyme is completely saturated with substrate, then its **apparent V_{max}** value will be less than that in free solution.

20.2.3 Applications of Immobilised Enzymes: General Principles

The immobilised enzyme system chosen for a given application should fit the requirements in terms of stability, activity, pH optimum and other characteristics: the factors discussed in Chapter 20.2.2 should all be considered.

Sometimes the changes in properties which take place when an enzyme is immobilised may be used to advantage. For example, if an enzyme-catalysed reaction cannot be linked directly to another because of incompatible pH-activity ranges, it may be possible to immobilise the enzymes in such a way that their pH-activity ranges now overlap, thus allowing them to be used in a single-rather than a two-step process.

Regardless of this, component enzymes of a coupled system may be immobilised together (e.g. entrapped in a gel) or, if immobilised separately, may later be placed in close proximity. This increases the efficiency of the coupled process by ensuring that the concentration of the substance which is the product of one of the enzyme-catalysed reactions and the substrate of the other is kept at a higher concentration in the neighbourhood of the enzyme than in the bulk solution. This may be used in an industrial application, or as a model system to investigate the properties of enzymes located closely together in the living cell (e.g. the **malate dehydrogenase/citrate synthase** system of mitochondria). Another example of the value of immobilised enzymes as model systems is the use of liposomes to investigate *in vitro* the effect of a lipid environment on the activity of enzymes which are associated with membranes *in vivo*.

In general, the property of immobilised enzymes which is of greatest industrial importance is the ease with which they can be separated from reaction mixtures. Hence, in contrast to systems involving soluble enzymes, the reaction can be stopped by physical removal of the immobilised enzyme, without requiring such procedures as heat inactivation which might affect the products of the reaction. Furthermore, the enzyme will still be active and largely uncontaminated, so can be used again. For these reasons, immobilised enzymes are ideal for use in continuously operated processes. At the present time, continuous industrial processes involving immobilised enzymes are usually carried out in simple **stirred-tank reactors** or in **packed-bed reactors** (Fig. 20.2). The direction of flow

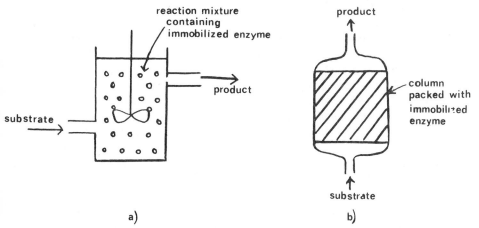

Fig. 20.2 Diagrammatic representation of (a) a continuously operated stirred-tank reactor (CSTR), and (b) a continuously operated packed-bed, or plugged-flow, reactor with an upwards direction of flow.

in packed-bed reactors may be in an upwards or a downwards direction, but the former is preferable in applications where a downward flow would cause compression or clogging of the packed-bed. A more recent development is the **fluidised-bed reactor**, which incorporates some of the features of both stirred-tank and packed-bed reactors: this is an upward flow system which is ideal for processes where the substrate solution is highly viscous and the product is gaseous. Finally, as mentioned in Chapter 20.2.1, some processes now make use of enzymes immobilised in membranes, hollow-fibres or tubes.

20.3 ENZYME UTILISATION IN INDUSTRY

20.3.1 Applications in Food and Drink Industries

Some applications of enzymatic analysis in industry were discussed in Chapter 19.2. Here we will consider some industrial processes where enzymes have a more direct involvement.

The traditional use of yeasts (e.g. *Saccharomyces carlsbergensis*) in the baking and brewing industries arose because they contain the enzymes for alcoholic fermentation: in common with other organisms, they metabolise hexose sugars to produce pyruvate, but, whereas animals convert this to lactate under anaerobic conditions, the anaerobic end-product in yeasts is ethanol, with carbon dioxide being evolved:

$$CH_3.CO.CO_2H \xrightarrow[\text{Mg}^{2+},\text{ TPP}]{\substack{\text{pyruvate} \\ \text{decarboxylase}}} CH_3.CHO \xrightarrow{\substack{\text{alcohol} \\ \text{dehydrogenase}}} CH_3.CH_2OH + CO_2.$$

$$\text{NADH} \quad \text{NAD}^+$$

pyruvic acid　　　　　　　　acetaldehyde　　　　　　　　ethanol

In the **baking of bread**, the preliminary process involves the mixing of wheat flour (mainly starch and protein) with yeast and water. Starch consists of D-glucose units linked by α-1,4 glycosidic bonds, with α-1,6 bonds at branching points; the enzymes α-**amylase** and β-**amylase** present in the flour cleave some of the α-1,4 bonds, the eventual products being glucose, maltose (a disaccharide) and some oligosaccharides which cannot be broken down further because of the presence of α-1,6 bonds. Glucose and maltose can then be metabolised by the yeast, and carbon dioxide is formed which distends the protein framework of the dough, ready for baking. However, wheat flour often has a low α-amylase content, so it may be supplemented with malt flour (see below) or, a more recent development, by fungal α-amylase (from *Aspergillus oryzae*). Wheat α-amylase is very heat stable, so it continues to act for a time during the baking process; this may lead to too much starch breakdown taking place, and cause the bread to be somewhat soggy. For this reason, wheat α-amylase is sometimes inactivated by a brief treatment with superheated steam prior to supplementation of the flour with the more heat-labile fungal α-amylase. **Proteases** from *Aspergillus oryzae* may be introduced to cause strictly-limited breakdown of the wheat protein (gluten), thus shortening mixing time and enabling a smoother, more uniform dough to be obtained. Malt flour also contains proteases, but in variable amounts, so its use might lead to excessive gluten breakdown.

In **brewing**, the main starting material is malt, produced by allowing barley seed to germinate in moist conditions. The reserve starch is broken down by the amylase present (as in wheat, discussed above and in Chapter 19.2) to give, among other products, glucose and maltose. The grains may then be roasted to prevent further growth and to add flavour, after which the soluble material present is extracted by water to produce the **wort**. This is further flavoured with hops, and then yeast is added to produce ethanol by alcoholic fermentation of the glucose and maltose.

Bacterial α-**amylase** (from *Bacillus subtilis*), which is even more heat stable than wheat α-amylase, is of increasing importance in the brewing industry. It may be used to produce a wort from unmalted barley (and maize), although some malt (a minimum of 20%) is usually added. Fungal **exo-1,4-α-glucosidase**, better known as **amyloglucosidase** or **glucamylase** (e.g. from *Aspergillus niger*) may also be added, since this can cleave α-1,6 as well as α-1,4 glycosidic bonds and so increase the yield of glucose and maltose from starch. Malt contains some **protease** activity, so bacterial proteases are usually added when a wort is being prepared from unmalted barley.

In the industrial production of **glucose** from starch, the latter is first solubilised and partly degraded by bacterial α-**amylase** (or, but less commonly than hitherto, by HCl), and then treated with fungal **amyloglucosidase**: as mentioned above, this can cleave both types of glycosidic bond found in starch, so gives a good yield of glucose. Glucose may also be obtained from cellulose-containing waste products by treatment with **cellulase** (e.g. from *Aspergillus niger*); as a further possibility, it may be produced, together with galactose, by the action of β-**galactosidase** (**lactase**) on lactose, which is present in whey and so is a major by-product of cheese manufacturing.

Invert sugar, a mixture of glucose and fructose, is produced from sucrose by the action of yeast β-**fructofuranosidase**, better known as **invertase** (an enzyme which can only be extracted by disruption of the yeast cell wall). Sucrose crystallises from solution more readily than invert sugar at the high concentrations required for jam making, so the latter is widely used for this purpose. Also, in confectionary, crystalline sucrose is coated with chocolate and invertase, causing the subsequent conversion of sucrose to invert sugar, and hence liquification to form a 'soft-centre'.

Invert sugar may also be produced from glucose by the action of **glucose isomerase** (bacterial or fungal), an enzyme now thought to be identical to **xylose isomerase**. The first commercial production of such a high-fructose syrup from glucose, using soluble glucose isomerase, was reported by Takasaki in 1966; one using glucose isomerase immobilised on DEAE-cellulose was described by Thompson and colleagues in 1974. At the present time, conversion of glucose to fructose is the biggest industrial application of immobilised enzymes. Fructose is much sweeter than glucose, so high-fructose syrup can be used to replace sucrose as a sweetener, where it is economical to do so. It may also find increased use in this respect because of fears about the adverse effects of sucrose on health, regardless of whether fructose can be proved to be any safer.

Certain **amino-acids** essential for animal growth (e.g. methionine) are found in low amounts in the proteins of some vegetable-based foodstuffs, so may have to be added to the diet. However, animals can only utilise L-amino-acids, whereas D- and L-amino-acids, in equal amounts, are produced by chemical processes. Dietary supplements usually consist of such mixtures, so half is wasted, even if no other harm is done. Hence the procedure introduced in 1969 by Chibata and colleagues for the production of L-amino-acids from racemic mixtures is of great significance: acetyl D- and L-amino-acids are passed through a system containing **aminoacylase** immobilised on a DEAE-Sephadex (or other) carrier, causing the deacylation of the L-form alone, which can then be separated; the unchanged D-form is then subjected to racemisation, and the process repeated.

The **clarification of cider, wines and fruit juices** (e.g. apple) is usually achieved by treatment with fungal **pectinases**. The pectins of fruit and vegetables play an important role in jam making and other processes by bringing about gel formation. However, they cause fruit drinks to be cloudy by preventing the flocculation of suspended particles. Pectinases are a group of enzymes including **polygalacturonases**, which break the main chains of pectins, and **pectinesterases**, which hydrolyse methyl esters. Their action releases the trapped particles and allows them to flocculate.

Cheese production involves the conversion of the milk protein, κ-casein, to para-casein by a defined, limited hydrolysis catalysed by **chymosin (rennin)**. In the presence of Ca^{2+}, para-casein clots and may be separated from the whey, after which the clot is allowed to mature under controlled conditions, still in the presence of chymosin, to form cheese. Since chymosin can only be extracted from calves killed before they are weaned (pepsin is produced instead of chymosin after weaning), the enzyme is in short supply, so there has been a large-scale search for an acceptable substitute. Proteases from animals (**pepsin**), plants (**ficin** and **papain**) and over a thousand micro-organisms have been tried, either on their own or mixed with calf chymosin; several have been largely, if not completely, successful (e.g. a protease from *Mucor rouxii*).

Papain is sometimes used as a **meat tenderiser**; some South American natives have traditionally wrapped their meat in leaves of papaya, the fruit from which papain is extracted. Papain (and other proteases) may also be used in the brewing industry to prevent **chill hazes**, caused by precipitation of complexes of protein and tannin at low temperatures.

20.3.2 Applications in Other Industries

The use of enzymes in medical diagnosis and therapy was discussed in Chapter 19.1. To give another example, immobilised enzymes may be components of artificial kidney machines, which are used to remove urea and other waste products from the body, where kidney disease prevents this being done by natural processes: urea enters the machine from the blood, by dialysis (termed **haemodialysis**), and is converted to CO_2 and NH_4^+ by immobilised **urease**; toxic NH_4^+ is then either trapped on ion exchange resins or incorporated into glutamate by the action of immobilised **glutamate dehydrogenase**, and subsequently adsorbed by active carbon, before the fluid is returned to the blood stream.

Some applications of enzymes in the chemical industry were discussed in Chapter 20.3.1, e.g. the production of L-amino-acids as possible supplements to foodstuffs.

Enzymes may also be of great value in the pharmaceutical industry, e.g. for the conversion of naturally-occurring penicillin G (benzyl penicillin) to 6–amino penicillanic acid (6-APA). This reaction, which is catalysed by bacterial **penicillin amidase**, cleaves the side chain of the substrate at mildly alkaline pH values:

benzyl penicillin   phenyl acetic acid   6-APA

Many commercial processes have been developed for the synthesis of 6-APA, including the use of *E. coli* immobilised in polyacrylamide gel, and the enzyme entrapped in fibres of cellulose acetate. The importance of 6-APA is that new side chains can be attached to give a variety of semisynthetic penicillins, e.g. ampicillin (α-methyl benzyl penicillin), a broad-spectrum antibiotic. This may be done by chemical processes or by the use of the enzyme under slightly acidic conditions, to favour the back reaction. Enzymes from different bacteria may be applied to catalyse the forward and back reactions, to make use of their slightly different characteristics and specificities: for example, a mutant of *Kluyvera citrophila* has been used for 6-APA production, and one of *Pseudomonas melanogenum* for ampicillin synthesis.

Washing powders incorporating bacterial proteases have been available for several years, although their commercial importance has waned due to fears about their effect on the respiratory system. The enzymes in question, **subtilisins** from *Bacillus subtilis* mutants, are stable to alkali, high temperature (e.g. 65°C), detergents and bleaches. They will attack blood and other protein stains.

Bacterial proteases are also used in the **leather and textile industries** to loosen hair (or wool) and enable it to be separated from hide.

SUMMARY OF CHAPTER 20

Enzymes are **extracted** on a large scale mainly from microbial sources, because a wide range of enzymes are available in micro-organisms, and because cells producing the required enzyme can be cultivated.

Most **fermentation** processes are carried out in submerged culture, where the conditions can be carefully controlled. The cells must be grown in the presence of suitable inducers, and in the absence of the approptiate repressors, if the required enzyme is to be produced. Most enzymes of commercial importance are extracellular ones: unlike intracellular enzymes, they leak out into the culture medium and so do not require extraction from the cells when the fermentation is finished. The degree to which the enzyme is then purified depends both on economic factors and on the intended application.

Enzymes may be **immobilised** by being bound to an insoluble carrier (by physical adsorption, ionic binding or covalent binding), by being entrapped in a gel or membrane, or by being cross-linked (often, but not always, in addition to being bound or entrapped). Immobilisation can effect the stability, pH optimum, apparent K_m and apparent V_{max} of an enzyme: this depends on the method of immobilisation and the nature of the carrier, so these factors must be considered when an immobilised enzyme system is being chosen for a particular application.

Immobilised enzymes are easily removed from a reaction mixture, thereby stopping the reaction and making the enzyme available to be used again. They are ideal for incorporation into continuous processes.

Enzymes have been used for centuries in the baking and brewing industries, as components of yeast cells and malt. More recently, many applications have been found in these and other industries for purified enzymes.

FURTHER READING

Abelson, P. H. (ed.) (1983), Biotechnology, *Science,* **219** (pages 609–746).

Bickerstaff, G. F. (1984), Applications of immobilised enzymes to fundamental studies on enzyme structure and function, in Wiseman, A. (ed.), *Topics in Enzyme and Fermentation Technology,* **9**, Ellis Horwood.

Bohak, Z. and Sharon, N. (1977), *Biotechnological Applications of Proteins and Enzymes,* Academic Press.

Chibata, I. (1978), *Immobilized Enzymes,* Kodansha, Halsted Press.

Chibata, I. Fukui, S. and Wingard, L. B. (eds.) (1982), *Enzyme Engineering,* **6**, Plenum.

Chibata, I. and Tosa, I. (1980), Immobilized microbial cells and their applications, *Trends in Biochemical Sciences,* **5** (pages 88–90);

Colowick, S. P. and Kaplan, N. O. (eds.), *Methods in Enzymology,* **44** (1976): Immobilized Enzymes; **58** (1979): Cell Culture; **68** (1980), **100** (1983): Recombinant DNA, Academic Press.

Emery, A. E. H. (1984), *An Introduction to Recombinant DNA,* Wiley.

Foster, R. L. (1980), *The Nature of Enzymology* (Chapter 7), Croom Helm.

Mosbach, K. (1980), Future trends: immobilized enzymes, *Trends in Biochemical Sciences,* **5** (pages 1–3).

Primrose, S. B., Derbyshire, P., Jones, I. M., Robinson, A. and Ellwood, D. C. (1984), The application of continuous culture to the study of plasmid stability, in Dean, A. C. R. Ellwood, D. C. and Evans, C. G. T. (eds.), *Continuous Culture,* **8**, Ellis Horwood.

Stryer, L. (1981), *Biochemistry,* 2nd edition (Chapter 31), Freeman.

Thomas, D. (1979), The future of enzyme technology, *Trends in Biochemical Sciences,* **4** (pages N207–209).

Wang, D. I. C., Cooney, C. L., Demain, A. L., Dunnill, P., Humphrey, A. E. and Lilly, M. D. (1979), *Fermentation and Enzyme Technology,* Wiley.

Wiseman, A. (ed.) (1975), *Handbook of Enzyme Biotechnology,* Ellis Horwood.

Answers to Problems

1.1 (a) acylcholine acyl-hydrolase (3.1.1)
  (b) carbamoyl phosphate : L-ornithine carbamoyl-transferase (2.1.3)
  (c) D-alanine : D-alanine ligase (ADP forming) (6.3.2)
  (d) glycerol : NAD⁺ 2-oxidoreductase (1.1.1)
       (Note – not 1.6.99. Oxidoreductases involving NAD⁺ are classified in
       the direction where NAD⁺ is the electron acceptor rather than NADH
       the electron donor, except where another redox catalyst is involved.
       Here, therefore, the reaction is classified in the opposite direction to
       that given.)
  (e) D-fructose-1,6-bisphosphate D-glyceraldehyde-3-phosphate-lyase
       (4.1.2)
  (f) NADH : ferricytochrome b_5 oxidoreductase (1.6.2)
  (g) UDP glucose 4-epimerase (5.1.3)

1.2 (a) 6.4.1.1
  (b) 1.8.1.2
  (c) 2.7.4.3
  (d) 5.3.1.1
  (e) 3.4.13.1
  (f) 3.2.1.1
  (g) 4.3.2.1

2.1 Gly-Ala-Thr-Arg-Ser-Phe-Val-Leu-Lys-Ala-Phe-Ser-Lys-Ile-Trp-*Glun*-
                                    -Thr-Ser

2.2 Probable structure:

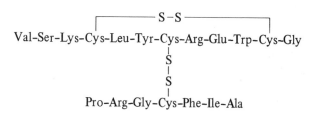

3.1 (a) (i) 4.74 (ii) 5.04

3.1 (b) Concentration of acetic acid in tube 4 = $(0.04 \times \frac{5}{6})$ mol.dm^{-3}.
Concentration of acetate in tube 4 = $(0.1 \times \frac{1}{6})$ mol.dm^{-3}.
pH of tube 4 (and hence approximate isoelectric pH of casein) = 4.44

3.2 (a) 99.9%, (b) 90.9%, (c) 9.09% and (d) 0.10%

6.1 (a)

Reaction	$\Delta G^{\oplus\prime}$(kJ.mol^{-1})
argininosuccinate + H$_2$O $\rightleftharpoons$ aspartate + citrulline	−34.4
arginine + fumarate $\rightleftharpoons$ argininosuccinate	−11.7
NH$_4^+$ + citrulline $\rightleftharpoons$ arginine + H$_2$O	+30.5
aspartate + H$_2$O $\rightleftharpoons$ malate + NH$_4^+$	+12.6
fumarate + H$_2$O $\rightleftharpoons$ malate	−3.0

$\Delta S^{\oplus\prime} = -43.9$ J.K^{-1}.mol^{-1}

(b) $\Delta G = +2.7$ kJ.mol^{-1} (Note - H$_2$O concentration is set at 1.0 mol.l^{-1}.)
$K'_{eq} = 3.2$; [fumarate$_{eq}$] = 2.4×10^{-4} mol.l^{-1} and
[malate$_{eq}$] = 7.6×10^{-4} mol.l^{-1}

6.2 (a) $E'_{\oplus}$(NAD$^+$|NADH + H$^+$) − $E'_{\oplus}$(malate|oxaloacetate) = −0.145 V

(b) $\Delta G = +31.4$ kJ.mol^{-1} ([H$^+$] = 10^{-7} mol.l^{-1} at pH 7, but this is not
required here since $\Delta G^{\oplus\prime}$ is being used and not $\Delta G^{\oplus}$.)

(c) $K'_{eq} = 1.3 \times 10^{-5}$; [oxaloacetate$_{eq}$] = [NADH$_{eq}$] = 0.0002 mol.l^{-1}

6.3 Activation energy for catalysed reaction = 43 kJ.mol^{-1} and for uncatalysed
reaction 69 kJ.mol^{-1}

6.4

[A$_0$] (mmol.l^{-1})	10.0	8.0	6.0	4.0	2.0
v_0 (absorbance units.minute^{-1})	0.107	0.085	0.064	0.042	0.021

Reaction is first order with respect to A.
(Note that graphs of absorbance against time do not necessarily pass through
the origin, even when no product is present at zero time. This could be due
to the instrument cells not matching, the reagent blank not being included
in the reference or to the inaccurate assessment of mixing time. Since it is
the slope which is being measured, this will not affect results provided the
same procedure is used throughout.)

6.5 (a)

Reaction	$\Delta G^{\circ\prime}(kJ.mol^{-1})$
glucose 6-P $\rightleftharpoons$ glucose 1-P	+7.2
UDP-glucose + 2P_i $\rightleftharpoons$ fructose 6-P + UTP	+21.7
glucose 1-P + UTP $\rightleftharpoons$ UDP-glucose + (PP)$_i$	0
(PP)$_i$ $\rightleftharpoons$ 2P_i	−26.8
glucose 6-P $\rightleftharpoons$ fructose 6-P	+2.1

$\Delta G = -1.8$ kJ.mol^{-1}.

(b) Activation energy for catalysed reaction = 43 kJ.mol^{-1} and for uncatalysed reaction = 68 kJ.mol^{-1}.

7.1 $K_m = 13$ mmol.l^{-1}; $V_{max} = 540\ \mu$mol.l^{-1}.minute^{-1}

7.2

NAD$^+$conc^n (mmol.l^{-1})	1.5	2.0	2.5	3.33	5.0	10.0
v_o (absorbance units.minute^{-1})	0.046	0.053	0.059	0.065	0.074	0.085

$K_m = 1.7$ mmol.l^{-1}; $V_{max} = 0.10$ absorbance units.minute^{-1}

7.3 $k_1 = 500$ mol^{-1}.l.s^{-1}; $k_2 = 1.5$ s^{-1}; $k_{-1} = 6$ s^{-1}

7.4 $$v_o = \frac{k_1k_2k_3[S_o][E_o]}{k_2k_3 + k_{-1}k_3 + k_{-1}k_2 + (k_1k_3 + k_1k_{-2} + k_1k_2)[S_o]}$$

8.1 With A: competitive inhibition.
With B: non-competitive inhibition.
Note that the 'uninhibited' reaction exhibits substrate inhibition

8.2 Non-competitive inhibition; $[I_o] = 1.3$ mmol.l^{-1}

8.3 Uncompetitive inhibition ($K_i = 2.1$ mmol.l^{-1}); $V_{max} = 0.04$ absorbance units.minute^{-1} in presence of 3.0 mmol.l^{-1} oxaloacetate

8.4 Non-competitive inhibition ($K_i = 2.0$ mmol.l^{-1}); $K'_m = 22$ mmol.l^{-1}, $V'_{max} = 360\ \mu$mol.l^{-1}.minute^{-1}, in presence of 3.0 mmol.l^{-1} inhibitor

8.5 *linear* competitive inhibition ($K_i = 4.5$ mmol.l^{-1})

8.6 Mixed inhibition (competitive–non-competitive); secondary plots are not linear, so inhibition is possibly partial

9.1 Primary plot excludes ping-pong bi-bi mechanism; inhibition studies with I show a competitive pattern, which excludes the compulsory-order mechanism where AX binds first; product inhibition studies with A similarly excludes the compulsory-order mechanism where B binds first.
Conclusion: probably a random-order ternary-complex mechanism

9.2 Primary plot excludes ping-pong bi-bi mechanism; product inhibition studies with oxaloacetate show a mixed pattern, which excludes a random-order mechanism and the compulsory-order mechanism where malate binds first. Equilibrium isotope exchange studies indicate the compulsory-order ternary-complex mechanism where NAD^+ binds first, so this is probably the mechanism which operates.

9.3 (a) Use of the dead-end inhibitor shows an uncompetitive pattern with respect to varying $AX(K_i = 2.0$ mmol.l^{-1}); (b) product inhibition by A gives a competitive pattern with respect to varying AX, which excludes the compulsory-order mechanism where B binds first; (c) the primary plot excludes a ping-pong bi-bi mechanism, and product inhibition by BX gives a mixed pattern with respect to varying B, which excludes a random-order mechanism, as well as the compulsory-order mechanism where B binds first; and (d) equilibrium isotope exchange studies also exclude the compulsory-order mechanism where B binds first.

Conclusion: probably the compulsory-order ternary-complex mechanism where AX binds first

9.4 $\phi_0 = \dfrac{k_{-3}k_5 + k_4k_5 + k_3k_5 + k_3k_4}{k_3k_4k_5}$; $\phi_{AX} = \dfrac{1}{k_1}$;

$\phi_B = \dfrac{k_{-2}k_{-3} + k_{-2}k_4 + k_3k_4}{k_2k_3k_4}$; $\phi_{AXB} = \dfrac{k_{-1}(k_{-2}k_{-3} + k_{-2}k_4 + k_3k_4)}{k_1k_2k_3k_4}$

9.5 There is a mixed inhibition pattern with respect to varying lysine, a competitive inhibition pattern with respect to varying 2-oxoglutarate and an uncompetitive pattern with respect to varying NADPH. Cleland's rules may be applied as for a two-substrate reaction, so saccharopine and 2-oxoglutarate compete for a site on the same enzyme form, and there is no reversible link between saccharopine and NADPH. The simplest mechanism consisten with these results is a compulsory-order one with 2-oxoglutarate binding first, then lysine and then NADPH, to form a 4-ary complex; $NADP^+$ is the first product to leave (as an irreversible step) and then saccharopine. For a more detailed discussion see Fjellstedt, T. A. and Robinson, J. C. (1975), *Archives of Biochemistry and Biophysics,* **168** (pages 536–548).

10.1 The variation of $\dfrac{V_{max}}{K_m}$ with pH at fixed $[E_0]$ indicates the involvement of two residues, of approximate pK_a values 4.2 and 8.2; the variation of V_{max} with pH at fixed $[E_0]$ indicates the involvement of a residue whose pK_a is approximately 4.1. Hence, two essential ionising residues are indicated: the degree of ionisation of the one of lower pK_a values is important in both the free enzyme and the enzyme–substrate complex, whereas the ionisation of the other is important only in the free enzyme (n.b. other types of investigation identify the two residues as histidine–159 and cysteine–25; the substrate forms a covalent link with cysteine in much the same way as with the essential serine residue in serine proteases)

12.1 Negative cooperativity

12.2 Positive cooperativity; $h = 1.8$

12.3 Negative cooperativity; four binding sites

13.1 Positive homotropic cooperativity is indicated: from the Hill plot, $h = 4$; from the sedimentation date, $n = 4$ (4 sub-units). Hence the Monod–Wyman–Changeux model may be operating

13.2 (a) Binding data show that there is no cooperativity; (b) the primary plot excludes a ping-pong bi-bi mechanism, and also shows there are some departures from linearity; product inhibition by NAD^+ gives a competitive pattern with respect to varying NADH, which excludes the compulsory-order mechanism where acetaldehyde binds first; and (c) equilibrium isotope exchange results exclude the compulsory-order mechanism where NADH binds first.

Conclusions: probably a random-order mechanism. (If this is the case, our hypothetical alcohol dehydrogenase is atypical, unless it can be assumed that a random-order mechansim does not prevent the *preferred* pathway being that where NADH binds first.)

The departures from linearity must be due to kinetic factors rather than to cooperative binding.

13.3 Positive homotropic cooperativity is indicated, with A an allosteric inhibitor. From the Hill plot, $h = 3$ for the inhibited reaction and $h = 2$ for the uninhibited reaction. The two molecular weights determined were presumably those for oligomer and monomer, so $n = 3$. Therefore $h = n$ when L is large (in presence of an allosteric inhibitor), so the Monod–Wyman–Changeux mechanism is indicated.

15.1 Specific activity $\simeq 11.5$ μmol.minute^{-1}.mg protein^{-1} in each case, the good agreement between the results suggesting that the assay procedure is valid. (Note that many students make the mistake of calculating specific activity as μmol.l^{-1}.minute^{-1}.mg protein^{-1}: the incubation volume (5 cm^3 in this problem) must be taken into account, since this is what is acted upon by the enzyme.)

16.1 Specific activity of precipitate = 1900 μmol.minute^{-1}.mg protein^{-1}.
Specific activity of supernatant = 2000 μmol.minute^{-1}.mg protein^{-1}.
Hence the purification step has *not* been successful

17.1 2.1 minutes

17.2 (a) Fig. 7.3a; (b) linear as long as substrate concentration is non-limiting; (c) Fig. 7.1 at fixed concentrations of other substrates (NADH being a co-substrate); (d) Fig. 6.3 (identical for an enzyme-catalysed reaction; see also Chapter 15.2.2); (e) Fig. 8.1a; (f) Fig. 8.12b; (g) Fig. 17.1; and (h) Fig. 17.2.

Abbreviations

ADP	adenosine-5′-diphosphate
AMP	adenosine-5′-monophosphate
ATP	adenosine-5′-triphosphate
DNA	deoxyribonucleic acid
$[E_o]$	initial (and usually total) enzyme concentration
FAD	flavin adenine dinucleotide (oxidised form)
$FADH_2$	flavin adenine dinucleotide (reduced form)
Hb	haemoglobin
K_m	Michaelis constant
Mb	myoglobin
NAD^+	nicotinamide adenine dinucleotide (oxidised form)
NADH	nicotinamide adenine dinucleotide (reduced form)
$NADP^+$	nicotinamide adenine dinucleotide phosphate (oxidised form)
NADPH	nicotinamide adenine dinucleotide phosphate (reduced form)

$$P_i \quad \text{inorganic orthophosphate} \left(\begin{array}{c} O \\ \parallel \\ HO-P-O^- \\ | \\ O^- \end{array} \right)$$

$$(PP)_i \quad \text{inorganic pyrophosphate} \left(\begin{array}{c} O \quad\quad O \\ \parallel \quad\quad \parallel \\ HO-P-O-P-O^- \\ | \quad\quad\quad | \\ O^- \quad\quad O^- \end{array} \right)$$

RNA	ribonucleic acid
$[S_o]$	initial substrate concentration
TPP	thiamine pyrophosphate
UDP	uridine diphosphate
v_o	initial (steady-state) reaction velocity
V_{max}	maximum possible v_o at fixed $[E_o]$

Index